AF618515

ISNM
International Series of
Numerical Mathematics
Vol. 116

Edited by
K.-H. Hoffmann, München
H. D. Mittelmann, Tempe
J. Todd, Pasadena

Multigrid Methods IV

Proceedings of the Fourth European Multigrid Conference, Amsterdam, July 6–9, 1993

Edited by

P.W. Hemker
P. Wesseling

Springer Basel AG

Editors

P.W. Hemker
CWI
Dept. of Numerical Mathematics
Kruislaan 413
1098 SJ Amsterdam
The Netherlands

P. Wesseling
Delft University of Technology
Faculty of Techn. Mathematics & Informatics
Mekelweg 4
2628 CD Delft
The Netherlands

A CIP catalogue record for this book is available from the Library of Congress, Washington D.C., USA

Deutsche Bibliothek Cataloging-in-Publication Data
Multigrid methods IV: proceedings of the Fourth European Multigrid Conference, Amsterdam, July 6–9, 1993. Ed. by P. W. Hemker ; P. Wesseling. – Basel ; Boston ; Berlin : Birkhäuser, 1994
(International series of numerical mathematics ; Vol. 116)
ISBN 978-3-7643-5030-7 ISBN 978-3-0348-8524-9 (eBook)
DOI 10.1007/978-3-0348-8524-9
NE: Hemker, Pieter W. [Hrsg.]; European Multigrid Conference <4, 1993, Amsterdam>; GT

Originally published by Birkhäuser Verlag, Basel, Switzerland in 1994

Camera-ready copy prepared by the editors
Printed on acid-free paper produced from chlorine-free pulp
Cover design: Heinz Hiltbrunner, Basel

ISBN 978-3-7643-5030-7

9 8 7 6 5 4 3 2 1

Contents

Preface

This volume contains a selection from the papers presented at the Fourth European Multigrid Conference, held in Amsterdam, July 6-9, 1993. There were 78 registered participants from 14 different countries, and 56 presentations were given.

The preceding conferences in this series were held in Cologne (1981, 1985) and in Bonn (1990). Also at the other side of the Atlantic special multigrid conferences are held regularly, at intervals of two years, always in Copper Mountain, Colorado, US. The Sixth Copper Mountain Conference on Multigrid Methods took place in April, 1993. Circumstances prevented us from putting a larger time interval between the Copper and Amsterdam meetings. The next European meeting is planned in 1996, a year later than the next Copper Meeting.

When the first multigrid conference was held in 1981 there was no doubt about the usefulness of a conference dedicated specially to multigrid, because multigrid was a new and relatively unexplored subject, still in a pioneering stage, and pursued by specialists. The past twenty years have shown a rapid growth in theoretical understanding, useful applications and widespread acceptance of multigrid in the applied disciplines. Hence, one might ask whether there is still a need today for conferences specially dedicated to multigrid. The general consensus is that the answer is affirmative. New issues have arisen that are best addressed or need also be addressed from a special multigrid point of view. Most prominent among these issues are parallel computing, adaptive computations and applications other than elliptic boundary value problems. Multigrid has much impact on computational fluid dynamics, but also in other fields profitable use of multilevel concepts is possible and starts to develop. In fact, in almost all areas in which intensive computing is a major tool, multilevel principles may bring improvements or even allow major breakthroughs. Hence, to exchange the experiences special multigrid conferences will continue to be useful in the foreseeable future.

Exchange of information on multigrid research is further aided by MGNet, in which papers and software are stored electronically, and may be retrieved by ftp. MGNet is maintained by C. Douglas of Yale University. Information on MGNet can be obtained by sending email to mgnet-requests@cs.yale.edu.

The papers in this volume are ordered alphabetically by author. The invited presentations are followed by a selection of contributed papers. Financial constraints put a page limit on this volume. Rather than severely limit the number of pages available for each contribution, reducing these more or less to technical abstracts, we preferred to give authors sufficient space to show interesting details, and to accept the consequence, that not all contributions could find a place in these pages. We made a selection, and the remaining contributions will be published by the Centre for Mathematics and Informatics (CWI) in Amsterdam in their CWI Tract series.

Several trends in the field that are discernible at present, are reflected in the papers presented at the conference. Maturing parallel computing technology has an increasing impact on scientific computing. This development is of prime concern to multigrid practitioners. After all, reduction of computing cost, measured in various norms, such as financial cost or elapsed wall clock time has from the start been the most (though not the only) appealing aspect of multigrid from a practical point of view. The holy grail of "just a few work units" takes on a new aspect in a parallel computing environment. Similar considerations of cost and quality lead to adaptive discretisation techniques for those applications that go beyond the realm of the smooth and continuous, to include sharp-edged features and discontinuities. Multigrid is especially suited here, because of the possibilities it offers for a-posteriori error estimation, and hence for the detection of special structures in solutions. Furthermore, multigrid for unstructured grids is actively pursued. In three-dimensional domains of complicated shape, unstructured grids are much more easily generated than structured grids and they give more flexibility when adapting the grid to the behaviour of the solution. However, there is still a long way to go before efficiency similar to that obtained for structured grids is obtained, especially for equations of second order. Algebraic multigrid, in which no reference is made at all to an underlying grid structure, shows progress, but also needs to be developed further.

Multigrid has become an indispensible tool in computational fluid dynamics. Significant new developments are seen in the treatment of evolution and hyperbolic problems. Steady progress is being made for the multifarious mathematical models that play a role in fluid dynamics. But other fields present huge computational challenges as well. A prime example is quantum chromodynamics. Aided by multigrid, significant breakthroughs seem in the offing. Another example is tribology, where computational models have improved significantly by application of multigrid. The reader will find papers about these topics in the present volume.

The conference was made possible by the Centre of Mathematics and Informatics (CWI), Amsterdam, and the University of Amsterdam. Financial support was provided by Akzo NV, IBM Nederland NV and the Royal Dutch Academy of Science (KNAW). We are also greatly indebted to Mr Frans Snijders and Ms Simone van der Wolff for their help in organising the conference in the historic setting of old Amsterdam.

Amsterdam / Delft, October 1993
P.W. Hemker
P. Wesseling

Part I

Invited Papers

1

On Robust and Adaptive Multi-Grid Methods

P. Bastian and G. Wittum[1]

ABSTRACT In the present paper we discuss the development and practical application of robust multi-grid methods to solve partial differential equations on adaptively refined grids. We review several approaches to achieve robust multi-grid methods and describe two special new strategies for anisotropic and convection diffusion problems. The performance of these algorithms is investigated for three selected test problems.

1 Introduction

In the present paper we discuss the development and practical application of robust multi-grid methods to solve partial differential equations on adaptively refined grids. Since a couple of years multi-grid methods are well established as fast solvers for large systems of equations arising from the discretization of differential equations. However, it is still a substantial unresolved question to find robust methods, working efficiently for large ranges of parameters e.g. in singularly perturbed problems. This applies to diffusion-convection-reaction equations, arising e.g. from modelling of flow through porous media, the basic equations of fluid mechanics and plate and shell problems from structural mechanics.

Multi-grid methods are known to be of optimal efficiency, i.e. the convergence rate κ does not depend on the dimension of the system, characterized by a stepsize h. Following [28] we call a multi-grid method robust for a singularly perturbed problem, if

$$\kappa(h,\varepsilon) \leq \kappa_0 < 1, \ \ \forall\varepsilon > 0, \ \ h > 0, \tag{1}$$

ε denoting the singular perturbation parameter. Up to now multi-grid methods satisfying (1) have been studied in the literature only for special model cases using structured grids, see [25], [26], [15], [27], [28], [29].

[1]Interdisziplinäres Zentrum für Wissenschaftliches Rechnen (IWR), Universität Heidelberg, Im Neuenheimer Feld 368, 69120 Heidelberg, Federal Republic of Germany, email: wittum@iwr.uni-heidelberg.de

Problems of the type mentioned, typically show degenerations in hyperplanes. To resolve these zones special dynamic grid adaptation techniques are necessary. Here it is necessary to rethink standard multi-grid techniques. In §2 we classify several multi-grid approaches for adaptively refined grids. On the one hand adaptively refined grids can substantially weaken the robustness requirement (1) as outlined in §3. On the other hand the unstructured grids generated by adaptive refinement require special numbering techniques so that the smoother does a good job on the problem. It is the main objective of the present paper to present a strategy to combine the techniques of robust multi-grid and adaptivity.

The techniques have been implemented within the software package *ug*, which will be shortly described in §4. Results of numerical tests for several practical problems are given in §5.

2 Multi-Grid Strategies

2.1 Basic Multi-Grid Techniques

Let the linear boundary-value problem

$$\begin{aligned} Ku &= f \text{ in } \Omega \\ u &= u_R \text{ on } \partial\Omega \end{aligned} \tag{2}$$

with a differential operator $K : U \to F$ between some function spaces be given on a domain $\Omega \subseteq \mathbf{R}^d$. Let (2) be discretized by some local discretization scheme on a hierarchy of admissible grids (cf. [13])

$$\begin{aligned} \Omega_l \quad &, \quad l = 0, \ldots, l_{max} \\ \Omega_l \;&\subseteq\; \Omega_{l+1} \subseteq \Omega \quad . \end{aligned} \tag{3}$$

We use nested grids only for ease of presentation. Most of the methods discussed below can readily be applied to general loosely coupled grids violating (3). The discretized equations on Ω_l are denoted by

$$\begin{aligned} K_l u_l &= f_l \text{ in } \Omega_l, \text{ for } l = 1, \ldots, l_{max} \quad , \\ u_l &= u_{R,l} \text{ on } \partial\Omega_l \end{aligned} \tag{4}$$

with

$$K_l : U_l \to F_l \quad , \tag{5}$$

U_l, F_l denoting the discrete analoga of U and F with finite dimension n. We assume that the discretized equations are sparse. Further let some "smoother"

$$S_l : U_l \to U_l \text{ for } l = 0, \ldots, l_{max} \quad , \tag{6}$$

and "grid transfer operators"

$$p_{l-1} : U_{l-1} \to U_l, \quad r_{l-1} : F_l \to F_{l-1}, \text{ for } l = 1, \ldots, l_{max} \quad , \tag{7}$$

be given.

Multi-grid methods are fast solvers for problem (4). We basically distinguish between additive and multiplicative multi-grid methods. The multiplicative method is the well-known classical multi-grid (cf. [12]) as given in algorithm 2.1:

Algorithm 2.1 *Multiplicative multi-grid method.*

```
mmgm(l, u, f)
integer l; grid function u, f;
{ grid function v, d; integer j;
    if (l = 0) u := K_l^{-1} f;
    else {
        u := S_l^{ν1}(u, f);
        d := r_{l-1}(K_l u - f);
        v := 0;
        for j:=1 step 1 to γ do mmgm(l - 1, v, d);
        u := u - p_{l-1} v;
        u := S_l^{ν2}(u, f);
    }
}
```

The additive multi-grid method is given by the following algorithm.

Algorithm 2.2 *Additive multi-grid method.*

```
amgm(l, u, f)
integer l; grid function v[l], d[l];
{ integer j;
    d[l] := K_l u - f; v[l] := 0;
    for j:=l step -1 to 1 do { d[j - 1] := r_{j-1} d[j]; v[j - 1] := 0;
    for j:=1 step 1 to l do v[j] := S_j^ν(v[j], d[j]);
    v[0] := K_0^{-1} d[0];
    for j:=1 step 1 to l do v[j] :=v[j] + p_{j-1} v[j - 1];
    u := u - v[l];
}
```

The structure of both algorithms can be seen from Figs. 1(a) and 1(b). The main difference between these two variants is that in the multiplicative method

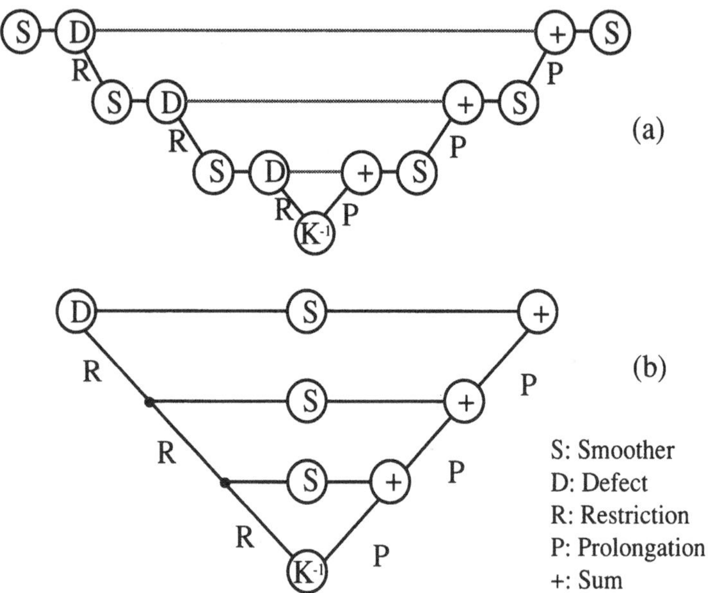

FIGURE 1. Outline of the V-cycle multiplicative multigrid algorithm mmgm (a) and of the additive multigrid algorithm amgm (b).

smoothing and restriction of the defect to the next coarser level are performed on one level after the other sequentially, while in the additve method smoothing on the different levels can be performed in parallel. Restriction and prolongation, however, are sequentially in the additive method too. Usually, the additive methods are applied as preconditioners, since acceleration methods like cg directly pick an optimal damping parameter, the multiplicative methods are used as solvers and as preconditioners. According to [31], these methods can be formulated as additive Schwarz methods.

Applying multi-grid methods to problems on locally refined grids one has to think about the basic question, how to associate grid-points with levels in the multi-grid hierarchy. Consider the hierarchy of grids $\{\Omega_l, l = 0, \ldots, l_{max}\}$ from (3). Early multi-grid approaches smooth all points in Ω_l. This may cause a non-optimal amount of work and memory of $O(n \log n)$ per multi-grid step. This problem was the starting point for Yserentant , [32], and Bank-Dupont-Yserentant, [1], to develop the method of hierarchical bases (HB) and the hierarchical basis multi-grid method (HB/MG). These were the first multi-grid methods with optimal amount of work per step for locally refined grids. This is due to the fact that on level l only the unknowns belonging to points in $\Omega_l \setminus \Omega_{l-1}$ are treated by the smoother. However, the convergence rate deteriorates with $\log n$. For the first time this problem was solved by the introduction of the additive method by Bramble, Pasciak and

TABLE 1. Multi-grid methods for locally refined grids.

smoothing pattern	basic structure	
	additive	multiplicative
(1) new points only	*HB* *Yserentant, 1984, [32]*	*HBMG* *Bank, Dupont, Yserentant, 1987, [1]*
(2) refined region only	*BPX* *Bramble, Pasciak, Xu, 1989, [8]*	*local multi-grid, [20], [9], [5]*
(3) all points	*parallel multigrid* *Greenbaum, 1986, [11]*	*classical multi-grid, [10]*

Xu, [8], (BPX). There on level l the smoother treats all the points in $\Omega_l \setminus \Omega_{l-1}$ and their direct neighbours, i.e. all points within the refined region.

Table 1 gives an overview of the multi-grid methods used for the treatment of locally refined grids and classifies the variant we call "local multi-grid". The methods mentioned above differ in the smoothing pattern, i.e. the choice of grid points treated by the smoother. The methods in the first two lines are of optimal complexity for such problems. The amount of work for one step is proportional to the number of unknowns on the finest grid. However, only the methods in the second line, BPX and local multi-grid converge independently of h for scalar elliptic problems. The basic advantage of the multiplicative methods is that they do not need cg-acceleration and thus can be directly applied to unsymmetric problems, further they show a better convergence rate and on a serial computer the additive process does not have any advantage. The local multi-grid scheme is the natural generalization of the classical multi-grid method to locally refined grids, since in case of global refinement, it is identical with the standard classical multi-grid method.

The local multi-grid has first been analyzed in 1991 by Bramble, Pasciak, Wang and Xu, [9]. They considered it as a multiplicative variant of their so-called BPX-method, [8]. However, they did not consider robustness. Further there exist predecessors of this method since a couple of years in some implementations (pers. communication by J.-F. Ma^ıtre and H. Yserentant). Without knowledge of this, the authors developed this method as a variant of standard multi-grid based on the idea of robustness (cf. [5]). The main advantage of this approach is that the application to unsymmetric and non-linear problems is straightforward (cf. [5]). Robustness for singularly perturbed problems is achieved by combining local multi-grid with robust smoothers (cf. [5]), as explained in the next section.

3 Robustness Strategies

3.1 Robust Smoothing

Already in 1981, Wesseling suggested the first robust multi-grid method for singularly perturbed problems discretized on structured grids [25], [26]. The main idea is to apply a smoother which solves the limit case exactly. This is possible e.g. for a convection-diffusion equation using a Gauß-Seidel smoother and numbering the unknowns in convection direction. Wesseling however, suggests to use an incomplete LU-smoother, since this handles the convection dominated case as well as the anisotropic diffusion (cf. [15], [28]). Main ingredients, however, are the use of structured grids and a lexicographic numbering.

A simple analysis of the hierarchical basis methods (HB, HB/MG) shows that the smoothing pattern is too poor to allow robust smoothing.

Remark 3.1 *The hierarchical basis method and the hierarchical basis multigrid method do not allow robust smoothing for a convection-diffusion equation. The smoothing pattern used in these methods does not allow the smoother to be an exact solver for the limit case. This holds for uniformly as well as for locally refined grids.*

Based on this observation, we extended the smoothing pattern, adding all neighbours of points in $\Omega_l \setminus \Omega_{l-1}$. This allows the smoother to solve the limit case exactly, provided the grid refinement is appropriate. This is confirmed by numerical evidence given in Chapter 5.

Up to now some theory is contained in [28],[29] and the new papers by Stevenson [21], [22] for uniformly refined grids. This theory shows that the basic requirement that the smoother is an exact solver in the limit case is not sufficient to obtain robustness. Additionally it must be guaranteed that the spectrum of the smoother is contained in $[-\vartheta, 1]$ for $0 \le \vartheta < 1$. This can be achieved by modification (cf. [28], [22]).

3.2 A Robust Smoother for Convection-Diffusion Problems

The construction of a robust smoother, which is exact or very fast in the limit, is the kernel of a robust multigrid method and makes up the main problem when applying this concept to unstructured grids. Here we need special numbering strategies.

In the following we present a strategy for the convection-diffusion equation

$$-\varepsilon\Delta u + c \cdot \nabla u = f \ , \tag{8}$$

with the convection vector c, and $\varepsilon > 0$. Discretizing the convection term by means of an upwind method, we can assign a direction to each link in the graph of the stiffness matrix. If the directed graph generated by this process is cycle-free, it defines a partial ordering of the unknowns. This partial ordering can be used to construct an algorithm for numbering of the unknowns, which brings the convective part of the stiffness matrix to a triangular form. The following numbering algorithm performs such an ordering on general unstructured grids, provided the convection graph is cycle-free.

Algorithm 3.1 *downwind_numbering.*

1. Assign the downwind direction from the discretization of the convective term to each link in the stiffness martix graph. Indifferent links are marked by 0.

2. Put $n =$ number of unknowns.

3. Find all vertices with minimal number of incoming links and put them in a fifo F.

4. Derive a total order from the directed acyclic graph

```
For all vertices L initialize Index(L) = 0;
While (F not empty) do
        get E from F;
(4a)    Put Index(E) := 1; Put E in fifo FP; i := 1;
(4b)    While (FP not empty) and (i < n) do
                 Get K from FP;
                 For all neighbors L of K do
                     If (L downwind from K) and (Index(L)≤ Index(K))
                       i := Index(L);
                       Index(L) := Index(K)+1;
                       Put L in FP;
```

5. Call quicksort with the vertex list and the criterion $Index(L) < Index(K) \Rightarrow L < K$. Output: Ordered vertex list.

Remark 3.2 *If the edge graph is cycle-free, loop (4b) terminates in $O(n)$-steps with $FP = \emptyset$. Loop (4) has complexity $O(q \cdot n)$ where q is the number of minimal elements in the edge graph, which is small. Because of calling quicksort in (5) the complexity of the whole algorithm equals $O(q \cdot n \ln n)$.*

If loop (4b) terminates with $FP \neq \emptyset$ and $i \geq n$, the edge graph contains a cycle.

This method has been used for the computations described in Section 5 . Meanwhile it has been improved by Bey (cf. [6]). Cycles in the matrix graph may occur, if there are vortices in the convection c. If c is vortex-free, cycles can occur if several triangles with sharp angles are neighbouring each other and are almost

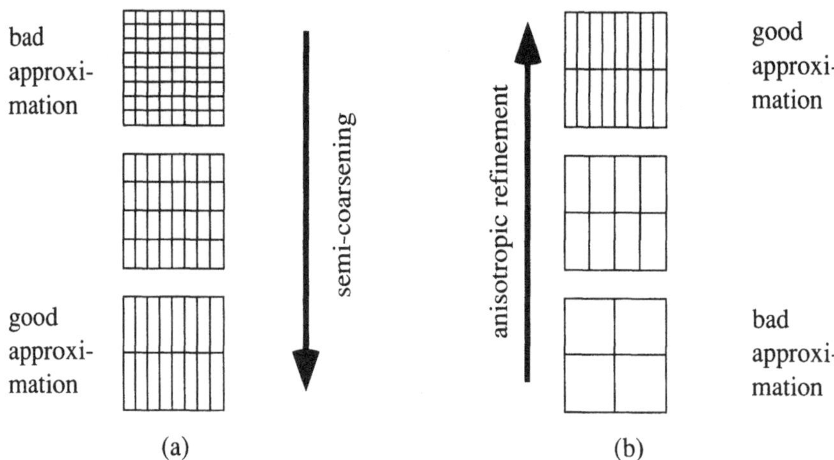

FIGURE 2. Illustration of semi-coarsening (a) and anisotropic refinement (b).

perpendicular to the flow direction (cf. [6]). These numerically caused cycles, however, can be simply eliminated by finding and cutting elementwise cycles. This is possible with $O(n)$ work count.

3.3 Semi-coarsening

Another strategy to obtain a robust multi-grid method is the so-called semi-coarsening approach (cf. [26]). The basic idea is to improve the coarse grid correction instead of the smoother. Starting with a fine and structured grid, coarsening is performed only in those co-ordinate directions, in which the scale of the equation is already resolved. E.g. for the anisotropic model problem

$$-\left(\varepsilon\partial_{xx}+\partial_{yy}\right)u=f\ ,\qquad \text{in}\Omega=(0,1)\times(0,1) \tag{9}$$

with corresponding boundary conditions one would coarsen an equidistant cartesian grid in case of small ε as shown in Figure 2(a).

Remark 3.3 *Such a sequence of coarse grids yields a robust multi-grid method for the anisotropic model problem (9) without using a special smoother, since the coarse grid resolves the scale in the direction where the smoother does not work.*

This semi-coarsening approach, however, is based on the use of fine grids which do not resolve the differential scale, otherwise there would be no semi-coarsening. Consequently this approach is not applicable as soon as the finest grid resolves the problem scale, which is crucial when solving differential equations.

This does not apply to so-called multiple semi-coarsening approaches, since these methods are able to construct sequences of coarse grids from any structured fine one, no matter if the scale is resolved. Thus we mainly have to look for an approach which allows to adapt the grid to the differential scale by adaptive refinement and to solve efficiently on the hierarchy of grids generated this way.

3.4 Anisotropic Refinement

Instead of starting with a fine grid and constructing the grid hierarchy by coarsening we start with a coarse grid and refine that anisotropically in order to resolve the scale successively. Such a refinement process is given e.g. by the "blue refinement strategy" due to Kornhuber, [16]. The basic idea is just to refine quadrilaterals with a "bad aspect ratio" by halving the longer edge. Bad aspect ratios can be introduced either by element geometry or by anisotropic coefficients in the equation. This is shown for the anisotropic model problem (9) in Fig. 2(b). Note that the discretization error is balanced on the *coarsest* grid for semi-coarsening, while it is balanced on the *finest* grid for the anisotropic refinement approach. Kornhuber described how to generalize this approach to triangular unstructured grids. Following this process we finally obtain a grid Ω_l which resolves the scale of the problem.

From this grid on we refine regularly and so the multi-grid process will obviously work without problems.

Remark 3.4 *A proof of robust multi-grid convergence is straightforward since the asymptotic behaviour is determined by the isotropic problem. So we need a robust method only for a finite sequence of grids up to a fixed $h > 0$, weakening the robustness requirement (1) to*

$$\kappa(h,\varepsilon) \leq \kappa_0 < 1, \quad \forall \overline{\varepsilon} \geq \varepsilon \geq \underline{\varepsilon} > 0,\ \forall h \geq \underline{h} > 0\ , \tag{10}$$

which makes the job much easier. Thus it is sufficient in many cases to use just a lexicographically numbered ILU_β, since we do not need the property that the smoother is exact in the limit case. It is sufficient that it reasonably accounts for the "main connections" up to a fixed range of $\varepsilon > 0$ and for finite h.

Since this process improves the approximation of the differential problem at the same time, this will be the appropriate approach to follow.

An example of that type is the skin problem described in §5.

3.5 Algebraic Multi-grid

Another approach yielding robustness is the family of algebraic multi-grid methods, see e.g. [24] and the references there. A new algebraic multi-grid approach is described by Reusken, [23], which shows to be fairly robust in practice. The basic idea of algebraic multi-grid is to decompose the stiffness matrix K into

$$K_l = \begin{pmatrix} K_{ff} & K_{fc} \\ K_{cf} & K_{cc} \end{pmatrix} \tag{11}$$

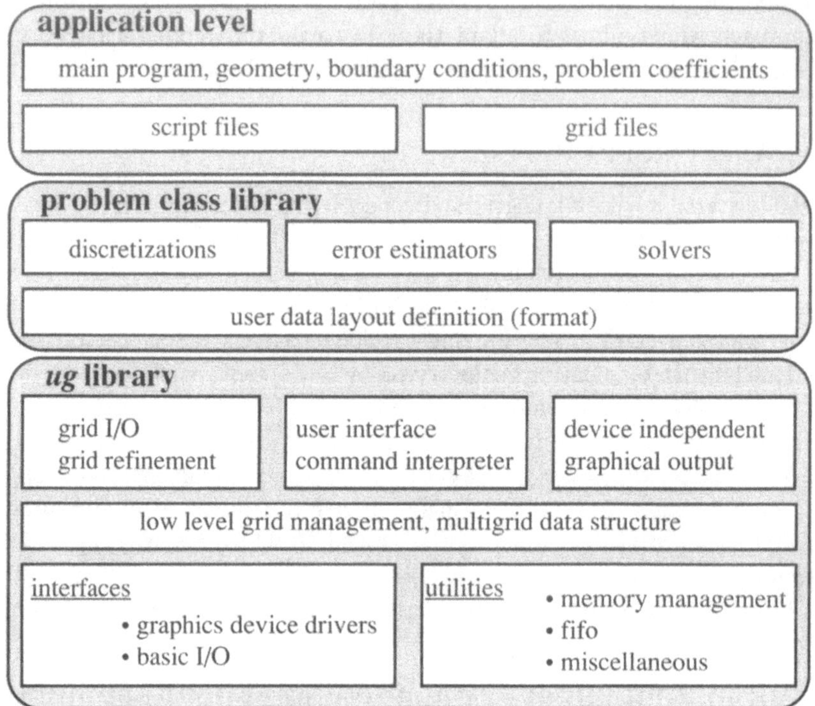

FIGURE 3. Overview of the internal structure of the ug code.

where K_{ff} denotes the part of K_l acting on the grid points which belong to the finest grid only, K_{cc} the part of K_l acting on coarse grids points only and the off-diagonal blocks represent the coupling between coarse and fine grid. The approximation of K_{cc} and the off-diagonal blocks within the multi-grid cycle have to be such, that it yields robustness. This is also satisfied for the frequency-decomposition multi-grid method, [14], and other multiple correction schemes, see ([18], [19]). However, these methods typically work only on structured grids and do also not provide a strategy to improve the approximation of the differential equation.

4 The Software Toolbox *ug*

The code *ug* ("unstructured grids") is used as a test-bed for the robustness strategies mentioned above and has been designed as problem independent as possible in order to allow reuse of its components for many different applications. It is a layered construction of several libraries, see Fig. 3 for an overview. The bottom layer contains all components that are totally independent of the PDE to be solved, e. g. grid I/O, grid refinement, device independent graphical output and the user interface. The next layer is the so-called problem class library that implements discretization, error estimators and solvers for a whole class of PDEs, e. g. a scalar conservation law. On top of that resides the user's application that provides the domain, boundary conditions and problem coefficients to the lower layers.

The relative code size of these layers indicates that the proper abstractions (interfaces) have been chosen: The *ug* layer typically makes about 75% of the executable, the problem class layer takes 20% in the convection-diffusion case (with many different solvers) and a main program typically is only 5%. This means in practice:

- 75% of the code can be reused *without any change* when switching to more complicated equations. This has been proved already for incompressible Navier-Stokes equations.
- The user interested in implementing new numerical algorithms (a problem class library) will never be concerned with low level programming.
- As a consequence of that his code is portable since machine dependencies typically arise only in the *ug* layer.

The concept of code reuse becomes even more important in a parallel environment, see [4] for a parallel implementation of *ug*.

5 Numerical Results

In the following we discuss the application of the above-mentioned robustness strategies to three problems, serving as paradigms for typical singularly perturbed problems.

5.1 The Skin Problem

As a first test problem we take the following one which is used to model the penetration of drugs through the uppermost layer of the skin (stratum corneum). The stratum corneum is made up of corneocytes which are embedded in a lipid layer. The diffusion is described by the diffusion equation

$$\begin{aligned} -\nabla(D(x,y)\nabla u) + \frac{\partial u}{\partial t} &= 0 \quad \text{in}\Omega \\ u &= 1 \quad \text{on}\Gamma_u \\ u &= 0 \quad \text{on}\Gamma_o \\ \frac{\partial u}{\partial n} &= 0 \quad \text{on}\Gamma_r \cup \Gamma_l \end{aligned} \qquad (12)$$

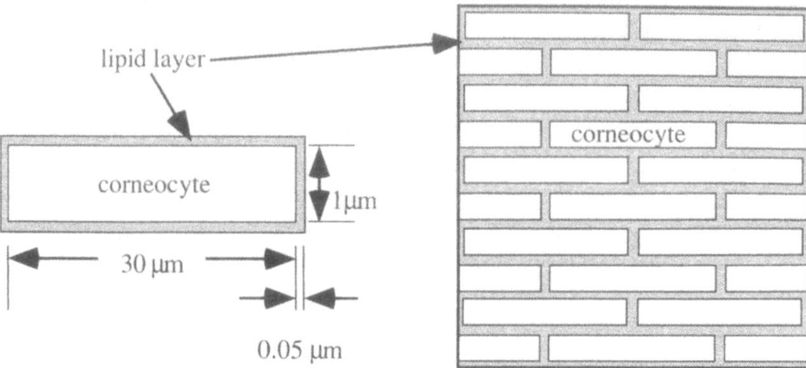

FIGURE 4. Right hand side: Structure of skin made up from corneocytes (white) and lipid layers (gray/black). The considered block of stratum corneum is $11\mu m$ by $60.2\mu m$. Left hand side: Elementary cell consisting of a corneocyte surrounded by one half of the lipid layer.

where Ω is the unit square and the diffusion coefficient $D(x, y)$ is given by

$$D(x,y) = \begin{cases} D_1 & \text{if } (x,y) \in \text{lipid} \\ D_2 & \text{if } (x,y) \in \text{corneocyte} \end{cases},$$

i.e. it may jump by some orders of magnitude across the corneocyte edges. The corneocytes are very flat and wide cells which in a two-dimensional cross-section are approximated by thin rectangles as shown in Fig. 4.

From Fig. 4 we see that the lipid layer is $0.1\mu m$ thick while the corneocytes are 1 by $30\mu m$ of size. Since the permeability may jump by some orders of magnitude between lipid and corneocyte, we must align the coarse-grid lines with the interfaces. So we just take the corners of the corneocytes as points for the coarse grid connecting them to form a tensor product grid. Thus we get rid of the problems induced by jumping coefficients. However, we obtain highly anisotropic grid cells in the lipid layer with an aspect ratio of approx. 1:150. Since such an aspect ratio makes the approximation strongly deteriorate and the multi-grid method as well, we use the anisotropic ("blue") refinement strategy to derive a robust multi-grid method and to create a grid which after 5 levels of blue refinement has elements not exceeding an aspect ratio of 1:5. Above that level we refine uniformly. To obtain a robust method on the coarser grids we use an ILU_β-smoother, cf. [28]. Average convergence factors for a (1,1,V)-cycle are given in Table 2. For more details on this problem see [17].

5.2 Convection-Diffusion Equation

As a second example we show results for the convection-diffusion equation

TABLE 2. Convergence rate of a (1,1,V)-mmgm applied to the stationary skin problem for various values of D_2 ($D_1 = 1$). The number of unknowns was 54385 on level 5 (6 grid levels).

D_2	1	10^{-1}	10^{-2}	10^{-3}	10^{-4}	10^{-5}	10^{-6}
ρ	0.08	0.22	0.39	0.41	0.45	0.45	0.43

TABLE 3. Robustness of a (1,1,V)-mmgm with ILU-smoother and downwind numbering. The method used 8 locally refined grids to discretize problem the convection-diffusion problem with over 10.000 unknowns on level 8. The convergence rate $\kappa(10)$ is averaged over 10 steps and refers to the finest grid.

ε	1	10^{-1}	10^{-2}	10^{-3}	10^{-4}	10^{-5}	10^{-6}	10^{-7}
$\kappa(10)$	0.068	0.067	0.075	0.102	0.092	0.068	0.033	0.018

$$-\epsilon\Delta u + c \cdot \nabla u = f \tag{13}$$

in the unit square with Dirichlet boundary conditions. We choose c as follows

$$c = \left(1 - \sin(\alpha)\left[2\left(x + \frac{1}{4}\right) - 1\right] + 2\cos(\alpha)\left[y - \frac{1}{4}\right]\right)^4 (\cos(\alpha), \sin(\alpha))^T \tag{14}$$

where α is the angle of attack. The boundary conditions are: $u = 0$ on $\{(x,y) : x = 0, 0 \leq y \leq 1\} \cup \{(x,y) : 0 \leq x \leq 1, y = 1\} \cup \{(x,y) : x = 1, 0 \leq y \leq 1\} \cup \{(x,y) : 0 \leq x < 0.5, y = 0\}$ and $u = 1$ on $\{(x,y) : 0.5 \leq x \leq 1, y = 0\}$. The jump in the boundary condition is propagated in direction α. We have $\text{div}c = 0$ and c varies strongly on Ω such that the problem is convection dominated in one part of the region and diffusion dominated in another part. As discretization we use a finite volume scheme with first order upwinding for the convective terms on a triangular grid. The grid is refined adaptively using a gradient refinement criterion. As smoother we took a Gauß-Seidel scheme with downwind numbering using algorithm 3.1 in a (1,1,V)-cycle mmgm. It is important to note that the smoother itself is not an exact solver. Thus we should see the benefit of multi-grid in the diffusion dominated part and of the robust smoother in the convection dominated one. This is confirmed by the results given in Table 3. There we show the residual convergence rate averaged over 10 steps for problem (13) on adaptively refined unstructured grids versus ε.

For the same problem with $\varepsilon = 10^{-7}$ the same mmgm but without downwind numbering shows a convergence rate of 0.95 averaged over 40 steps and taking the smoother with downwind numbering but without coarse grid correction as a solver,

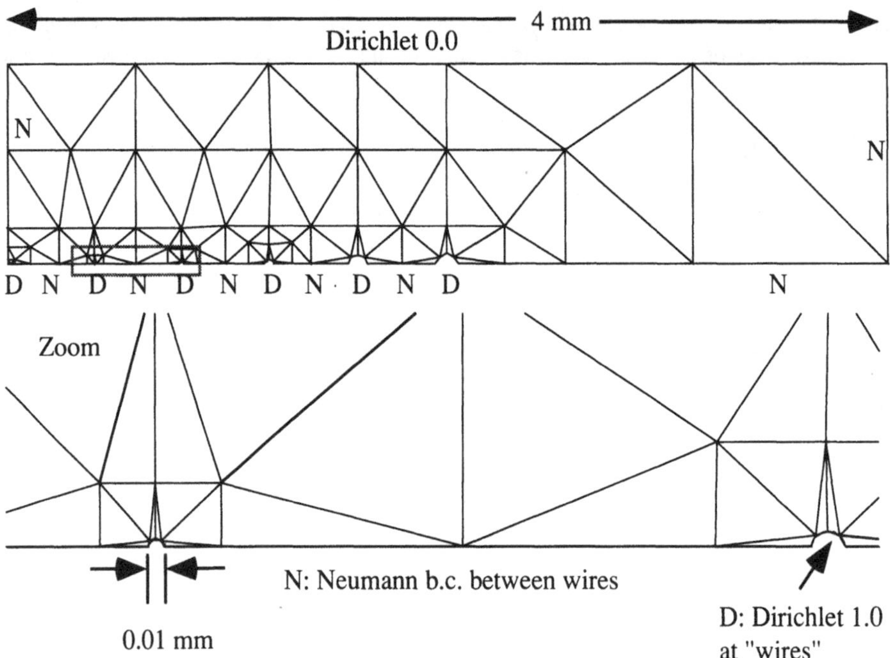

FIGURE 5. Problem definition, coarse grid and zoom for the drift chamber problem.

we end up with a convergence rate of 0.949 as well. This confirms the outlined concept of robust multi-grid. Results of 3d computations can be found in [6].

5.3 Drift Chamber

This problem solves the Laplacian $-\Delta u = 0$ in the domain given by Fig. 5. The boundary conditions are of Dirichlet and Neumann type as indicated in the figure. The feature of this problem are the small wires with Dirichlet boundary conditions that must be resolved on the coarse grid. The smallest wire has a radius of 0.005 mm, while the whole chamber is 4 mm wide and 1 mm thick. So one has to trade off between a coarse grid with few unknowns but a large aspect ratio in grid cells and a coarse grid with equal sized triangles but a large number of unknowns. The grid in Fig. 5 is a reasonable compromise with 85 nodes and 112 triangles but still aspect ratios are large and a robust smoother is required.

Table 4 shows the results of multiplicative and additive multigrid with several different smoothers applied after 3, 4, 5 and 6 levels of uniform refinement. Specifically the smoothers were damped jacobi with $\omega = 2/3$ (djac), (symmetric) Gauß-Seidel (gs, sgs) and ILU without modification and with $\beta = 0.35$ (ILU, ILU_β). We make the following remarks:

TABLE 4. Results for different solver/smoother combinations for the drift chamber problem. Multigrid data: (2,2,V) cycle for jacobi smoother, $\nu = 1$ for amgm, (2,2,V) cycle for all other smoothers, initial solution $u = 0$, numbers are iterations for a reduction of the residual by 10^{-6} in the euclidean norm. The grid nodes have been ordered lexicographically, iteration numbers exceeding 100 are marked with an asterisk, diverging iterations are marked with $\uparrow$.

highest level		3	4	5	6
grid nodes		3809	14785	58241	231169
mmgm	djac	*	*	*	*
	gs	79	99	*	*
	sgs	48	59	66	70
	ILU	33	$\uparrow$	$\uparrow$	$\uparrow$
	ILU_β	9	9	9	9
mmgm+cg	djac	31	38	43	43
	sgs	13	16	17	18
	ILU	10	$\uparrow$	$\uparrow$	$\uparrow$
	ILU_β	6	6	6	6
amgm+cg	djac	74	99	*	*
	sgs	36	46	53	57
	ILU	62	$\uparrow$	$\uparrow$	$\uparrow$
	ILU_β	20	24	25	26

1. h independent convergence is only achieved with the ILU_β smoother. The optimal value was $\beta = 0.35$ but the choice is not very sensitive and good results are achieved with values between 0.2 and 0.5. This corresponds nicely with the theory in [28].

2. The additive method shows qualitatively the same behaviour as the multiplicative multi-grid method but has worse numerical efficiency.

3. Multiplicative multi-grid with a symmetric Gauß-Seidel smoother used as preconditioner in a conjugate gradient method is the only combination giving also relatively satisfactory results, being only a factor 3 slower in computation time than the ILU_β smoother.

4. The diverging iteration for ILU without modification can be explained by accumulating roundoff errors. Since the global stiffness matrix is symmetric positive definite but *not* an M-matrix due to obtuse angles the diagonal elements in the ILU decomposition can become very small which leads to instabilities. The modification helps in this case too, since it enlarges the diagonal.

References

[1] R. E. BANK, T. F. DUPONT, H. YSERENTANT: *The Hierarchical Basis Multigrid Method* , Numer. Math., **52**, 427-458 (1988).

[2] R. E. BANK: *PLTMG: A software package for solving elliptic partial differential equations. Users Guide 6.0.* SIAM, Philadelphia, 1990.

[3] P. BASTIAN, G. WITTUM: *Adaptivity and Robustness.* In: Adaptive Methods, Proceedings of the Ninth GAMM Seminar, Notes on Numerical Fluid Mechanics, Vieweg Verlag, Braunschweig, 1993, to appear.

[4] P. BASTIAN: *Parallel Adaptive Multigrid Methods.* IWR Report, Interdisziplinäres Zentrum für Wissenschaftliches Rechnen, Universität Heidelberg, 1993, to appear.

[5] P. BASTIAN: *Locally Refined Solution of Unsymmetric and Nonlinear Problems.* In: Hackbusch, W., Wittum, G. (eds.): Incomplete Decompositons - Theory, Algorithms, and Applications, NNFM, vol. 41, Vieweg, Braunschweig, 1993.

[6] J. BEY, G. WITTUM: *A Robust Multigrid Method for the Convection-Diffusion Equation on locally refined grids.* In: Adaptive Methods, Proceedings of the Ninth GAMM Seminar, Notes on Numerical Fluid Mechanics, Vieweg Verlag, Braunschweig, 1993, to appear.

[7] A. BRANDT: *Guide to Multigrid Development.* in Hackbusch W., Trottenberg U. (eds.): Multigrid Methods. Proceedings Köln-Porz, 1981. Lecture Notes in Mathematics, Bd. 960, Springer, Heidelberg, 1982.

[8] J. H. BRAMBLE, J. E. PASCIAK, J. XU: *Parallel Multilevel Preconditioners*, Math. Comput., **55**, 1-22 (1990).

[9] J. H. BRAMBLE, J. E. PASCIAK, J. WANG, AND J. XU, *Convergence estimates for multigrid algorithms without regularity assumptions*, Math. Comp., **57**, (1991), pp. 23–45.

[10] R. P. FEDORENKO: *Ein Relaxationsverfahren zur Lösung elliptischer Differentialgleichungen.* (russ.) UdSSR Comput Math Math Phys 1,5 1092-1096 (1961).

[11] A. GREENBAUM: *A Multigrid Method for Multiprocessors.* Appl. Math. Comp., **19**, 75-88 (1986).

[12] W. HACKBUSCH: *Multi-grid methods and applications.* Springer, Berlin, Heidelberg (1985).

[13] W. HACKBUSCH: *Theorie und Numerik elliptischer Differentialgleichungen.* Teubner, Stuttgart, 1986.

[14] W. HACKBUSCH: *The Frequency Decomposition Multi-grid Method.* Part I: Application to Anisotropic Equations. Numer. Math., 1989.

[15] R. KETTLER: *Analysis and comparison of relaxation schemes in robust multi-grid and preconditioned conjugate gradient methods.* In: Hackbusch,W., Trottenberg,U. (eds.): Multi-Grid Methods, Lecture Notes in Mathematics, Vol. 960, Springer, Heidelberg, 1982.

[16] R. KORNHUBER, R. ROITZSCH: *On Adaptive Grid Refinement in the Presence of Boundary Layers.* Preprint SC 89-5, ZIB, Berlin, 1989.

[17] R. LIECKFELDT, G. W. J. LEE, G. WITTUM, M. HEISIG: *Diffusant concentration profiles within corneocytes and lipid phase of stratum corneum.* Proceed. Intern. Symp. Rel. Bioact. Mater., 10 (1993) Controlled Release Society, Inc.

[18] W. A. MULDER: *A New Multigrid Approach to Convection Problems.* J. Comp. Phys., **83**, 303-323 (1989).

[19] C. W. OOSTERLEE, P. WESSELING: *Multigrid Schemes for Time-Dependent Incompressible Navier-Stokes Equations.* Report 92-102, Fac. Technical Math. and Inf., Delft University of Technology, Delft (1992), to appear in Impact of Comp. in Science and Eng.

[20] M. C. RIVARA, *Design and data structure of a fully adaptive multigrid finite element software*, ACM Trans. on Math. Software, 10 (1984), pp. 242–264.

[21] R. STEVENSON: *On the robustness of multi-grid applied to anisotropic equations: Smoothing- and Approximation-Properties.* Preprint Rijksuniversiteit Utrecht, Wiskunde, 1992.

[22] R. STEVENSON: *New estimates of the contraction number of V-cycle multi-grid with applications to anisotropic equations.* In: Hackbusch, W., Wittum, G. (eds.) : Incomplete Decompositions, Algorithms, theory, and applications. NNFM, vol 41, Vieweg, Braunschweig, 1993.

[23] A. REUSKEN: see this volume.

[24] K. STÜBEN: *Algebraic Multigrid (AMG): Experiences and Comparisons.* Appl. Math. Comput., **13**, 419-451 (1983).

[25] P. WESSELING: *A robust and efficient multigrid method.* In: Hackbusch, W., Trottenberg, U. (eds.): Multi-grid methods. Proceedings, Lecture Notes in Math. 960, Springer, Berlin (1982).

[26] P. WESSELING: *Theoretical and practical aspects of a multigrid method.* SIAM J. Sci. Statist. Comp. **3**, (1982), 387-407.

[27] G. WITTUM: *Filternde Zerlegungen - Schnelle Löser für große Gleichungssysteme.* Teubner Skripten zur Numerik Band 1, Teubner, Stuttgart, 1992.

[28] G. WITTUM: *On the robustness of ILU-smoothing.* SIAM J. Sci. Stat. Comput., **10**, 699-717 (1989).

[29] G. WITTUM: *Linear iterations as smoothers in multi-grid methods.* Impact of Computing in Science and Engineering, **1**, 180-215 (1989).

[30] J. XU: *Multilevel theory for finite elements.* Thesis, Cornell Univ., 1988.

[31] J. XU: *Iterative Methods by Space Decomposition and Subspace Correction: A Unifying Approach*, SIAM Review, **34**(4), 581-613, (1992).

[32] H. YSERENTANT: *Über die Aufspaltung von Finite-Element-Räumen in Teilräume verschiedener Verfeinerungsstufen.* Habilitationsschrift, RWTH Aachen, 1984.

2

A Generalized Multigrid Theory in the Style of Standard Iterative Methods

Craig C. Douglas[1]

ABSTRACT A basic error bound for multigrid methods is given in terms of residuals on neighboring levels. The terms in this bound derive from the iterative methods used as solvers on each level and the operators used to go from a level to the next coarser level. This bound is correct whether the underlying operator is symmetric or nonsymmetric, definite or indefinite, and singular or nonsingular. We allow any iterative method as a smoother (or rougher) in the multigrid cycle. One of the advantages of this theory is that all of the parameters are available during execution of a computer program. Hence, adaptively changing levels can be achieved with certainty of success. This is particularly important for solving problems in which there is no known useful convergence analysis. Two problems arising in modeling combustion problems (flame sheets and laminar diffusion flames with full chemistry) are discussed.
While this theory is quite general, it is not always the correct approach when analyzing the convergence rate for a given problem. A discussion of when this theory is useful and when it is hopelessly nonsharp is provided.

1 Introduction

In this paper, linear problems

$$Au + f = 0, \quad u, f \in \mathcal{M},\ A \in \mathcal{L}(\mathcal{M}) \tag{1}$$

are solved using a nested space multigrid iterative method. The operator (matrix) A is typically the discretized (by finite elements, differences, or volumes) version of a partial differential equation.

Many multigrid papers begin by narrowing their scope just to problems which are symmetric and positive definite, symmetric and indefinite, or nonsymmetric

---

[1]Department of Computer Science, Yale University, P. O. Box 2158, New Haven, CT 06520-2158. and Mathematical Sciences Department, IBM Research Division, Thomas J. Watson Research Center, P. O. Box 218, Yorktown Heights, NY 10598-0218 E-mail: *na.cdouglas@na-net.ornl.gov*.

and indefinite. In each case, these papers assume the problem is nonsingular, a set of smoothers is defined, and one or more specific multigrid algorithms are defined (e.g., a V, W, or F cycle). Finally, analysis is provided, usually in only one particular norm. For excellent traditional multigrid theoretical treatments of problems, see [1], [8], [17], and [21].

The analysis in this paper is correct whether the underlying operator is symmetric or nonsymmetric, definite or indefinite, and singular or nonsingular. Any iterative method is allowed as a smoother or rougher in the multigrid cycle. Any multigrid cycle is allowed, including adaptively chosen ones. Finally, the analysis is not dependent on any specific norm. In fact, different norms can be used on different levels (though doing this can produce misleading convergence rates).

The purpose of this paper is to provide a discussion on when to use the theoretical tool in [12] for analyzing nested space multilevel algorithms that are applied to any problem with any set of properties. The approach is simple enough to implement in computer programs without adding an excessive amount of overhead. There are similar procedures, known as aggregation-disaggregation methods (see [6]) when A is not derived from partial differential equations; the theory in this paper applies directly to these methods.

The basic correction multigrid algorithm is defined in the traditional recursive style in §2. This is then rephrased into a nonstandard form in §3. This leads to the two flavors of analysis in §4, one quite simple (and rarely sharp) and the other somewhat more complicated (and sharper). Examples and the practicality of this analysis are given in §5.

The theory in §4 depends on three sets of parameters which are available either dynamically or in advance. The basic convergence (divergence) result is not stated in a "nice" closed form, as is usual in multigrid papers, but in terms of the convergence rate of the next coarser level's rate.

2 A standard multilevel formulation

Suppose that there is a set of solution spaces $\{\mathcal{M}_k\}_{k=1}^{j}$, which approximate $\mathcal{M}=\mathcal{M}_j$ in some sense, and that $\dim(\mathcal{M}_k) \leq \dim(\mathcal{M}_{k+1})$. In the partial differential equation case, the $\mathcal{M}_k$ correspond to discrete problems on given grids (which are not necessarily nested). Then the multigrid approximation to (1) requires solving a sequence of problems of the form

$$A_k u_k + f_k = 0, \quad u_k, f_k \in \mathcal{M}_k, \; A_k \in \mathcal{L}(\mathcal{M}_k). \tag{2}$$

That there exist mappings between the neighboring spaces is assumed:

$$R_k : \mathcal{M}_k \to \mathcal{M}_{k-1} \quad \text{and} \quad P_{k-1} : \mathcal{M}_{k-1} \to \mathcal{M}_k$$

as well as mappings

$$Q_k : \mathcal{M}_k \to \mathcal{M}_{k-1} \quad \text{such that} \quad A_{k-1} = Q_k A_k P_{k-1}.$$

For partial differential equations, there are natural definitions of Q_k depending on the discretization method and the grids. See [12] for a more complete discussion of natural choices for Q_k.

Since for most applications, $\dim(\mathcal{M}_k) < \dim(\mathcal{M}_{k+1})$, Q_k cannot be inverted. However, the theory in §4 uses Q_k^{-1}. Thus, the interpretation of Q_k^{-1} must be explained. For finite element methods commonly used in practice, $\mathcal{M}_k$ represents a refinement of $\mathcal{M}_{k-1}$ and

$$Q_k = \begin{cases} I & \text{on } \mathcal{M}_{k-1}, \\ 0 & \text{on } \mathcal{M}_k - \mathcal{M}_{k-1}; \end{cases}$$

this is true for both the h-version and the p-version of the methods.

The same relation holds for refinements in the finite difference case. Hence, Q_k^{-1} can be taken to be injection of $\mathcal{M}_{k-1}$ into $\mathcal{M}_k$ in each of the cases described; otherwise Q_k^{-1} should be taken as a pseudoinverse. Note that a Moore-Penrose type pseudoinverse may not be the best choice; a Drazin type pseudoinverse may be better.

For $k \geq 1$, assume there are iterative methods, represented by M_k and N_k, and possibly dependent upon the data (e.g., conjugate gradients), which are used as smoothers (or roughers) on level k before and after, respectively, the residual correction step (on level 1, note that there is never a residual correction step nor, usually, a smoother N_1).

In the multigrid literature, the term smoother has become synonymous with the direct or iterative methods M_k and N_k. The term was used in [4] to describe the effect of one or more iterations of a relaxation method on each of the components of the error vector. For many relaxation methods (e.g., SSUR and Gauss-Seidel), the norm of each error component is reduced each iteration; hence, the term smoother. For many other iterative methods (e.g., SSOR or conjugate gradients), while the norm of the error vector is reduced each iteration, the norm of some of the components of the error may grow each iteration; hence, the term rougher. For some iterative methods (e.g., Bi-CGSTAB), the norm of the error vector does not necessarily decrease each iteration, much less smooth all of the error components. The term smoother in the traditional multigrid sense will be used, even though it is technically wrong.

Standard multigrid analysis assumes the smoothers have the form

$$B_k(w_k^{\ell+1} - w_k^{\ell}) = f_k + A_k w_k^{\ell}, \quad \ell = 0, 1, \cdots, \ell_k,$$

where B_k corresponds to some scaled iterative method on each level k (e.g., symmetric Gauss-Seidel or conjugate gradients). This frequently leads to an analysis which assumes a fixed ℓ_k throughout the multigrid iterations. Neither assumption is required in §4.

There are two principal variants of multigrid algorithms. One variant is composed of correction schemes, which start on some level j and only use the coarser levels k, $k < j$, for solving residual correction problems. The other variant is composed of nested iteration schemes, which begin computation on level 1 and work their way to some level j, using each level k, $k < j$, both to generate an initial guess for level $k+1$ and for solving residual correction problems. Analysis of nested iteration algorithms in the context of this paper can be found in [12]; more traditional analyses can be found in [2], [7], [8], and [17].

In this paper, only correction schemes are considered. Define a k-level (*standard*) correction multigrid scheme by

ALGORITHM $\mathrm{MG}(k, \{\mu_\ell\}_{\ell=1}^{k}, x_k, f_k)$

(1) If $k = 1$, then solve $A_1 x_1 = f_1$ exactly or by smoothing

(2) If $k > 1$, then repeat $i = 1, \cdots, \mu_k$:

(2a) Smoothing: $x_k \leftarrow M_k^{(i)}(x_k, f_k)$

(2b) Residual Correction:

$$x_k \leftarrow x_k + P_{k-1}\ \mathrm{MG}(k-1, \{\mu_\ell\}_{\ell=1}^{k-1}, 0, R_k(A_k x_k + f_k))$$

(2c) Smoothing: $x_k \leftarrow N_k^{(i)}(x_k, f_k)$

(3) Return x_k

This definition requires that $\mu_1 = 1$. Steps (2a) and (2b) are sometimes referred to as *pre-smoothing* and *post-smoothing*, respectively, in the literature.

Symmetric multigrid schemes assume that $M_k = N_k$. *Nonsymmetric* multigrid schemes usually assume that $N_k = I$, where I is the identity. However, it is computationally more efficient to assume $M_k = I$ since the residual on level $k-1$ is f_{k-1} and does not need to be recomputed. Only rarely is the complete algorithm analyzed.

The standard V and W cycles correspond to Algorithm $\mathrm{MG}(j, \{1, \cdots, 1\}, \cdot, \cdot)$ and Algorithm $\mathrm{MG}(j, \{1, 2, \cdots, 2, 1\}, \cdot, \cdot)$, respectively (the definition of the W cycle frequently causes confusion). The F cycle [5] corresponds to something "in between" the V and W cycles.

3 A nonstandard multilevel formulation

In this section, a subtle change is made to Algorithm MG, which produces a simplified analysis for multigrid methods.

To make the notation of this section consistent, a fake (extra) level $j+1$ is introduced. Define

$$\mathcal{M}_{j+1} = \mathcal{M}_j, \qquad P_j = R_{j+1} = Q_{j+1} = I, \qquad A_{j+1} = A_j,$$

and the initial residual on level $j+1$, z_{j+1}, by

$$A_{j+1} x_{j+1}^{(-1)} + f = z_{j+1}.$$

This transforms the problem on all computational levels to one of solving a residual correction problem instead of the real problem on the finest grid and residual correction problems on the coarser grids.

Associated with each level k is a norm $\|\cdot\|_k$, which can be arbitrary. The norms can be different on each level, though the usefulness of this is unclear. For simplicity, the subscript from the norm symbol will be dropped.

Define a k-level (*nonstandard*) correction multigrid scheme using parameters z_{k+1} (the residual on level $k+1$ at some step) and $x_k^{(-1)}$ (the initial guess for level k, which is normally 0, except at the finest level) by

ALGORITHM NSMG$(k, z_{k+1}, x_k^{(-1)})$

(1) Initial residual: $R_{k+1}z_{k+1} \in \mathcal{M}_k$

(2) Pre-Smoothing: $x_k^{(0)} = M_k^{(1)} x_k^{(-1)}$ such that
$$A_k x_k^{(0)} + R_{k+1}z_{k+1} = z_k^{(0)}, \text{ where } \|z_k^{(0)}\| \le \rho_k^{(1)} \|z_{k+1}\|$$

(3) Let $\hat{x}_k^{(1)} = x_k^{(0)}$, $\hat{z}_k^{(1)} = z_k^{(0)}$, and $\gamma_1^{(1)} = 0$

(4) Repeat $i = 1, \cdots, \mu_k$

(4a) If $i > 1$, then

(4a1) Residual: $A_k x_k^{(i-1)} + R_{k+1}z_{k+1} = \hat{\theta}_k^{(i)}$

(4a2) Pre-Smoothing: $\hat{x}_k^{(i)} = M_k^{(i)} x_k^{(i-1)}$ such that
$$A_k \hat{x}_k^{(i)} + R_{k+1}z_{k+1} = \hat{z}_k^{(i)},$$
where
$$\|\hat{z}_k^{(i)}\| \le \rho_k^{(i)} \|\hat{\theta}_k^{(i)}\|$$

(4b) If $k > 1$, then

(4b1) Correction: $\gamma_k^{(i)} = P_{k-1}\bar{x}_{k-1}^{(i)}$, where
$$\bar{x}_{k-1}^{(i)} = \text{NSMG}(k-1, \hat{z}_k^{(i)}, 0)$$
and
$$A_{k-1}\bar{x}_{k-1}^{(i)} + R_k \hat{z}_k^{(i)} = \bar{z}_{k-1}^{(i)}$$

(4c) Residual: $A_k(\hat{x}_k^{(i)} + \gamma_k^{(i)}) + R_{k+1}z_{k+1} = \theta_k^{(i)}$

(4d) Post-Smoothing: $x_k^{(i)} = N_k^{(i)}(\hat{x}_k^{(i)} + \gamma_k^{(i)})$ such that
$$A_k x_k^{(i)} + R_{k+1}z_{k+1} = z_k^{(i)},$$
where
$$\|z_k^{(i)}\| \le \epsilon_k^{(i)} \|\theta_k^{(i)}\|$$

(5) Return $x_k^{(\mu_k)}$

Algorithm MG was defined in §2 in an intentionally imprecise manner. Algorithm NSMG is a precise, but nonstandard definition of Algorithm MG. The first smoothing reduces the norm of the residual on level k by a factor involving the norm of the residual on level $k+1$, which is nonstandard. For subsequent smoothings, this factor involves the norm of the residual on level k instead. The parameters $\{\mu_\ell\}$, which determine how many iterations of the multilevel algorithm to do on each level, can be considered either fixed or adaptively chosen during the course of computation.

Standard multigrid theory analyzes the case when a certain number of smoothing steps are used. This may be explicitly stated (e.g., [1]), or it may be phrased as to require the choice of a constant number of smoothing iterations such that some error reduction condition is satisfied (e.g., [7]). This is worst case analysis and rarely models the behavior seen in practice. However, it allows the proof of certain complexity results of optimal order.

The nonstandard formulation allows two interpretations of smoothing: first as the standard form, and second as fixing the factors $\epsilon_k^{(i)}$ and $\rho_k^{(i)}$ and letting the number of smoothing steps vary per iteration.

4 Analysis

In this section, assume that $\{\mathcal{M}_k\}$ is nested and analyze $z_j^{(i)}$ under minimal assumptions. Two flavors of analysis are considered. The first is a trivial analysis that should not be used when anything is really known about the problem. The second is an affine space decomposition analysis that is somewhat sharper than the first treatment.

The first result assumes only a simple property about each of the restrictions R_k: there exists a constant, $\delta_k \in \mathbb{R}$, such that

$$\|(I - Q_k^{-1} R_k)u\| \leq \delta_k \|u\|, \quad u \in \mathcal{M}_k. \tag{3}$$

Since normally $\dim(Range(Q_k^{-1})) < \dim(\mathcal{M}_k)$, $\delta_k \geq 1$. In many cases it is possible to choose norms for which $\delta_k = 1$ and which are meaningful for the underlying elliptic problem.

The problem is to determine conditions for $\{\rho_k^{(i)}, \epsilon_k^{(i)}\}$ in order to guarantee convergence of Algorithm NSMG. The results do not depend directly on properties of the A_k and f_k.

The basic theorem is as follows.

Theorem 1 *Assume that z_{j+1} is the residual on level $j+1 \geq 2$ and that the prolongation operators P_k, $1 \leq k \leq j$, are imbeddings and the inverse of the operator restrictions Q_k^{-1}, $2 \leq k \leq j+1$, are embeddings:*

$$P_k \equiv i_{\mathcal{M}_k \to \mathcal{M}_{k+1}} \quad \text{and} \quad Q_k^{-1} \equiv i_{\mathcal{M}_{k-1} \to \mathcal{M}_k}. \tag{4}$$

Let

$$E_1^{(1)} = \epsilon_1^{(1)} \rho_1^{(1)} \quad \text{and} \quad E_k^{(\mu_k)} = \prod_{i=1}^{\mu_k} \left(\epsilon_k^{(i)} \rho_k^{(i)} \left[\delta_k + E_{k-1}^{(\mu_{k-1})} \right] \right), \quad k > 1.$$

Then,

$$\|Q_j^{-1} z_j^{(\mu_j)}\| \leq E_j^{(\mu_j)} \|z_{j+1}\|.$$

The proof of Theorem 1 is a double induction argument and can be found in [12].

Remark 1 *In some instances, different restriction operators $R_k^{(i)}$ are used during a multigrid cycle. Substituting $\delta_k^{(i)}$ for δ_k covers this case.*

Remark 2 *For the V cycle with $\epsilon_j^{(i)} = \epsilon_j$ and $\rho_j^{(i)} = \rho_j$, $j = 1, \cdots, k$, the definition of $E_k^{(1)}$, $k > 1$, simplifies to*

$$E_k^{(1)} = \sum_{\ell=1}^{k} \left(\prod_{m=0}^{\ell-1} \epsilon_{k-m} \rho_{k-m} \right) \delta_{k-\ell} + \rho_1 \prod_{m=2}^{k} \epsilon_m \rho_m.$$

Remark 3 *When adaptively choosing when to change levels, the error term for the coarser level will be different each time a correction step is performed. Substituting $E_k^{(\mu_k^{(i)})}$ for $E_k^{(\mu_k)}$ covers this case.*

Remark 4 *For numerous problems, $\delta_k \geq 1$ guarantees that Theorem 1 is not sharp nor even realistic. See §5 for another interpretation of δ_k that is computationally useful since for specific residual vectors u in (3), δ_k can be much less than 1.*

Remark 5 *Many papers have been written analyzing multigrid using a variational point of view instead of an algebraic one. Rewrite (2) as*

$$\textit{find } u_k \in \mathcal{M}_k \textit{ such that } \quad a_k(u_k, v_k) + f_k(v_k) = 0, \quad \forall v_k \in \mathcal{M}_k.$$

Then Theorem 1 can be rewritten in a variational form.

Now consider an affine space analysis. Each space $\mathcal{M}_j$ is decomposed approximately into the parts which are corrected by the residual correction steps, and the parts which are relatively unaffected. This theory is considerably more complicated, but sharper than that in Theorem 1.

Each space $\mathcal{M}_j$ is assumed to be decomposable into a smooth part $\mathcal{S}_j$ and a rough part $\mathcal{T}_j$, e.g.,

$$\mathcal{M}_j = \mathcal{S}_j \oplus \mathcal{T}_j, \quad \text{where} \quad \mathcal{T}_j = \mathcal{M}_{j-1} \quad \text{and} \quad \mathcal{S}_j = \mathcal{M}_{j-1}^{\perp} \cap \mathcal{M}_j. \tag{5}$$

So, $\mathcal{S}_j$ contains the high frequency components and $\mathcal{T}_j$ contains the low frequency ones. Note that other definitions for $\mathcal{S}_j$ and $\mathcal{T}_j$ can be used.

Let $1 \leq k \leq j$. Assume that $v_k \in \mathcal{M}_k$. Let

$$|||v_k||| \equiv |||v_k|||_k \equiv \|v_k|_{\mathcal{S}_k}\|$$

and

$$< v_k > \equiv < v_k >_k \equiv \|v_k|_{\mathcal{T}_k}\|.$$

If v_k are w_k are the residuals before and after a post-smoothing iteration using N_k, and $\|w_k\|^2 = \epsilon_k^2\|v_k\|^2$, then there exist $\epsilon_{k,S}$ and $\epsilon_{k,T}$ such that

$$\|w_k\|^2 = \epsilon_{k,S}^2|||v_k|||^2 + \epsilon_{k,T}^2 < v_k >^2. \tag{6}$$

Similarly, if v_k are $\bar{w}_k$ are the residuals before and after a pre-smoothing iteration using M_k, and $\|\bar{w}_k\|^2 = \rho_k^2\|v_k\|^2$, then there exist $\rho_{k,SS}$, $\rho_{k,ST}$, $\rho_{k,TT}$, and $\rho_{k,TS}$ such that

$$\begin{aligned} |||\bar{w}_k|||^2 &= \rho_{k,SS}^2|||v_k|||^2 + \rho_{k,ST}^2 < v_k >^2 \quad \text{and} \\ < \bar{w}_k >^2 &= \rho_{k,TS}^2|||v_k|||^2 + \rho_{k,TT}^2 < v_k >^2. \end{aligned} \tag{7}$$

As was noted at the end of §3, these parameters will probably only be bounded with estimates of some form.

The result here requires more precise knowledge than (3), namely that for any $u \in \mathcal{M}_k$, there exist constants $\delta_{k,S}$ and $\delta_{k,T} \in \mathbb{R}$ such that

$$|||(I - Q_k^{-1}R_k)u|||^2 \le \delta_{k,S}^2|||u|||^2 \quad \text{and} \quad < (I - Q_k^{-1}R_k)u >^2 \le \delta_{k,T}^2 < u >^2.$$

The problem is to determine conditions for $\{\rho_{k,XY}^{(i)}, \epsilon_{k,X}^{(i)}\}$, $X, Y \in \{S, T\}$, in order to guarantee convergence of Algorithm NSMG. As before, the results do not depend directly on properties of the A_k and f_k.

A sharper convergence result than Theorem 1 is as follows.

Theorem 2 *Assume that z_{j+1} is the residual on level $j+1 \ge 2$ and that P_k, $1 \le k \le j$, and Q_k^{-1}, $2 \le k \le j+1$, satisfy (4). Let*

$$E_1^{(1)} = \epsilon_{1,S}^{(1)}\rho_{1,S}^{(1)} \equiv E_{1,SS}^{(1)} \quad \textit{and} \quad E_{1,ST}^{(1)} = E_{1,TS}^{(1)} = E_{1,TT}^{(1)} = 0.$$

For $1 < k \le j$, let

$$E_{k,SS}^{(i)} = \epsilon_{k,S}^{(i)}\left[\left(\delta_{k,S} + E_{k-1,SS}^{(\mu_{k-1})}\right)\rho_{k,SS}^{(i)} + E_{k-1,ST}^{(\mu_{k-1})}\rho_{k,ST}^{(i)}\right],$$

$$E_{k,TS}^{(i)} = \epsilon_{k,T}^{(i)}\left[\left(\delta_{k,T} + E_{k-1,TT}^{(\mu_{k-1})}\right)\rho_{k,TS}^{(i)} + E_{k-1,TS}^{(\mu_{k-1})})\rho_{k,SS}^{(i)}\right],$$

$$E_{k,TT}^{(i)} = \epsilon_{k,T}^{(i)}\left[\left(\delta_{k,T} + E_{k-1,TT}^{(\mu_{k-1})}\right)\rho_{k,TT}^{(i)} + E_{k-1,TS}^{(\mu_{k-1})})\rho_{k,ST}^{(i)}\right],$$

and

$$E_{k,ST}^{(i)} = \epsilon_{k,S}^{(i)}\left[\left(\delta_{k,S} + E_{k-1,SS}^{(\mu_{k-1})}\right)\rho_{k,ST}^{(i)} + E_{k-1,ST}^{(\mu_{k-1})})\rho_{k,TT}^{(i)}\right].$$

Then,

$$\|Q_j^{-1}z_j^{(\mu_j)}\| \le \prod_{i=1}^{\mu_j} \max\left\{E_{j,SS}^{(i)} + E_{j,TS}^{(i)},\ E_{j,ST}^{(i)} + E_{j,TT}^{(i)}\right\}\|z_{j+1}\|. \tag{8}$$

The proof of (8) is a double induction argument and can be found in [12].

Remark 6 *For a symmetric multilevel algorithm (see §2), all of the terms in Theorem 2 exist. It is possible to to see that whenever an individual term is large, there is another term multiplying it that is small.*

Remark 7 *For nonsymmetric multilevel algorithms, the expressions simplify since some of the individual terms are either* 0 *or* 1.

Remark 8 *For simple enough the* $\delta_{j,T} \approx 0$ *and* $\delta_{j,S} \approx 1$.

Special care is required when using this theory since it is, in some sense, too general. It is quite easy to calculate various terms in the two theorems using incompatible norms, resulting in nonsensical results.

5 Examples

In this section, δ_k is computed for several examples. The first is for Dirichlet problems on $\mathbb{R}^2$ with simple, but not entirely trivial meshes. While the estimates are rather pessimistic, some advice is offered on practical uses of the simple theory in §4. Next, an example is presented where Theorem 1 is sharp. Finally, two problems arising in attempting to numerically simulate flames are examined.

Assume that for each k, $k = 1, \cdots, j$, the spaces $\mathcal{M}_k$ has a bilinear hat function basis over uniform squares of side length h_k. This does not imply that the domain Ω is either rectangular or convex, just polygonal (possibly with holes) with boundary segments either parallel to the axes or inclined 45° to the axes (which requires appropriate modifications to some of the basis functions).

Set

$$D_{ij} = \{(i+1,j),(i-1,j),(i,j+1),(i,j-1)\}$$

and

$$\hat{D}_{ij} = \{(i+1,j+1),(i+1,j-1),(i-1,j+1),(i-1,j-1)\}.$$

Let ${R_k}^{(9)} v_{ij}$ be the following weighted sum of v_{ij} and its eight neighbors from level k:

$${R_k}^{(9)} v_{ij} = \frac{1}{4}\left[v_{ij} + \frac{1}{2}\sum_{(k,\ell)\in D_{ij}} v_{k\ell} + \frac{1}{4}\sum_{(k,\ell)\in \hat{D}_{ij}} v_{k\ell}\right]$$

We approximate $\delta_k^{(9)} = \delta_k({R_k}^{(9)})$ using a piecewise bilinear hat function v on level $k-1$ which is centered at some point $(i+1, j+1)$ on level k. Note that, if $v_{ij} = (-1)^{i+j}$, then $R_k v_{ij} = 0$ at any interior point of the $(k-1)$-grid. Thus, $\delta_k \geq 1$; since R_k satisfies a maximum principal, it then follows that

$$\|(I - Q_k^{-1}{R_k}^{(9)})v\|_{\ell^\infty} \leq \|v\|_{\ell^\infty}$$

and that

$$\delta_k^{(9)} = 1.$$

Let ${R_k}^{(5)} v_{ij}$ be the following weighted sum of v_{ij} and its four neighbors from level k:

$${R_k}^{(5)} v_{ij} = \frac{1}{4} \left[2v_{ij} + \frac{1}{2} \sum_{(k,\ell) \in D_{ij}} v_{k\ell} \right]$$

Again, the same argument shows that, with respect to the ℓ^∞,

$$\delta_k^{(5)} = 1.$$

If there are boundary elements associated with the edges at 45° to the axes, ${R_k}^{(9)}$ and ${R_k}^{(5)}$ can be mixed to form R_k.

Besides motivating the affine space analysis, the theory of this section can actually be used in computer programs to adaptively change the parameter choices on coarser levels k (μ_k and the number of iterations in the smoothers). Consider Laplace's equation on the unit interval, two levels, a uniform mesh, a central difference discretization, linear interpolation and projection, and one Jacobi iteration as the smoother. Sharp theory says that the convergence rate is bounded by 0.5. In a strictly nonrigorous exercise, 5000 randomly chosen problems were generated. In theory, $\delta_2^{(3)} = 1$, where $\delta_2^{(3)}$ is derived using a three point restriction operator R_2. However, for individual residual vectors v, the following was calculated:

$$\delta(v) = \frac{\|(I - Q_2^{-1} R_2)v\|}{\|v\|}.$$

The following was observed.

Statistic	$\delta(v)$
Minimum	0.3444
Maximum	0.9312
Average	0.7126

Further, there was a direct correlation between the size of the estimated $\delta(v)$ and the actual error reduction produced by one multigrid iteration.

Now consider the affine space analysis. Assume that only post smoothing is performed; this causes many of the terms in Theorem 2 to be either 1 or 0. In this case, Theorem 2 predicts that the convergence rate is bounded by 0.5, which is sharp. Unfortunately, Theorem 2 predicts an overly pessimistic convergence rate when two post smoothing steps are used (c.f., [1] which gets the right bound in both cases).

For some problems, multigrid with particular smoothers is known to be a terrible method. For example, let $q \geq 5$ in

$$\begin{cases} -10^q u_{xx} - 10^{-q} u_{yy} & = f \text{ in } (0,1)^2, \\ u & = 0 \text{ on } \partial(0,1)^2, \end{cases}$$

and choose a central difference discretization on a uniform mesh and Jacobi as the smoother. Then the coarse grid corrections do not necessarily improve the approximation to the solution. In this case, Theorem 1 actually is sharp. (The fix to making multigrid work well for this problem is to use either a line relaxation or a conjugate gradient method as the smoother or rougher.)

The examples given so far were not the of interest to the authors of [12] when this theory was developed, however. Two problems which are currently being studied arise in numerical simulation of flames. These are complicated nonlinear coupled partial differential equations which are amenable to solution by multigrid methods provided that the right solvers are used on each level. The first is a flame sheet model (see [13]) while the second is a laminar, axisymmetric diffusion flame model (see [16]).

In the flame sheet model, the chemical reactions are described with a single one step irreversible reaction corresponding to infinitely fast conversion of reactants into stable products. This reaction is assumed to be limited to a very thin exothermic reaction zone located at the locus of stoichiometric mixing of fuel and oxidizer, where temperature and products of combustion are maximized. To further simplify the governing equations, one neglects thermal diffusion effects, assumes constant heat capacities and Fick's law for the ordinary mass diffusion velocities, and takes all the Lewis numbers equal to unity. With these approximations, the energy equation and the major species equations take on the same mathematical form and by introducing Schvab-Zeldovich variables, one can derive a source free convective-diffusive equation for a single conserved scalar. Although no information can be recovered about minor or intermediate species in the flame sheet limit, the temperature and the stable major species profiles in the system can be obtained from the solution of the conserved scalar equation coupled to the flow field equations. Further, the location of the physical spatially distributed reaction zone and its temperature distribution can be adequately predicted by the flame sheet model for many important fuel-oxidizer combinations and configurations. Since being studied as a means of obtaining an approximate solution to use as an initial iterate for a one dimensional detailed kinetics computation in [19], flame sheets have been routinely employed to initialize multidimensional diffusion flames.

A schematic of the physical configuration is given in Figure 1 (though not drawn to scale). It consists of an inner cylindrical fuel jet (radius R_I =0.2cm), an outer co-flowing annular oxidizer jet (radius R_O =2.5cm) and a dead zone extending to R_{max} =7.5cm. The inlet velocity profile of the fuel and oxidizer are a plug flow of 35cm/s. This yields a typical value for the Reynolds number of 550. Further, the flame length is approximately L_f =3cm and the length of the

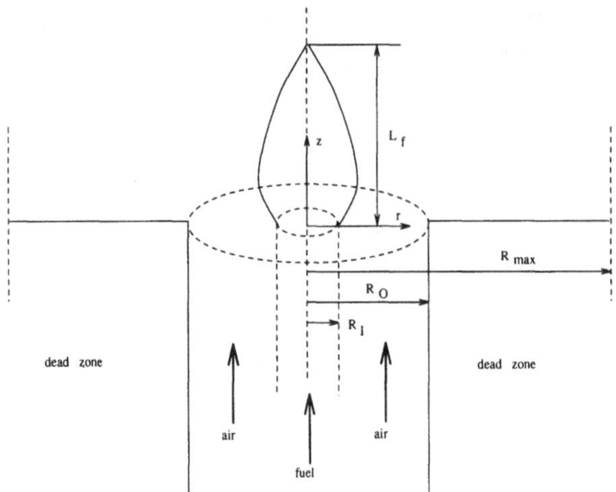

FIGURE 1. Flame sheet physical configuration

computational domain is set to L =30cm. Although the fuel and oxidizer reservoirs are at room temperature (300°Kelvin), we need to assume, in the flame sheet model, that the temperature already reaches the peak temperature value along the inlet boundary at $r = R_I$. This peak temperature is estimated for a methane-air configuration to be 2050°K. Hence, the inlet profile of the conserved scalar, $S^0(r)$, is specified in such a way that the resulting temperature distribution blends the room temperature reservoirs and the peak temperature by means of a narrow Gaussian centered at R_I. The narrowness of the Gaussian profile has a relevant influence on the calculated flame length, so that its parameters have to be determined appropriately.

A damped Newton multilevel solver is used (see [3] and [18]). Due to the model used, nonstaggered grids can be used, though they are tensor product grids with quite variable mesh spacings. The linear problems solved on each level are 36 point operators. We found that GMRES with a Gauss-Seidel preconditioner was a very good solver for each level. The code uses a left preconditioned residual norm to determine when the solutions are adequate. In calculating $\delta_k^{(i)}$ in this norm, we found it to be in the interval $[10^6, 10^8]$ frequently. This required that the ϵ's and ρ's be quite small in order to achieve convergence. However, $\delta_k^{(i)} \ll \|z_{k+1}\|$ so that this is not really an imposition. Even so, we saw speed ups of a factor of 10.5 on an IBM RS6000-560 workstation over the unigrid solution approach (see [13]).

While $\delta_k^{(i)}$ was reduced dramatically by using a semi-coarsening approach, the overall run time increased by 50% over the traditional multigrid approach.

We used a damped Newton multilevel approach instead of a full approximation scheme (see [20]) because experiments us to believe that in the full chemistry case, FAS will be too expensive.

The second flame numerical simulation is of a laminar, axisymmetric, methane-air diffusion flame using nonlinear damped Newton multigrid (see [16]). The physical configuration is based on an inner cylindrical fuel stream surrounded by a coflowing oxidizer jet and the inlet velocities are high enough to produce a lifted flame with a triple flame ring structure at its base. Computationally, we solve the total mass, momentum, energy, and species conservation equations with complex transport and finite rate chemistry submodels. The velocity field is predicted using a vorticity-velocity formulation and the governing partial differential equations are discretized on a nonstaggered grid. The numerical solution involves a pseudo transient process and a steady-state Newton iteration combined with nonlinear damped Newton multigrid. Coarse grid information is used to provide initial starting estimates for the Newton iteration on the finest level and also to form correction problems, thus yielding significant savings in the execution times.

The physical configuration consists of an inner methane-nitrogen jet (with radius 0.2cm), an air coflow (with radius 2.5cm), and the computational domain is $[0, 7.5] \times [0, 30]$ (all units are centimeters). The temperature and species mass fractions values for the surrounding air are the same as the ones for the dead zone. This physical configuration was chosen because experimental data and a numerical solution using primitive variables were already available for this problem.

Once again, a variable width tensor product set of grids was used. Due to the high number of chemical species in the calculation, the discrete Jacobians were 270 point operators. In the left preconditioned norm, $\delta_k^{(i)}$ was frequently in the interval $[10^6, 10^{10}]$. However, $\delta_k^{(i)} \ll \|z_{k+1}\|$ so that this is not really an imposition. Still, a factor of 9.7 speed up was achieved on a 57×73 fine grid over a unigrid approach. In this example, $\delta_k^{(i)}$ was not reduced dramatically by using a semi-coarsening approach.

6 Multiple coarse grid methods

In [12], the Theorems 1 and 2 are extended to a multiple coarse space model. In this case, there are multiple δ's for each level, the quantity depending on the number of coarse level correction problems that are associated with each level.

While the theorems of §4 may not be satisfactory for simple problems, the multiple coarse space theory is for these problems. This style of analysis is much more accurate due to the fact that we can show that the δ's can be quite small, including being 0 for the case of the domain reduction method (see [9], [14], and [15]).

7 Conclusions

It is possible to prove a convergence result for multigrid and aggregation-disaggregation methods with minimal knowledge about the problem. By treating multigrid as a simple iterative method, almost nothing needs to be known about the grids, solution spaces, linear systems of equations, iterative methods used as smoothers (or roughers), restriction and prolongation operators, or the norms used on each level.

Being able to prove such a result is much easier than showing that it is useful all of the time. In fact, this theory is normally not sharp enough to satisfy theoreticians. It should be used in computational settings in which almost nothing is known about the convergence rate a priori.

One of the advantages of this theory is that all of the parameters are available during execution of a computer program. Hence, adaptively changing levels can be achieved with certainty of success.

Code availability

A series of codes, Madpack (see [11] and its references), are available from MGNet [10] which are compatible with the philosophy applied here and with the earlier theory in [8].

Acknowledgements

I am indebted to Professor Jim Douglas, Jr., Alexandre Ern, and Professor Mitchell Smooke for helpful discussions.

References

[1] R. E. Bank and C. C. Douglas. Sharp estimates for multigrid rates of convergence with general smoothing and acceleration. *SIAM J. Numer. Anal.*, 22:617–633, 1985.

[2] R. E. Bank and T. Dupont. An optimal order process for solving elliptic finite element equations. *Math. Comp.*, 36:35–51, 1981.

[3] R. E. Bank and D. J. Rose. Analysis of a multilevel iterative method for nonlinear finite element equations. *Math. Comp.*, 39:453–465, 1982.

[4] A. Brandt. Multi–level adaptive solution to boundary–value problems. *Math. Comp.*, 31:333–390, 1977.

[5] A. Brandt. Guide to multigrid development. In W. Hackbusch and U. Trottenberg, editors, *Multigrid Methods*, pages 220–312. Springer–Verlag, New York, 1982.

[6] F. Chatelin and W. L. Miranker. Acceleration by aggregation of successive approximation methods. *Lin. Alg. Appl.*, 43:17–47, 1982.

[7] C. C. Douglas. *Multi–grid algorithms for elliptic boundary–value problems.* PhD thesis, Yale University, May 1982. Also, Computer Science Department, Yale University, Technical Report 223.

[8] C. C. Douglas. Multi–grid algorithms with applications to elliptic boundary–value problems. *SIAM J. Numer. Anal.*, 21:236–254, 1984.

[9] C. C. Douglas. A tupleware approach to domain decomposition methods. *Appl. Numer. Math.*, 8:353–373, 1991.

[10] C. C. Douglas. MGNet: a multigrid and domain decomposition network. *ACM SIGNUM Newsletter*, 27:2–8, 1992.

[11] C. C. Douglas. Implementing abstract multigrid or multilevel methods. Technical Report YALEU/DCS/TR-952, Department of Computer Science, Yale University, New Haven, 1993. In the Proceedings of the Sixth Copper Mountain Conference on Multigrid Methods, N. D. Melson (ed.), NASA CP 3224, Langley, VA, 1993, pp. 127-141.

[12] C. C. Douglas and J. Douglas. A unified convergence theory for abstract multigrid or multilevel algorithms, serial and parallel. *SIAM J. Numer. Anal.*, 30:136–158, 1993.

[13] C. C. Douglas and A. Ern. Numerical solution of flame sheet problems with and without multigrid methods. *submitted to Advances in Comp. Math.*, 1993. Also available as Yale University Department of Computer Science Report YALEU/DCS/TR-955, New Haven, CT, 1993.

[14] C. C. Douglas and J. Mandel. A group theoretic approach to the domain reduction method. *Computing*, 48:73–96, 1992.

[15] C. C. Douglas and W. L. Miranker. Constructive interference in parallel algorithms. *SIAM J. Numer. Anal.*, 25:376–398, 1988.

[16] A. Ern, C. C. Douglas, and M. D. Smooke. Numerical simulation of laminar diffusion flames with multigrid methods. Technical Report YALEU/DME/TR-xxx, Department of Mechanical Engineering, Yale University, New Haven, 1993. In preparation.

[17] W. Hackbusch. *Multigrid Methods and Applications.* Springer–Verlag, Berlin, 1985.

[18] W. Hackbusch and A. Reusken. Analysis of a damped nonlinear multilevel method. *Numer. Math.*, 55:225–246, 1989.

[19] D. E. Keyes and M. D. Smooke. Flame sheet starting estimates for counterflow diffusion flame problems. *J. Comput. Phys.*, 73:267–288, 1987.

[20] C. Liu, Z. Liu, and S. F. McCormick. Multigrid methods for numerical simulation of laminar diffusion flames. *AIAA*, 93-0236:1–11, 1993.

[21] P. Wesseling. *An Introduction to Multigrid Methods.* John Wiley & Sons, Chichester, 1992.

3

Turbulence Modelling as a Multi-Level Approach

L. Fuchs[1]

ABSTRACT Large Eddy Simulation (LES) of turbulent incompressible flows is shown to be directly related to the concept of multi-level approximation of a differential problem on a sequence of successively refined grids. We discuss how a "Dynamic"-LES (D-LES) turbulence model can be modified so that it fits naturally into an adaptive Multi-Grid scheme. The modified model assumes that the functional form of relationship between the (time- and/or space-) averaged second moments and the averaged variables, is grid independent. This assumption, and the other assumptions on the averaging operator, are satisfied also by several artificial viscosity forms.

1 Introduction

In many situations the fine structures of a flow field are not of interest. In most cases, the level of spatial and temporal resolution, that is of practical interest is by several orders of magnitudes larger than the Kolmogorov (i.e. smallest) scales of the turbulent field. Thus, we face a rather intricate situations: we need information only on the "large eddies", but these depend also on the smaller ones that we are completely uninterested in. It is interesting to note that these small scales, cannot be computed anyway for engineering problems, due to the enormous (and non existing) computational power required for such calculations. Thus, the natural approach has always been to use a "model" that functionally simulates the interaction between the fine- (unresolved) and the large- (resolved) structures of the flow field. There are several "main-stream" models for this purpose: Eddy viscosity type models (using commonly the two equation, e.g. $k - \epsilon$ model), the Reynolds Stress Model (RSM) and models that use the so called Large Eddy Simulation (LES) approach. The eddy-viscosity concept is introduced into the LES approach, by estimating an equivalent turbulent viscosity from the resolved field (see below). A review of these different models, their background and applicability, can be found in many papers (c.f. [1-5]).

[1]Department of Mechanics/Applied CFD,
Royal Institute of Technology, S-100 44 Stockholm

This paper considers the so called "Dynamic LES" turbulence model. In this model one does not assume a "calibrated" value of a model paramter, but rather computes the modal parameter locally in each instant (time step). To be able to do so, one has to assume that the form of the model is filter independent. If the spatial filter that is used, is a local polynomial (similar to the one used to approximate the derivatives by finite-differences), one may compute the model "constant" from a double filtered relation, that is equivalent to the defect that is computed in the Multi-Grid (MG) process. We also discuss how one may simplify the "eddy viscosity" model, by applying the basic LES assumption on the generalized second moments of the mean components of the velocity vector. Furthermore, we point out how the basic LES assumption can be integrated into an adaptive MG solver.

2 Averaged Equations

Consider incompressible flows:

$$u_{k,k} = 0 \; , \tag{1}$$

$$u_{i,t} + (u_i u_k)_{,k} = -p_{,i} + \sigma_{ik,k} \; , \tag{2}$$

where

$$\sigma_{i,j} = 2\nu s_{i,j} \quad ; \quad s_{i,j} = \frac{1}{2}(u_{i,j} + u_{j,i}) \; . \tag{3}$$

These equations can be solved in principle, numerically. Accurate numerical solutions require that all spatial and temporal scales are resolved. Even if we neglect the potential difficulties due to solution multiplicity and bifurcations, one still cannot solve the discrete system with adequate resolution, for Reynolds numbers (Re = U L $/\nu$), that are not small enough. Furthermore, for most practical applications the fine details of the flow are not of interest and it suffices to obtain time-averaged values of the dependent variables at some discrete set of points in space. Thus, for these reasons it is natural to seek the space- and/or time-averaged values of dependent variables. Averaging the governing equation is not new and in fact it has already been carried out by Reynolds about hundred years ago. Incidentally, Reynolds considered the space averaged quantities and not the time-averaged quantities that one often associates with "Reynolds averages". In general terms, Germano [6] carried out a formal averaging of the Navier-Stokes equations. The highlights of this derivations are given here for the sake of discussion.

Consider averaging dependent variables:

$$< u_i(\mathbf{x},t) >_{l,\theta} = \int u_i(\mathbf{x}',t') \; G(\mathbf{x}-\mathbf{x}',t-t';l,\theta) \; d\mathbf{x}' \; dt' \; , \tag{4}$$

with

$$\int G(\mathbf{x}-\mathbf{x}',t-t';l,\theta) \; d\mathbf{x}' \; dt' = 1 \; , \tag{5}$$

l and θ are characteristic space- and time-scales which are the smallest to be of interest. Examples for such averaging are pure time-averages (often called "Reynolds averages"), where one "filters-out" all smaller time-scales. Local spatial averaging (a "hat-function" or a Gaussian filter) is a corresponding averaging that is used in so called "Large Eddy Simulations (LES)" [3-5]. The classical decomposition of the dependent variables into the sum of the average and a fluctuating component, is not always convenient. This is so whenever the average of the fluctuations and the average of the mean times the fluctuations do not vanish. When such averaging is applied to the Navier-Stokes equations, one ends-up with additional (cross-correlation terms) that are unknown and therefore such averaging does not contribute to the solution of the equations. If on the other hand, the above mentioned averages vanish, one ends-up with the "Reynolds averaged" Navier-Stokes equations. These equations are identical to the non-averaged equation with an additional term in the momentum equations (the divergence of the "Reynolds-stresses"). Germano [6] generalized this concept by requiring the averages (spatial and/or temporal) have the following properties:

$$< f + g >=< f > + < g > , \tag{6}$$

$$< \alpha f >= \alpha < f > \quad \text{for } \alpha = \text{ constant} , \tag{7}$$

and

$$< f_{,t} >=< f >_{,t} \ ; \ < f_{,k} >=< f >_{,k} \ . \tag{8}$$

One may define a "generalized central moments" (second and third, respectively) by:

$$\tau(f, g) =< fg > - < f >< g > , \tag{9}$$

$$\tau(f, g, h) \ = \ $$

$$< fgh > - < f > \tau(g, h) - < g > \tau(f, h) - < h > \tau(f, g) - < f >< g >< h > \ . \tag{10}$$

By applying the average $< * >$ to the Naver-Stokes equations one obtains:

$$< u_k >_{,k} = 0 , \tag{11}$$

$$< u_i >_{,t} + (< u_i >< u_k >)_{,k} = - < p_{,i} > + < \sigma_{ik} >_{,k} - [\tau(u_i, u_k)]_{,k} \ . \tag{12}$$

This definition of $\tau(u_i, u_j)$ leads to "averaging invariance" of the Navier-Stokes equations. In this formulation, one still has to compute $\tau(u_i, u_j)$, or otherwise model it. One may compute $\tau(u_i, u_j)$ by deriving transport equations for the second moments. However, these equations contain in turn third-moments. This "closure" problem is the prime difficulty in modelling turbulent flows.

Our aim here is to show how certain models (the one that we refer to here as "Dynamic Large Eddy Simulations - D-LES) can be related to a hierarchical modelling and how this model can be consider as a natural element of a multi-level approach.

3 Turbulence Modelling via Eddy-Viscosity

The closure problem of turbulence has been handled traditionally by different models. An often used model is the concept that the "Reynolds stress" can be modeled as being proportional to the rate of shear of the averaged flow field $< s_{i,j} >$. This relation is also known as Boussinesq's assumption. The underlying assumption is that turbulence can be modeled by an equivalent viscosity (the "eddy viscosity"); typically:

$$\tau_{i,j} - \frac{1}{3}\delta_{i,j}\tau_{k,k} = -2\nu_t < s_{i,j} > , \tag{13}$$

where

$$\tau_{i,j} \equiv \tau_\Delta(u_i, u_j) . \tag{14}$$

Δ is the characteristic (spatial) length and $< * >$ is the corresponding space averaging. By dimensional analysis, and by assuming a balance between production and dissipation of turbulent fluctuating energy, one finds that the eddy-viscosity is proportional to the square of the characteristic length-scale. A typical eddy-viscosity model based upon spatial averaging is the Smagorinsky [3-5] model:

$$\nu_t = c\Delta^2 (2 < s_{l,m} > < s_{l,m} >)^{\frac{1}{2}} . \tag{15}$$

The model "constant" c in the Smagorisky model, is unfortunately not universal and even it is not a constant within a given flow field. Thus, the numerical value of c has to be "calibrated" for each type of problem. An alternative approach has been proposed recently [6-13]. In this approach one leaves the numerical value of the constant, as a problem dependent parameter to be computed as part of the solution. Furthermore, the value of c may vary both in space and time.

In the following we assume that $< * >$ corresponds to a polynomial averaging. We use the following notation:

$$< f >_\Delta \equiv \overline{f} \quad \text{and} \quad < f >_{\widehat{\Delta}} \equiv \widehat{f} , \tag{16}$$

Then, by definition

$$\tau_{i,j} = \overline{u_i u_j} - \overline{u_i}\,\overline{u_j} , \tag{17}$$

Smagorisky's model can be written as:

$$\tau_{i,j} - \frac{1}{3}\delta_{i,j}\tau_{k,k} = -2\nu_t \overline{s_{i,j}} \quad ; \quad ; \nu_t = c\Delta^2 |\overline{s_{i,j}}| . \tag{18}$$

The main point of the so called *"dynamic"* LES model is that one assumes that the same functional relation exists between the second central moment and the rate of shear of the averaged velocity field. The form of this relation should be *independent* of the filter size Δ. Thus, we apply a second filter (with $l = \widehat{\Delta}$) after the first filter ($l = \Delta$)

Denote the generalized turbulent stress by $T_{i,j}$ as:

$$T_{i,j} = \widehat{\overline{u_i u_j}} - \widehat{\overline{u_i}}\,\widehat{\overline{u_j}}\ , \tag{19}$$

Germano [G2] noticed the following identity:

$$L_{i,j} = T_{i,j} - \widehat{\tau_{i,j}} = \widehat{\overline{u_i}\,\overline{u_j}} - \widehat{\overline{u_i}}\,\widehat{\overline{u_j}}\ , \tag{20}$$

or by using the eddy viscosity model:

$$L_{i,j} - \frac{1}{3}\delta_{i,j}L_{k,k} = -2c\widehat{\Delta}^2|\widehat{\overline{s_{i,j}}}|\,\widehat{\overline{s_{i,j}}} + 2c\Delta^2\widehat{|\overline{s_{i,j}}|\,\overline{s_{i,j}}}\ . \tag{21}$$

It is evident that $L_{i,j}$,$\overline{s_{i,j}}$ and $\widehat{\overline{s_{i,j}}}$ can be computed explicitly. Since all components in (21) above are known, with the exception of c, the model parameter can be computed locally. However, since (21) is a tensorial relation with six independent components (symmetric tensor), one has to make further assumptions. One may assume that c itself is a symmetric tensor, or else to assume a scalar approximation (times the identity tensor) to this tensor. In recent two years several approaches have been reported: Germano et al [8] used an approximate local value of c which was later averaged to filter out high frequency fluctuations in c. Lilly [10] has suggested a least square approximation of the six components:

$$c\Delta^2 = -\frac{L_{m,n}M_{m,n}}{M_{k,l}M_{k,l}}\ , \tag{22}$$

where

$$M_{m,n} = 2\left\{(\frac{\widehat{\Delta}}{\Delta})^2|\widehat{\overline{s_{i,j}}}|\,\widehat{\overline{s_{i,j}}} - \widehat{|\overline{s_{i,j}}|\,\overline{s_{i,j}}}\right\}\ . \tag{23}$$

From the relation above it is obvious that c may vary in space and time. It is also clear that the dissipation rate (of fluctuation energy) is proportional to c. Negative values of c imply that energy is being transferred into the system. This effect is called "back-scatter" since normally energy is transferred from the large scales to the small ones, where they dissipate. In the back-scatter mechanism, energy is being transferred from the small scales to the large scales. This back-scatter process is now accepted as being physically valid [7]. A third important property of the model is that c vanishes for laminar flows. Thus, the averaged equations become identical to the original equations when the flow is not turbulent. This property is not found in most other turbulence models that require an external (often manual) switching mechanism to turn the turbulence model on and off.

It should be noted that the "eddy-viscosity" model is not the only one that can be derived by pure dimensional considerations. It is easy to show that very conventional terms that are often called (or implicitly act as) artificial viscosity, may function as closure models for the second moments. Such terms can often be expressed as:

$$[\tau_{i,j}]_{,j} = c\Delta^n|u_k|\partial_k\nabla^n u_i\ . \tag{24}$$

With $n = 1$ one obtains the classical upwind scheme (or the "hybrid" scheme for larger cell Re, when combined with central differences). For $n = 3$ one recovers Kuwahara's [14] third order upwind scheme (a non-linear form of a fourth order artificial viscosity as is often used in compressible flow calculations; c.f. Jameson [16]).

4 Multi-Level Adaptive Modelling

As noticed above, one has to apply at least two filtering functions. From the two filtered values, one may deduce the local value of the second moments (provided that the functional behavior of the moments is the same for both filters). This is exactly the way we compute the coarse grid solution in the Multi-Grid (MG) process. The defect that is added to the governing equation on a coarser grid, is the difference in the residual on the coarse grid (double filtered function; corresponding to $T_{i,j}$) and the filtered residual on the finer grid (corresponding to $\widehat{\tau_{i,j}}$). That is, the defect is corresponding to $L_{i,j}$ in (21). Thus, turbulence modelling can be considered as solving the differential equation on a rather coarse grid with a local mean value at the computational points. In terms of MG process this means that one has to correct the equation by a defect as is usually done in the MG process. Thus, the LES-process described in the previous section is nothing else but a standard coarse grid solution with a so called τ-correction. Bearing this in mind, one may now skip altogether the need to rely on the "eddy viscosity" concept. Instead one may compute directly the divergence of the second moment tensor ($[\tau_{i,j}]_{,j}$).

With this interpretation one may compute directly the defect

$$(L_{i,j})_{,j} = c\Delta^2 g_i \ , \tag{25}$$

where g_i is the divergence of the difference in the second moments in (20). To compute c one may use a root-mean-square fitting, yielding simply that

$$c\Delta^2 = (L_{i,j})_{,j} g_i / g_i g_i \ . \tag{26}$$

The only possible source for difficulty here, is when $g_i \equiv 0$; In such cases however, one has also that $c = 0$. This type of situation occurs for laminar regions in the flow field!

The suggested approach has another advantage over the "eddy viscosity" model, when the model coefficient, c, varies in space. Large spatial variations cause large "spikes" in the defect and requires therefore smoothing as is done by Germano at al [8]. With this type of "dynamic-LES" it is natural to consider artificial viscosity as a "high-frequency spatial filter". The second moments can then be identified as the (explicit) high-order artificial viscosity terms. This would also supports the observations of Kuwahara that LES-like results are obtained without adding an explicit Sub-Grid-Scale (SGS) model [14,15]. A more closer comparison of the above mentioned D-LES models (based on the "eddy viscosity" assumption or on the defect of the second moments) is given in [13].

The other issue associated with the above described procedure is the question when can the model be applied. That is, one has to ensure that the underlying assumption is valid. This happens when the spatial resolution is high enough so that the smallest resolved scales belong to the "inertial sub-range" of the turbulent spectrum. This scale is not known in advance, though it may be estimated (to its order of magnitude). For practical purposes one should rely on adaptive methods, that will introduce (local) grid refinements as long as the asymptotic behavior of the model is not reached. As MG processes yield data required for adaptive refinements as part of the solution procedure, it is natural that the D-LES scheme should be tightly connected to a MG-process.

5 Concluding Remarks

The Multi-Level philosophy can be directly extended to the modelling of turbulent flows in the frame work of Large Eddy Simulations (LES). The basic "dynamic" LES turns out to be identical to the calculation of the defects in the MG procedure. The main issue of this type of modelling is whether or not one may assume the same functional relationship, used for modelling the second moments on the different grids. This assumption is presumably correct once one has entered to the "inertial subrange" of the turbulent spectrum. To ensure this, one has to use an "adaptive" scheme that refines the spatial discretization, locally, to the level required by the particular flow. All these aspects fit naturally into the concept of Multi-level computation of turbulence.

References

[1] Rogallo, R.S. and Moin, P.: "Numerical simulation of turbulent flows", *Annual Review Fluid Mechanics*, **16**, pp. 99-137, 1984.

[2] Speziale, C.G.: "Analytical methods for the development of Reynolds-Stress closures in turbulence",*Annual Review Fluid Mechanics*, **23**, pp. 107-157, 1991.

[3] Smagorinsky J., "General Circulation Experiments with the Primitive Equations". *Mon. Weather Rev.*, vol **91**, No 3, pp.99, March 1963.

[4] Deardorff J.W., "A Numerical study of three-Dimensional Turbulent Channel Flow at Large Reynolds Numbers" *J.Fluid Mech.*, vol **41**, part 2, pp. 453, 1970.

[5] Bardina, J., "Improved Subgrid scale Models for Large Eddy Simulation". AIAA Paper No. 80-1357, 1980.

[6] Germano, M. "Turbulence; the filtering approach". *J. Fluid Mech.*, vol **238**, pp. 325, 1992.

[7] Piomelli, U., "Subgrid-scale backscatter in turbulent and transient flows". *Phys. Fluids A*, vol **3**, No 7, pp. 1766, July 1991

[8] Germano, M., Piomelli, U., Moin, P. & Cabot, W., "A Dynamic Subgrid-Scale Eddy Viscosity Model".*Phys. Fluids A*, vol **3**, No 7, pp. 1760, July 1991.

[9] Germano, M., Piomelli, U., Moin, P. & Cabot, W., Erratum: "A Dynamic Subgrid-Scale Eddy Viscosity Model". *Phys. Fluids A*, vol **3**, No 12, pp. 3128, 1991.

[10] Lilly, D.K., "A Proposed modification of the Germano subgrid-scale closure method".*Phys. Fluids A*, vol **4**, No 3, pp. 633, 1992.

[11] Moin, P., et. al, "A Dynamic Subgrid-Scale Model for Compressible Turbulence and Scaler Transport". *Phys. Fluids A*, vol **3**, No 11, pp. 2746, 1991.

[12] Moin, P., "A New Approach for Large Eddy Simulation of Turbulence and Scalar Transport" Proc. Monte Verita Coll. on Turbulence, Sept. 91-09-13.

[13] Olsson, M. & Fuchs, L. "Simulation of a co-annular swirling jet using a dynamic SGS-model". AIAA Paper 94-654, 1994.

[14] Kawamura, T., & Kuwahara, K., "Computation of High Reynolds Number Flow around a Circular Cylinder with Surface Roughness". AIAA Paper No. 84-0340, 1984.

[15] Naitoh, K. & Kuwahara, K., "Large Eddy Simulation and Direct Simulation of compressible turbulence and combusting flows in engines based on the BI-SCALE method". *Fluid Dynamics Research*, **10**, pp. 299-325, 1992.

[16] Jameson, A. "Numerical solution of the Euler equations for compressible inviscid fluids". Proc INRIA Workshop on Numerical Methods for the Euler Equations of Fluid Dynamics. Rocquenfort, Eds F. Angrand et al, SIAM, 1983.

4

The Frequency Decomposition Multi-Grid Method

Wolfgang Hackbusch[1]

ABSTRACT The FD (frequency decomposition) multi-level algorithm is presented. It uses multiple coarse-grid corrections with particularly associated prolongations and restrictions. We discuss the construction of the algorithm and the proof of robustness by means of the techniques from domain decomposition methods.

1 Introduction: Smoothing versus Coarse Grid Correction

Multi-grid methods are known as very fast solvers for a large class of discretised partial differential equations. However, often, the components of the multi-grid algorithm have to be adapted to the given problem and sometimes the problems are modified in order to make them acceptable for multi-grid methods. In particular, singular perturbation problems require special care. In latter case, the problem depends on the discretisation parameter and an additional parameter of the differential equation. An iteration is called *robust*, if the convergence speed is uniform for all of these parameters.

The traditional remedy is the choice of a special smoothing iteration, while the coarse-grid correction is the standard one. For instance, block-versions of the Gauß-Seidel iteration, the alternating line-Gauß-Seidel smoothing (cf. Stüben - Trottenberg [9]), the incomplete LU-decomposition (ILU) and its block version (cf. Hemker [11], Kettler [13], Wittum [21]) were introducted for this purpose.

The simplest but typical example of this kind is the anisotropic equation

$$-\alpha u_{xx}(x,y) - \beta u_{yy}(x,y) = f(x,y) \tag{1.1}$$

with non-negative coefficients. As soon as the ratio α/β approaches 0 or ∞, the multi-grid method with pointwise Gauß-Seidel converges very slowly. The obtained convergence rates are independent of the discretisation parameter but not of the

[1]Institut für Informatik und Praktische Mathematik,
Christian-Albrechts-Universität zu Kiel, D-24098 Kiel, Germany

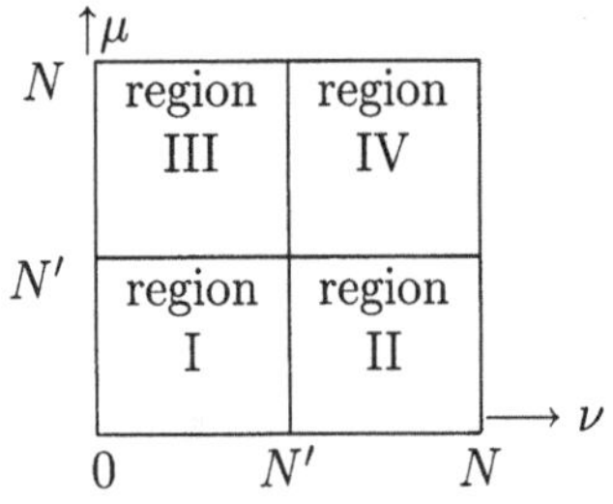

Fig. 1.1. Frequency diagram, $N' = N/2$

ratio α/β. Using line-Gauß-Seidel smoothing, one obtains good convergence for $\alpha/\beta \leq 1$ or $\alpha/\beta \geq 1$ depending on the line direction, while alternation line-relaxation as well as modified ILU version yields uniform convergence for all α/β (cf. Hackbusch [3], Wesseling [19]).

The Fourier analysis leads to the following explanation. Region I of Fig. 1.1 consists of the "low frequencies". Components of this part are reduced by the coarse-grid correction. All other (high frequency) components must be eliminated by the smoothing process. In the case of $\alpha/\beta \ll 1$ and pointwise relaxation, the frequencies in region II are poorly reduced, while for $\alpha/\beta \gg 1$ region III fails.

A slight modification of the standard coarse-grid correction is the semi-coarsening (cf. Brandt [1], Hackbusch [3 §3.4.1]).

In the *frequency decomposition multi-grid algorithm* we follow an alternative approach. We perform a more complicated coarse-grid correction using multiple coarse grids, while the smoothing iteration may be as simple as possible (e.g., damped Jacobi method). In the one-dimensional case, the standard coarse grid $\Omega_0^{\ell-1}$ consists of the grid points indexed by even numbers. The grid consisting of the odd points represents a second coarse grid $\Omega_1^{\ell-1}$. In the two-dimensional case, there are four coarse grids as indicated in Fig. 1.2.

The subscripts ij of $\Omega_{ij}^{\ell-1}$ indicate that the grid is obtained from the standard one by a shift of ih in x direction and jh in y direction. We will see that together with the special choice of prolongations each coarse-grid will contribute a correction to one of the four regions in the frequency diagram of Fig. 1.1.

There are also other multi-grid approaches with multiple coarse-grid correction. Mulder's [14] method uses a coarse-grid correction involving three grids: the standard one and the semi-coarsened grids w.r.t. both directions (cf. Fig. 1.3). The related prolongations are the usual ones. The aim of the method is the robustness.

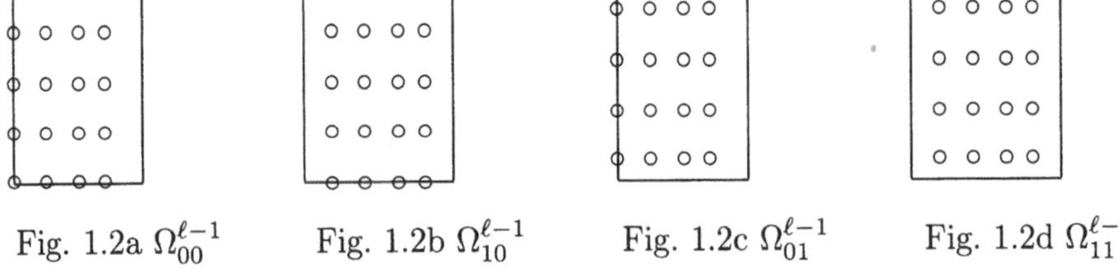

Fig. 1.2a $\Omega_{00}^{\ell-1}$ Fig. 1.2b $\Omega_{10}^{\ell-1}$ Fig. 1.2c $\Omega_{01}^{\ell-1}$ Fig. 1.2d $\Omega_{11}^{\ell-1}$

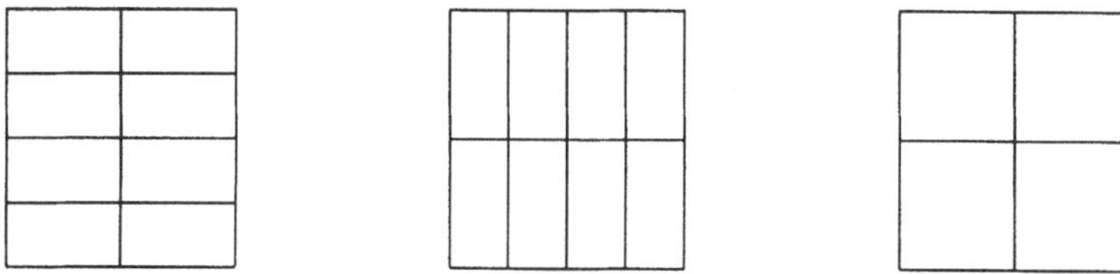

Fig. 1.3 The three coarse grids of Mulder's coarse-grid correction

This is discussed in the recent papers of Naik - Rosendale [15] and Oosterlee - Wesseling [16].

The same coarse grids as in Fig. 1.2a-d are used in the PSMG method of Frederickson and McBryan [2]. The name PSMG (parallel superconvergent multigrid) indicates the aim of the method. The convergence of standard problems (not singular perturbation problems) is accelerated by tuning the parameters of the nine-point prolongation, which are the same for all four coarse grids.

The frequency decomposition multi-grid method will be defined in Section 2 and analysed in Section 3. It has first been presented in Hackbusch [4]. For details we refer to Hackbusch [6,7].

2 Construction of the Frequency Decomposition Multi-Grid Method

Let the fine grid correspond to level ℓ. The four different coarse grids $\Omega_{00}^{\ell-1}$, $\Omega_{10}^{\ell-1}$, $\Omega_{01}^{\ell-1}$, $\Omega_{11}^{\ell-1}$ at level $\ell-1$ are defined as in Fig. 1.2a-d. The set of these four indices is denoted by

$$J = \{(0,0),(0,1),(1,0),(1,1)\}. \tag{2.1}$$

REMARK 2.1. The fine grid Ω^ℓ is the union of the grids $\Omega_\kappa^{\ell-1}$, $\kappa \in J$.

Each grid is associated with a prolongation

$$p_\kappa : \Omega_\kappa^{\ell-1} \to \Omega^\ell \qquad (\kappa \in J). \tag{2.2}$$

For $\kappa = (0,0)$, p_κ is the standard one: for the other indices, p_κ represents a nonstandard prolongation:

$$p_{00} = \frac{1}{4}\begin{bmatrix} 1 & 2 & 1 \\ 2 & 4 & 2 \\ 1 & 2 & 1 \end{bmatrix}, \quad p_{10} = \frac{1}{4}\begin{bmatrix} -1 & 2 & -1 \\ -2 & 4 & -2 \\ -1 & 2 & -1 \end{bmatrix}, \tag{2.3a}$$

$$p_{01} = \frac{1}{4}\begin{bmatrix} -1 & -2 & -1 \\ 2 & 4 & 2 \\ -1 & -2 & -1 \end{bmatrix}, \quad p_{11} = \frac{1}{4}\begin{bmatrix} +1 & -2 & +1 \\ -2 & 4 & -2 \\ +1 & -2 & +1 \end{bmatrix}. \tag{2.3b}$$

The use of completely different prolongations for each coarse grid is characteristic for the frequency decomposition multi-grid algorithms. The other variants with multiple coarse grids mentioned in Section 1 involve a standard interpolation. The range of the prolongations p_κ, $\kappa \neq (0,0)$, contains also high frequencies. $p_{10}u_{\ell-1}$ is oscillatory in x direction, $p_{01}u_{\ell-1}$ in y direction, and $p_{11}u_{\ell-1}$ is oscillatory in both directions.

REMARK 2.2. The span of range (p_κ) for all $\kappa \in J$ is the space of all fine-grid functions.

REMARK 2.3. For different indices $\iota, \kappa \in J$, range (p_ι) and range (p_κ) are orthogonal.

As in standard multigrid methods, the restriction is defined as the adjoint of the prolongation.

$$r_\kappa = p_\kappa^H \qquad (\kappa \in J). \tag{2.4}$$

Remark 2.3 is equivalent to

REMARK 2.4. $r_\iota p_\kappa = 0$ holds for different indices $\iota, \kappa \in J$.

In each coarse grid $\Omega_\kappa^{\ell-1}$ we will have to solve a coarse-grid equation with a coarse-grid matrix $A_\kappa^{\ell-1}$. As in the standard case, these matrices are defined by the Galerkin product

$$A_\kappa^{\ell-1} = r_\kappa A^\ell p_\kappa \qquad (\kappa \in J) \tag{2.5}$$

from the fine-grid matrix A^ℓ. Note that positive definiteness of A^ℓ implies positive definiteness of all coarse-grid matrices.

The new part of the algorithm is the multiple coarse-grid correction. Let J_0 be a subset of the index set J. The simplest choice would be $J_0 := J$. At least J_0 has to contain the index $(0,0)$. Using all coarse grids $\Omega_\kappa^{\ell-1}$ with $\kappa \in J_0$, we are lead to the multiple coarse-grid correction

$$u_\ell \mapsto u_\ell - \sum_{\kappa \in J_0} p_\kappa (A_\kappa^{\ell-1})^{-1} r_\kappa (A^\ell u_\ell - f_\ell). \tag{2.6}$$

If $(0,0)$ is the only index in J_0 (2.6) represents the standard coarse-grid correction. The purpose of the additional terms in (2.6) is to correct also oscillatory errors from the regions II to IV of Fig. 1.1. More precisely, the index $\kappa = (1,0)$ corresponds to region II $\kappa = (0,1)$ to III, and $\kappa = (1,1)$ to IV. Since the different parts of the spectrum should be corrected by the different coarse-grid corrections, the resulting method is called the frequence decomposition (FD) two-level iteration.

In order to demonstrate that the robustness is a consequence of the multiple coarse-grid correction (2.6) and not of a suitably chosen smoothing iteration, we choose the damped Jacobi iteration. Of course, the choice of more sophisticated smoothing processes can only improve the convergence.

The FD two-level method consists of pre-smoothing (ν steps) followed by the multiple coarse-grid correction (2.6). The two-level method depends on the choice of J_0. For $J_0 = \{(0,0)\}$ one obtains the standard two-grid method.

As for the standard multi-grid method, the straightforward version of the FD multi-grid algorithm is obtained from the FD two-level algorithm by replacing the exact solution in the coarse-grid correction (2.6) by the recursive application of the same method. Note that the treatment of the different coarse-grid equations can be performed in parallel.

If the FD multi-grid algorithm is performed with a *fixed* subset J_0, the following operation count holds in the two-dimensional case.

REMARK 2.5. If $\#J_0 < 4$, the V-cycle requires $O(n_l)$ operations. For $\#J_0 = 4$ the V-cycle takes $O(n_l \log n_l)$ operations.

The W-cycle with $\#J_0 > 2$ requires a too large amount of work. Therefore, a modification will be discussed in §4. The statement in three dimensions is similar: One has to replace $\#J_0 < 4$, $\#J_0 = 4$ by $\#J_0 < 8$, $\#J_0 = 8$, respectively.

Numerical results are reported in Hackbusch [6]. In addition, we show the convergence for the three-dimensional anisotropic equation $\alpha u_{xx} + \beta u_{yy} + \gamma u_{zz} + u = f$ in $[0,1]^3$ with $\alpha = 0.05$, $\beta = 0.001$, and $\gamma = 1$. The involved coarse grids carry three subscripts indicating the shift into the x, y, and z direction. Let J_0 consist of $(0,0,0)$, $(1,0,0)$, $(0,1,0)$, $(1,1,0)$, and $(0,0,1)$. Choose smoothing by the Jacobi iteration damped by $\omega = 1/2$. Then the FD V-cycle produces the following rates depending on the number ν of smoothing steps.

ν	1	2	3	5	10
rate	0.524	0.275	0.235	0.226	0.203

Using all coarse grids ($J_0 = J$), we obtain rates which are only weakly dependent on ν.

ν	1	2	3	5	10
rate	0.245	0.240	0.235	0.226	0.203

From these results we can draw the following conclusions. It is not true that the convergence rate behaves like $1/\nu$ as for standard multi-grid methods (cf. Hackbusch [3]). Furthermore, the smoothing may play a minor role if $J_0 = J$. Consequently, in the next chapter we will study the convergence of the FDE two-level method without any smoothing (i.e., $\nu = 0$).

3 Convergence Analysis

We will formulate the FD-multigrid method as a special variant of the additive Schwarz iteration. Then the convergence analysis for domain decomposition methods applies to the frequency decomposition multi-grid method and yields robust convergence. Since the convergence analysis for Schwarz-type iterations is restricted to positive definite problem, we have to restrict our considerations to the symmetric and positive case, although the FD multi-grid method is also applicable to nonsymmetric problems.

First, we remind the reader to the convergence theory of the additive Schwarz method (for more details compare Hackbusch [8]). Then, we raise the question of robustness, i.e., we want to obtain convergence estimates independent not only of the dimension of the problem but also of the inherent parameters. It turns out that this question is easy to answer if the class consists of all non-negative linear combinations of positive semidefinite matrices. Finally, we make use of the orthogonality of the subspaces (cf. Remark 2.3).

We consider a general system

$$Ax = b \tag{3.1}$$

with a Hermitian and positive definite matrix A of the size $n \times n$. We denote the linear space of the vectors x, b by X. The characteristic feature of the subspace iteration is the (approximate) solution of smaller subproblems, which we denote by

$$A_\kappa y^\kappa = c^\kappa \qquad (\kappa \in J). \tag{3.2}$$

Here J is an index set. The size of A_κ is $n_\kappa \times n_\kappa$. The vectors x, b from (3.1) and y^κ, c^κ from (3.2) belong to the respective vector spaces X, $X_\kappa (\kappa \in J)$. The connection between X_κ and X is given by an injective prolongation

$$p_\kappa : X_\kappa \mapsto X \qquad (\kappa \in J). \tag{3.3}$$

Endowing X and X_κ with the standard Euclidean scalar product $\langle \cdot, \cdot \rangle$, we are able to define the transposed (Hermitian) mapping

$$r_\kappa = p_\kappa^H : X \mapsto X_\kappa \qquad (\kappa \in J). \tag{3.4}$$

By assumption, r_κ is surjective. The (positive definite) matrices A_κ from (2.2) are defined by the Galerkin product

$$A_\kappa := r_\kappa A p_\kappa. \tag{3.5}$$

In the case of the FD multi-grid method, p_κ from (3.3) are the prolongations (2.3a,b), (3.4) describes the restrictions to the coarse grids, and (3.5) coincides with (2.5).

The corresponding additive Schwarz iteration

$$x^m \mapsto x^{m+1} := x^m - \omega \sum_{\kappa \in J} p_\kappa A_\kappa^{-1} r_\kappa (A x^m - b). \tag{3.6}$$

is a damped variant of the FD two-level method (2.6).

Using the second and third normal forms $x^{m+1} = x^m - N(Ax^m - b)$ and $W(x^{m+1} - x^m) = b - Ax^m$ of the linear iteration (cf. Hackbusch [7, §3]), one can represent the additive Schwarz iteration by the corresponding matrices

$$N = \omega \sum_{\kappa \in J} N_\kappa \quad \text{with} \quad N_\kappa := p_\kappa A_\kappa^{-1} r_\kappa, \qquad W := N^{-1}. \tag{3.7}$$

The convergence is described by the constants γ and Γ in

$$\gamma W \leq A \leq \Gamma W \qquad\qquad (\gamma > 0). \tag{3.8a}$$

Here, $B \leq C$ means that $C - B$ is positive semi-definite. If γ and Γ are the best possible bounds in (2.8a),

$$\kappa = \Gamma / \gamma \tag{3.8b}$$

is the condition number of $W^{-1}A$ (cf. Hackbusch [7, §8.3]).
Choosing the optimal damping factor $\omega_{\text{opt}} := 2/(\gamma + \Gamma)$ in (3.6), we obtain the convergence rate

$$\|M\|_A := \|A^{1/2} M A^{-1/2}\| = (\Gamma - \gamma)/(\Gamma + \gamma) = (\kappa - 1)/(\kappa + 1), \tag{3.9}$$

where $M = I - NA$ is the iteration matrix. Using the additive Schwarz iteration (3.6) as basic iteration of the Chebyshev or cg-method, we obtain an asymptotic rate equal to $(\sqrt{\kappa} - 1)/(\sqrt{\kappa} + 1)$.

Concerning the construction or estimation of the bounds γ and Γ in (2.7), two wellknown lemmata are applied (cf. Widlund [20], Hackbusch [7, §11],).

LEMMA 3.1. *$\Gamma := \#J$ satisfies (3.8a). Since for the FD two-level methods $\#J$ is the number of coarse grids, $\Gamma \leq 4$ is a bound independent of any parameter.*

LEMMA 3.2. *Assume that for any $x \in X$ there is a decomposition $x = \Sigma_\kappa p_\kappa x^\kappa$ such that*

$$\sum_{\kappa \in J} \langle A_\kappa x^\kappa, x^\kappa \rangle \leq C \langle Ax, x \rangle. \tag{3.10}$$

Then, (2.8a) holds with $\gamma = 1/C$.

Next, we study the uniform convergence for a class $\mathcal{U}$ of matrices $A = A(\alpha_1, \alpha_2, \ldots)$ depending on parameters $\alpha_1, \alpha_2, \ldots$. The set $\mathcal{U}$ is defined as follows:

$$\mathcal{U} := \{\mathcal{A} = \sum_{\nu \in \mathcal{I}} \alpha_\nu \mathcal{A}^{(\nu)} > 0 \quad \text{with} \quad \alpha_\nu \geq 0\}, \tag{3.11}$$

where $A^{(\nu)} (\nu \in I)$ are positive semi-definite matrices. In the case of the anisotropic equation (1.1), the linear combination $\alpha A^{(1)} + \beta A^{(2)}$ of the one-dimensional second differences $A^{(1,2)} = [-1 \ 2 \ -1]$ in x and y direction yields the standard discretisation.

For the determination of $\gamma = 1/C$ by Lemma 3.2 one may use

LEMMA 3.3. *Let $A^{(\nu)} (\nu \in I)$ be the positive semi-definite matrices from (3.11) and assume that (3.10) holds for each $A^{(\nu)}, \nu \in I$, with a constant C_ν:*

$$\sum_{\kappa \in J} \langle A_\kappa^{(\nu)} x^\kappa, x^\kappa \rangle \leq C_\nu \langle A^{(\nu)} x, x \rangle \qquad (x \sum p_\kappa x^\kappa \quad \text{from Lemma 3.2, } \nu \in I), \tag{3.12}$$

where $A_\kappa^{(\nu)} := r_\kappa A^{(\nu)} p_\kappa$. Then (3.10) holds for all $A \in \mathcal{U}$ with the constant $C := \max\{C_\nu : \nu \in I\}$.

The hypothesis of this paper is the orthogonality as stated in Remark 2.3:

$$\text{range}(p_\iota) \text{ and range } (p_\kappa) \text{ are orthogonal for different indices } \iota, \kappa \in J. \tag{3.13}$$

The products $r_\kappa p_\kappa (\kappa \in J)$ are positive definite. Therefore, the orthogonal projection onto X_κ =range(p_κ)

$$Q_\kappa := p_\kappa (r_x p_\kappa)^{-1} r_\kappa \qquad (\kappa \in J) \tag{3.14}$$

are well-defined. The orthogonality (3.13) allows a unique and easily describable decomposition of vectors $\kappa \in X$. For all $x \in X$ the identities (3.15a.b) hold:

$$x = \sum_{\kappa \in J} (Q_\kappa x), \quad Q_\kappa x \in \text{ range } (p_\kappa), \tag{3.15a}$$

$$x = \sum_{\kappa \in J} p_\kappa x^\kappa \quad \text{with } x^\kappa := (r_\kappa p_\kappa)^{-1} r_\kappa x. \tag{3.15b}$$

In the following we derive sufficient conditions for inequality (3.12) of Lemma 3.3.

The (Euclidean) scalar product, on which the definition of the transposed matrices and the orthogonality are based, is denoted by $\langle \cdot, \cdot \rangle$. The corresponding norm is $\|x\| := \langle x, x \rangle^{1/2}$. We use identical symbols $\| \cdot \|$, $\langle \cdot, \cdot \rangle$ for elements from X (fine grid functions) and X_κ = range (p_κ). In addition, we define the energy norm on X:

$$|||x||| := \langle Ax, x \rangle^{1/2} \qquad \text{for } x \in X. \tag{3.16a}$$

For the space X_κ, we choose suitable positive definite matrices B_κ and define the (energy) norm

$$|||x^\kappa||| := \langle B_\kappa x^\kappa, x^\kappa\rangle^{1/2} \qquad \text{for } x^\kappa \in X_\kappa. \tag{3.16b}$$

In the sequel, we consider the following estimates:

$$|||p_\kappa||| \le C_{p,\kappa}, \tag{3.17a}$$

$$|||r_\kappa||| \le C_{r,\kappa}, \tag{3.17b}$$

$$|||(r_\kappa p_\kappa)^{-1}||| \le |||C_{rp,\kappa}. \tag{3.17c}$$

Here, $|||\cdot|||$ also denotes the induced matrix norm, e.g.,

$$|||p_\kappa||| := \max\{|||p_\kappa x^\kappa|||/|||x^\kappa||| : 0 \ne x^\kappa \in X_\kappa\}.$$

LEMMA 3.4. *Condition (3.17a) is equivalent to the inequality*

$$A_\kappa \le C_{p,\kappa}^2 B_\kappa \qquad \text{with } A_\kappa \text{ from (3.5).}$$

LEMMA 3.5. *Assume that the matrices B_κ and $r_\kappa p_\kappa$ commute. Then condition (3.17c) holds with the constant*

$$C_{rp,\kappa} := 1/ \text{ smallest eigenvalue of } r_\kappa p_\kappa.$$

From (3.17a-c) and (3.14) we conclude

$$|||Q_\kappa||| \le C_{Q,\kappa} \qquad \text{with } C_{Q,\kappa} := C_{p,\kappa} C_{r,\kappa} C_{rp,\kappa}. \tag{3.18}$$

LEMMA 3.6. *Let $C_{Q,\kappa}$ be the bound in (3.18). Then inequality (3.12) holds with*

$$C := \sum_{\kappa\in J} \dot{C}_{Q,\kappa}^2.$$

To obtain parameter and dimension independent convergence, one has to prove (3.17a-c) with dimension independent constants $C_{p,\kappa}^{(\nu)}$, $C_{r,k}^{(\nu)}$, $C_{rp,\kappa}^{(\nu)}$ for all $A^{(\nu)}$ ($\nu \in I$) instead of A. Note that A is involved in $|||x||| := \langle Ax, x\rangle^{1/2}$. Also the matrices $B_\kappa^{(\nu)}$ involved in $|||x^\kappa|||$ may depend on $\nu \in I$.

As mentioned after (3.11), the first matrix $A^{(1)}$ may represent the second difference $[-1\ \ 2\ \ -1]$ in x direction, which can be regarded as the tensor product of the one-dimensional stencil $[-1\ 2\ -1]$ w.r.t. x times the one-dimensional identity $[0\ 1\ 0]$ w.r.t. y. It is not astonishing that the analysis of $A = A^{(1)}$ reduces to the analysis of the one-dimensional cases (cf. Hackbusch [7]), where only two coarse grids (indexed by $\kappa = 0, 1$) and the associated prolongations

$$p_0 = [1/2\ \ 1\ \ 1/2], \quad p_1 = [-1/2\ \ 1\ \ -1/2]$$

are involved. To illustrate the proof technique, we demonstrate the estimation (3.18) for $A^{(1)}$ and $\kappa = 0, 1$. The results are stated in Lemmata 3.7 and 3.8.

Let $(N_x - 1) \times (N_x - 1)$ be the size of the matrix $A^{(1)}$. For simplicity, we assume N_x to be even. The energy norm $|||\xi||| = \langle A\xi, \xi\rangle^{1/2}$ from (3.16a) can also be represented by

$$|||\xi|||^2 = \sum_{j=1}^{N_x} |\xi_j - \xi_{j-1}|^2.$$

The index j refers to the x-location $x = jh$. For $j \geq N_x$ or $j \leq 0$, $\xi_j := 0$. We choose $B_0 := [-1 \;\; 2 \; -1]$. Then, the squared energy norm (3.,16b) related to the coarse grid $\Omega_{2h} = \{0, 2h, 4h, \ldots, 1\}$ equals

$$|||\xi^0|||^2 = \langle B_0\xi^0, \xi_0\rangle \;=\; \sum_{j=1}^{N_x/2} |\xi^0 - \xi^0_{j-1}|^2 \quad \text{for } \xi^0 \in X_0,$$

where the index j refers tot the grid point $x = 2jh \in \Omega_{2h}$. We prove

LEMMA 3.7. *In the given case, the inequalities (3.17a-c) hold with the constants*

$$C_{p,0} = 1/\sqrt{2}. \quad C_{r,0} = \sqrt{8}, \quad C_{rp,0} = 1. \tag{3.19}$$

PROOF. (i) Let $x \in X$ and $\xi = r_0 x \in X_0$. From

$$\begin{aligned}\xi_j - \xi_{j-1} &= (\tfrac{1}{2}x_{2j-1} + x_{2j} + \tfrac{1}{2}x_{2j+1}) - (\tfrac{1}{2}x_{2j-3} + x_{2j-2} + \tfrac{1}{2}x_{2j-1}) = \\ &= \tfrac{1}{2}(x_{2j-2} - x_{2j-3}) + \tfrac{3}{2}(x_{2j-1} - x_{2j-2}) + \tfrac{3}{2}(x_{2j} - x_{2j-1}) + \tfrac{1}{2}(x_{2j+1} - x_{2j})\end{aligned}$$

one concludes that

$$\begin{aligned}&|\xi_j - \xi_{j-1}|^2 \leq (1^2 + \sqrt{3}^2 + \sqrt{3}^2 + 1^2)/2^2 \times \\ &\times \{|x_{2j-2} - x_{2j-3}|^2 + 3|x_{2j-1} - x_{2j-2}|^2 + \\ &\quad +3|x_{2j} - x_{2j-1}|^2 + |x_{2j+1} - x_{2j}|^2\} = \\ &= 8\{\tfrac{1}{4}|x_{2j-2} - x_{2j-3}|^2 + \tfrac{3}{4}|x_{2j-1} - x_{2j-2}|^2 + \\ &\quad +\tfrac{3}{4}|x_{2j} - x_{2j-1}|^2 + \tfrac{1}{4}|x_{2j+1} - x_{2j}|^2\}\end{aligned}$$

for $2 \leq j \leq N_x/2 - 1$ (use the Schwarz inequality). For $j = 1$ one has to note that $\xi_{j-1} = 0$. Then

$$\begin{aligned}&|\xi_1 - \xi_0|^2 = |\xi_1|^2 = \\ &= |\tfrac{1}{2}x_1 + x_2 + \tfrac{1}{2}x_3|^2 = [\tfrac{1}{2}(x_3 - x_2) + \tfrac{3}{2}(x_2 - x_1) + 2(x_1 - x_0)]^2 \leq \\ &\leq [(\tfrac{1}{2})^2 + (\tfrac{3}{2})^2 + 2]\,[(x_3 - x_2)^2 + (x_2 - x_1)^2 + 2(x_1 - x_0)^2].\end{aligned}$$

The index $j = N_x/2$ is treated similarly. Summing over $1 \leq j \leq N_x/2$, we obtain $|||\xi|||^2 \leq 8|||x|||^2$, which proves $C_{r,0} = \sqrt{8}$.

(ii) The product $A_0 := r_0 A p_0$ equals $A_0 = \frac{1}{2}[-1 \;\; 2 \; -1]$. Hence, $A_0 \leq \frac{1}{2}B_0$ and Lemma 3.1 prove $C_{p,0} = 1/\sqrt{2}$.

(iii) The product $r_0 p_0$ equals $\frac{1}{4}[1\ \ 6\ \ 1] = 2I - \frac{1}{4}[-1\ \ 2\ \ -1]$. Since it commutes with B_0, Lemma 3.5 can be applied. The smallest eigenvalue of $\frac{1}{4}[1\ \ 6\ \ 1]$ is bounded from below by $\frac{1}{4}(6\ -1\ -1) = 1$ proving $C_{rp,0} = 1$. □

For the case of $\kappa = 1$, i.e., $p_1 = [-1/2\ \ 1\ \ -1/2]$, choose $B_1 = I$ in (3.16b).

LEMMA 3.8. *In this case, the inequalities (3.17a-c) hold with*

$$C_{p,1} = \sqrt{8}, \qquad C_{r,1} = 1/\sqrt{2}, \qquad C_{rp,1} = 1. \tag{3.20}$$

PROOF. (i) Let $x \in X$ and $\xi = p_1^H x \in X_1$. Summing

$$\begin{aligned} |\xi|^2 &= |\tfrac{1}{2}(x_{2j} - x_{2j-1}) - \tfrac{1}{2}(x_{2j+1} {-}_{2j})|^2 \\ &\le \tfrac{1}{2}\left(|x_{2j} - x_{2j-1}|^2 + |x_{2j+1} - x_{2j}|^2\right) \end{aligned}$$

over j, one obtains $||\xi||^2 = |||\xi|||^2 \le |||\xi|||_x^2 = \frac{1}{2}|||x|||_x^2$ proving $C_{r,1} = 1/\sqrt{2}$.

(ii) The product $A_1 = r_1 A p_1$ yields the tridiagonal matrix

$$A_1 = \frac{1}{2}\begin{bmatrix} 7 & 3 & & & \\ 3 & 10 & 3 & & \\ & \ddots & \ddots & \ddots & \\ & & 3 & 10 & 3 \\ & & & 3 & 7 \end{bmatrix}.$$

Its largest eigenvalue is bounded by $\frac{1}{2}(10 + 3 + 3) = 8$ proving $A_1 \le 8I$.
From Lemma 3.1 the estimate $C_{p,1} = \sqrt{8}$ follows.

(iii) The product $r_1 \rho_1$ equals

$$\frac{1}{4}\begin{bmatrix} 5 & 1 & & & \\ 1 & 6 & 1 & & \\ \ddots & & \ddots & & \ddots \\ & & 1 & 6 & 1 \\ & & & 1 & 5 \end{bmatrix}$$

Its smallest eigenvalue is 1. Hence, Lemma 3.5 yields $C_{rp,1} = 1$. □

The bounds (3.19) and (3.20) result in $C_{Q,0} = C_{Q,1} = 2$. Since the same estimates hold for the second difference w.r.t. y, Lemmata 3.6 and 3.3 prove uniform convergence of the anisotropic equation discretised by

$$A = \begin{bmatrix} a & c & a \\ b & e & b \\ a & c & a \end{bmatrix}$$

with $b = \alpha$, $c = \beta$, $a = e = 0$. Inequality (3.12) can also be proved for $A^{(3)}=$ identity and the rotated difference star

$$A^{(4)} = \begin{bmatrix} -1 & 0 & -1 \\ 0 & 4 & 0 \\ -1 & 0 & -1 \end{bmatrix}.$$

According to Lemma 3.3, the nine-point star (4.2): $A = bA^{(t)}+cA^{(2)}+eA^{(3)}+aA^{(4)}$ with $a, b, c, e \geq 0$ leads to uniform convergence of the additive Schwarz method and hence of the FD two-level method.

A diagonally second difference is not allowed in this context, but another FD approach including this term is described by Katzer [12].

4 The Multi-Level Version

In the following, we discuss the generalisation of the two-level iteration (2.6) to a multi-level iteration corresponding to the well-known V- and W-cycles of the multi-grid method (cf. Hackbusch [3]). Before analysing the W-cycle, we have to discuss the form of the auxiliary systems (3.2): $A_\kappa y^\kappa = c^\kappa$, which now are to be solved recursively by the same additive Schwarz method. One may check that the Galerkin products $A_\kappa^{(\nu)} := r_\kappa^H A^{(\nu)} p_\kappa$ belong again to the same class $\mathcal{U}$. The Galerkin product of a general matrix $A \in \mathcal{U}$ does not leave $\mathcal{U}$ and the convergence analysis of the two-level iteration is also true for the subproblems (3.2).

Let $\kappa = \kappa^{\text{two}}$ be the uniform condition number of the *two-level* iteration at all levels. As in the standard case, we can prove the following result (cf. Hackbusch [7]): If the two-level convergence is fast enough, it implies multi-grid convergence.

THEOREM 4.1. *If $\kappa^{two} < 4$, the convergence rate ρ_ℓ of the W-cycle is bounded uniformly: $\rho_\ell \leq \rho^* < 1$, where ρ^* is the solution of $\rho = (\kappa^{two} - 1 + \rho^2)/(\kappa^{two} + 1 - \rho^2)$.*

The unmodified W-cycle is unpractical because of the unfavourable (sequential) operation count. There are 4^i coarse-grid problems at level $\ell - i$ of dimension $N_{\ell-i} \approx N_\ell/4^i$ summing up to $N_\ell = \dim(X)$. But since the W-cycle induces 2^i recursive calls of the method at level $\ell - 1$, the total amount of computational work is of order $N_\ell + 2^1 N_\ell + 2^2 N_\ell + \ldots + 2^\ell N_\ell = O(N_\ell^2)$.

However, most of the coarse-grid matrices have a constant (parameter independent) condition number. Hence, the approximate solution of the coarse-grid equation by two W-cycles can be replaced by one (or few) steps of the Richardson iteration. The analysis (Hackbusch [7]) shows that only $O(2^i)$ of the 4^i coarse-grid problems can lead to a larger condition number. Together with the number of 2^i recursive calls at level $\ell - i$, we arrive at $O(2^i)2^i N_{\ell-i} = O(N_\ell)$ operations. Summing over all levels, we obtain a total amount of work (sequential version) in the order $O(\ell N_\ell) = O(N_\ell \log N_\ell)$.

Finally, we add that the corresponding V-cycle or F-cycle requires only a work of $O(N_\ell)$, while without modification an additional logarithmic factor appears as mentioned in Remark 2.5.

References

[1] BRAND., A.: Stages in developing multigrid solutions. In: Numerical Methods for Engineers (eds.: E. Absi, R. Glowinski, H. Veysseyre), Paris, Dunod, 1980, pages 23-43.

[2] FREDERICKSON, P.O. and O.A. MCBRYAN: Recent developments for the PSMG multiscale method. In Hackbusch - Trottenberg [10] 21-39.

[3] HACKBUSCH, W.: *Multi-grid methods and applications.* Springer, Heidelberg, 1985.

[4] HACKBUSCH, W.: A new approach to robust Multi-Grid methods. In: McKenna, J., Temam, R. (eds.) ICIAM '87: Proceedings of the First International Conference on Industrial and Applied Mathematics, SIAM, Philadelphia 1988.

[5] HACKBUSCH, W.: (ed.): *Robust Multi-Grid methods.* Proceedings, 4th GAMM-Seminar Kiel, Jan. 1988. Notes on Numerical Fluid Mechanics, vol. 23, Vieweg, Braunschweig, 1989.

[6] HACKBUSCH, W.: The frequence decomposition multi-grid method I. Application to anisotropic equations. *Numer. Math. 56* (1989) 229-245.

[7] HACKBUSCH, W.: The frequency decomposition multi-grid method II. Convergence analysis based on the additive Schwarz method. *Numer. Math. 63* (1992) 433-453.

[8] HACKBUSCH, W.: *Iterative Lösung großer schwachbesetzter Gleichungssysteme.* Teubner, Stuttgart 1991. English translation: *Iterative Solution of Large Sparse System of Equations.* Springer-Verlag, New York 1993.

[9] HACKBUSCH, W. and U. TROTTENBERG (eds.): Multi-Grid Methods, Proceedings, Lecture Notes in Mathematics 960. Springer Berlin-Heidelberg, 1982.

[10] HACKBUSCH, W. and U. TROTTENBERG (eds.): *Multi-Grid methods III.* Proceedings, Bonn, Oktober 1990. ISNM 98, Brikhäuser, Basel, 1991.

[11] HEMKER, P.W.: The incomplete LU-decomposition as a relaxation method in Multi-Grid algorithms. In: Miller, J.J.H. (ed.): *Boundary and interior layers - computational and asymptotic-methods,* Boole Press, Dublin, 1980, pages 306-311.

[12] KATZER, E.: A subspace decomposition two-grid method for hyperbolic equations. Contribution to this volume.

[13] KETTLER, R.: Analysis and comparison of relaxation schemes in robust multigrid and preconditioned conjugate gradient methods. In: Hackbusch - Trottenberg [1] 502-534.

[14] MULDER, W.: A new multigrid approach to convection problems. *J. Comp. Phys.* 83 (1989) 303-323.

[15] NAIK, N.H. and J. VAN ROSENDALE: The improved robustness of multigrid elliptic solvers based on multiple semicoarsened grids. *SIAM Num. Anal.* 30 (1993) 215-229.

[16] OOSTERLEE, C.W. and P. WESSELING: On the robustness of a multiple semi-coarsened grid method. To appear in ZAMM

[17] STÜBEN, K. and U. TROTTENBERG: Multi-grid methods: fundamental algorithms, model problem analysis and applications. In Hackbusch - Trottenberg [1] 1-176.

[18] WESSELING, P.: Theoretical and practical aspects of a multigrid method. *SIAM J. Sci. Statist. Comput. 3* (1982) 387-407.

[19] WESSELING, P.: *An introduction to multigrid methods.* Wiley, Chichester 1991.

[20] WIDLUND, O.: Optimal iterative refinement methods. In: Chan - Glowinski - Périaux - Widlund (eds.), *Domain decomposition methods.* Proceedings, SIAM Philadelphia 1989. Pages 114-125.

[21] WITTUM, G.: On the robustness of ILU-smoothing. In Hackbusch [3] 217-239.

5

Multiscale Methods for Computing Propagators in Lattice Gauge Theory

P. G. Lauwers[1]

ABSTRACT Gauge theories, a special kind of Quantum Field Theories, are the best mathematical framework to describe all known basic interactions in nature. In particular, the theory of the strong interactions (nuclear and subnuclear forces) is a four-dimensional $SU(3)$ gauge theory called Quantum Chromodynamics (QCD). In state-of-the-art QCD simulations, requiring massive amounts of computer time, more than 95% of the CPU-time is spent computing propagators, i.e., inverting the huge *fermion matrix*. Although a multiscale approach may be called for to speed up many aspects of QCD simulations, first real breakthroughs should be expected thanks to more efficient multiscale algorithms for inverting the fermion matrix. Several strategies, proposed recently by different groups, are presented and discussed.

1 Brief introduction to lattice gauge theories

During the last fifty years, one of the great achievements in the physical sciences has definitely been the development of Quantum Field Theory (QFT) as the description of basic interactions in nature. That quantum gauge theories, a particular kind of QFT, are now generally accepted to be the correct description of at least three out of the four known types of interactions, is an intriguing fact [1]. Electromagnetism is described by Quantum Electrodynamics (QED), a $U(1)$ gauge theory. The weak interactions, e.g. responsible for β-decay, have been unified with QED in the Glashow-Salam-Weinberg theory(GSW), an $SU(2) \times U(1)$ gauge theory. Finally, the strong interactions, e.g. binding protons and neutrons within the nuclei, are described by Quantum Chromodynamics (QCD), an $SU(3)$ gauge theory. Even gravitation, the fourth basic interaction, is generally assumed to be a gauge theory [2].

In the framework of QED, extremely precise predictions can be made by

[1]German National Research Center for Computing Science (GMD)
Institute I1.T, P.O.B. 1316, D-53731 Sankt Augustin, Germany

means of perturbation expansions. Many of them have been verified experimentally in high-precision measurements. In QCD the situation is completely different. Some of the most relevant aspects of the theory, e.g. the hadron spectrum, can not be investigated by perturbative methods. Although protons and neutrons, the main constituents of the nuclei, are generally believed to be composite states made up of three quarks bound together by QCD, no analytic method has been found enabling us to derive this simple fact from first principles. At present, the only tool for investigating this and many more equally essential properties of the theory is the Monte Carlo simulation of lattice QCD [3].

1.1 The Model

Lattice QCD is defined on a finite four-dimensional hypercubic lattice Λ, usually with periodic boundary conditions. L denotes the number of lattice points in each direction and h is the basic lattice distance. To obtain real physical predictions, a highly nontrivial double limit must be taken in the end: (i) $Lh \to \infty$ (the physical extent of the lattice goes to infinity) and simultaneously (ii) $h \to 0$ (the lattice distance goes to zero).

The first kind of variables in the theory are the link elements $U_\mu^{\alpha\beta}$, located on the links of the lattice. They are elements of the fundamental representation of the gauge group $SU(3)$, i.e., three-dimensional special unitary matrices. The upper indices α and β are $SU(3)$ indices, taking the values 1 through 3; the lower index μ takes the values 1 through 4 depending on the direction of the link. The second type of variables are the fermion fields ψ^α, located on the lattice points. They are anticommuting complex variables (Grassmann variables), transforming under the fundamental representation of the gauge group: the upper index α is the gauge group index, taking the values 1 through 3. A given state of the system, with values for all variables specified, is called a *configuration*.

A gauge transformation of a configuration is defined by means of a map g: $\Lambda \to G$, where G is the gauge group; in two dimensions this means that a group element $g(i,j)$ (fundamental representation) is assigned to each coordinate pair (i,j). Under the gauge transformation g, the fermion field $\psi(i,j)$ is transformed into $g(i,j)\psi(i,j)$. The link element $U_1(i+\frac{1}{2},j)$, i.e. the group element located on the link connecting the points (i,j) and $(i+1,j)$, is transformed into $g(i,j)U_1(i+\frac{1}{2},j)g^\dagger(i+1,j)$, where $g^\dagger$ denotes the hermitian conjugate of g. It is evident, that many structures can be built with the U's and the ψ's that are invariant under any gauge transformation. An important gauge invariant quantity is the *plaquette action* S_{plaq}, defined for every *plaquette* on the lattice; a plaquette is an elementary square consisting of four lattice links. In two dimensions a typical plaquette has the corners (i,j), (i+1,j), (i+i,j+1) and (i,j+1). For this plaquette the gauge invariant $SU(3)$ plaquette action is defined by

$$S_{plaq} = \frac{1}{3}\, Trace\, [U_1(i+\frac{1}{2},j)U_2(i+1,j+\frac{1}{2})U_1^\dagger(i+\frac{1}{2},j+1)U_2^\dagger(i,j+\frac{1}{2})] \quad . \quad (1)$$

Another gauge invariant quantity, containing fermion fields as well as link variables, is $\psi^\dagger(i,j)U_1(i+\frac{1}{2},j)\psi(i+1,j)$.

A very important entity in lattice gauge theories is the *fermion matrix* M, a huge $nV_\Lambda \times nV_\Lambda$ complex matrix, where V_Λ denotes the number of lattice points of the lattice Λ and n the dimension of the fundamental representation of the gauge group. In two dimensions, the explicit expression for the fermion matrix is

$$\begin{aligned} M^{\alpha\beta}_{(i,j)(k,l)} &\equiv \frac{1}{2h}[U_1^{\alpha\beta}(i+\frac{1}{2},j)\delta_{j,l}\delta_{i+1,k} - U_1^{\dagger\alpha\beta}(i-\frac{1}{2},j)\delta_{j,l}\delta_{i-1,k}] \\ &+ \frac{(-1)^i}{2h}[U_2^{\alpha\beta}(i,j+\frac{1}{2})\delta_{i,k}\delta_{j+1,l} - U_2^{\dagger\alpha\beta}(i,j-\frac{1}{2})\delta_{i,k}\delta_{j-1,l}] \\ &+ m_q\delta_{i,k}\delta_{j,l}\delta^{\alpha\beta} \\ &= D^{\alpha\beta}_{(i,j)(k,l)} + m_q\delta_{i,k}\delta_{j,l}\delta^{\alpha\beta} \quad , \end{aligned} \tag{2}$$

where m_q denotes the quark mass. The explicit form of the fermion matrix M is not unique; Eq. 2 is the two-dimensional version of the *staggered formulation* [4]. An important alternative is the so-called *Wilson formulation* [5]. Which version is better, is still an open question, which should be decided on physical grounds.

In lattice gauge theories, the content of the model is defined by the *action* S, a function of all the variables of the theory. In the case of lattice QCD, this action S consists of two terms: $S = S_G + S_F$. The gauge part S_G is a simple function of the plaquette action S_{plaq}, defined in Eq. 1:

$$S_G = \beta \sum_{plaq} [1 - Real(S_{plaq})] \, , \tag{3}$$

where the sum runs over all plaquettes of Λ. The fermionic part S_F, expressing the dynamics of the fermion fields as well as their interaction with the gauge fields, can be written as a function of the fermion matrix M and the fermion field ψ in the following way:

$$S_F \equiv (\bar{\psi}, M[U]\psi) \, ; \tag{4}$$

in two dimensions the explicit expression is

$$S_F = \sum_{(i,j)\in\Lambda} \sum_{(k,l)\in\Lambda} \sum_{\alpha,\beta=1}^{3} \psi^{\dagger\alpha}(i,j)\, M^{\alpha\beta}_{(i,j)(k,l)}\, \psi^\beta(k,l) \quad . \tag{5}$$

A fundamental property of gauge theories is that all physical quantities in the theory are gauge invariant. As a consequence, all physical quantities corresponding to a configuration can be expressed as gauge invariant combinations of ψ's and U's: e.g. the plaquette action S_{plaq} is directly related to the energy density. Predictions for a physical quantity $F[\psi^\dagger,\psi,U]$ come, as is the case in all quantum theories, as its *expectation value* $< F >$ given by the following expression:

$$< F >= \frac{1}{Z}\int [d\psi^\dagger]\,[d\psi]\,[dU]\,F[\psi^\dagger,\psi,U]\exp[-S[\psi^\dagger,\psi,U]] \, , \tag{6}$$

where the *partition function* Z, serving as a normalization factor, is defined by

$$Z = \int [d\psi^\dagger]\,[d\psi]\,[dU]\,\exp[-S[\psi^\dagger,\psi,U]]\,. \tag{7}$$

In these expressions the integration $\int [d\psi^\dagger]\,[d\psi]\,[dU]$ stands for the integration over all possible configurations of the system. The fermion fields, being Grassmann variables, can be integrated out. After this integration has been carried out, Eq. 6 and 7 take the following form:

$$< F >= \frac{1}{Z}\int [dU]\,F[M^{-1},U]\det(M[U])\exp(-S_G[U])\,, \tag{8}$$

with

$$Z = \int [dU]\,\det(M[U])\exp(-S_G[U])\,. \tag{9}$$

Notice the appearance in these expressions of both the determinant $\det(M[U])$ and the inverse $M^{-1}[U]$ of the fermion matrix. The expectation value $< F >$ can now be given a simple interpretation: it is the weighted average of $F[M^{-1},U]$ over all possible U-configurations with the expression $\det(M[U])\exp(-S_G[U])$ as weighting factor. Exactly for this type of problems, at least if the weighting factor is real and nonegative for all configurations, Monte Carlo simulation methods were developed many years ago [6]. For a more rigorous and complete introduction to lattice gauge theories I must refer to the literature [3].

1.2 Monte Carlo simulations: quenched and full

The ideas behind a Monte Carlo (MC) simulation are simple. Instead of computing the weighted average of $F[M^{-1},U]$ over all possible U-configurations – an impossible computational task for the large systems being studied – a *Markov chain* of sample configurations is generated with $\det(M[U])\exp(-S_G[U])$ as probability distribution. If all the rules of the game are carefully followed, then the regular average of $F[M^{-1},U]$ over the sample configurations is a good estimator for the expectation value $< F >$, at least for long enough Markov chains [6, 7].

For reasons having to do with the double limit required for physical predictions (see Section 1.1), reliable MC simulations of lattice gauge theories must be carried out on large systems - at least 24^4 and preferably much larger. This requirement turned the simulation of the full theory into a computational task that, until recently, was unmanageable on the available computer systems. If one wanted physical predictions from lattice gauge theories anyway, one was forced to use the *quenched* approximation, a drastic, but in many cases acceptable, approximation: $\det(M[U])$ is set equal to 1 in Eq. 8 and 9. In the framework of this approximation, one should clearly distinguish two computational tasks: (i) the generation of statistically independent gauge field configurations with probability distribution $\exp(-S_G[U])$ by means of a MC *update* algorithm; (ii) computation of the measurements $F[M^{-1}, U]$ for these configurations, requiring the computation of $M^{-1}[U]$. The traditional update algorithms for QCD are local and suffer from Critical Slowing Down (CSD). Much effort has gone into the search for efficient nonlocal update algorithms for lattice QCD and also for simpler, but physically equally relevant, models in statistical mechanics. This search was very successful for a whole series of models, leading to very efficient algorithms of two types: (i) Swendsen-Wang cluster algorithms [8] and (ii) Multigrid algorithms [9]. For QCD in four dimensions, however, the search still continues. The computation of the *measurements* $F[M^{-1}, U]$ requires for most relevant physical quantities the inversion of the fermion matrix M, a task that can be accelerated by multigrid methods. Because this inversion plays an even more important role in the simulation of the full model, it will be discussed in that context.

Over the years, as computer power increased, the interest in MC simulations of *full* QCD was renewed, especially after the invention of a much faster exact algorithm: the *Hybrid Monte-Carlo algorithm* [10]; in this context the word exact means introducing no additional approximations. Using huge amounts of computer time on some of the biggest computers available at present, this algorithm enables us to take a first glance at some of the most fundamental properties of QCD, albeit on lattices, barely large enough to guarantee reliable physical results [11]. The main computational obstacle preventing us from obtaining sufficient statistics for larger systems is the fact that the Hybrid Monte Carlo Algorithm requires the frequent inversion of the fermion matrix M as part of the updating process. In all recent large-scale full-QCD simulations, more than 95% of the CPU time was spent inverting the fermion matrix by means of standard algorithms as the conjugate gradient algorithm and the preconditioned minimal residual algorithm. A direct consequence is that any real breakthrough in inversion algorithms will almost certainly lead to a breakthrough in lattice QCD. Because of the nature of the problem, a multiscale approach looks very promising. Several groups are pursuing this approach with somewhat differing methods and goals. This variety of efforts will be the topic of the following sections.

2 Parallel-Transported Multigrid (PTMG)

Although gauge theories require extra precautions, PTMG is, in spirit, close to multigrid methods used for solving elliptic partial differential equations. *Naive PTMG*, was proposed and tested for inverting the fermion matrix M in the *massive Schwinger model*: two-dimensional $U(1)$ lattice gauge theory [12]. Afterwards a more stable version was developed – *standard PTMG*, – one order of magnitude faster than the conjugate gradient algorithm [13]. It was successfully generalized to $SU(2)$ [14] and $SU(3)$ in two dimensions [15].

2.1 The computational problem

The inversion of the fermion matrix M is achieved by solving the linear system

$$\sum_{(k,l)}\sum_{\beta} M[U]^{\alpha\beta}_{(i,j),(k,l)}\Phi[U]^{\beta}_{(k,l)} = \delta_{i,1}\delta_{j,1}\delta^{\alpha,1} . \tag{10}$$

For simplicity's sake, explicit formulae are given for the two-dimensional case; the generalization to four dimensions is almost always straightforward. The numerical solution $\Phi[U]$ is one of the columns of $M^{-1}[U]$. A measure for the accuracy of the approximate solution $\Phi[U]$ is the *residual* defined by

$$r^{\alpha}_{(i,j)} = \delta_{i,1}\delta_{j,1}\delta^{\alpha,1} - \sum_{(k,l)}\sum_{\beta} M[U]^{\alpha\beta}_{(i,j),(k,l)}\Phi[U]^{\beta}_{(k,l)} \tag{11}$$

or its norm $|r|$, given by

$$|r|^2 = \sum_{(i,j)}\sum_{\alpha} |r^{\alpha}_{(i,j)}|^2 . \tag{12}$$

If we denote the exact but unknown solution of Eq.10 by $\tilde{\Phi}[U]$ and define the error E, corresponding to the present approximation $\Phi[U]$, by

$$E^{\alpha}_{(i,j)} = \tilde{\Phi}[U]^{\alpha}_{(i,j)} - \Phi[U]^{\alpha}_{(i,j)} , \tag{13}$$

then Eq. 10 can be rewritten in a completely equivalent form:

$$\sum_{(k,l)}\sum_{\beta} M[U]^{\alpha\beta}_{(i,j),(k,l)} E^{\beta}_{(k,l)} = r^{\alpha}_{(i,j)} . \tag{14}$$

The method, used for solving the problem iteratively, is the *Kaczmarz local relaxation procedure* [16]. As is always the case for local algorithms, it suffers from Critical Slowing Down (CSD), caused by the existence of *Approximate Zero Modes* (AZM), eigenvectors of M with very small eigenvalues (absolute value). In general it can not compete with the conjugate gradient algorithm. The situation changes completely, if the Kaczmarz algorithm is used as the *smoother* of a multigrid algorithm.

2.2 The multigrid approach (MG)

The main idea of MG is trying to solve the problem on a coarser lattice Λ^1 with mesh size $h^{(1)}$, where the smooth AZM of Λ^0 are less smooth. For this purpose, we need a *restriction* (coarsening) operator I_0^1, which transforms the components of the residual on Λ^0 to the coarse grid Λ^1:

$$r^{(1)} = I_0^1 r^{(0)} . \tag{15}$$

On the coarse grid we then solve

$$M^{(1)}E^{(1)} = r^{(1)} , \tag{16}$$

where $M^{(1)}$ stands for the appropriate "translation" of the fermion matrix on the coarse grid Λ^1. Finally, we need the *interpolation* operator I_1^0, translating the solution of Eq. 16 back to Λ^0:

$$E^{(0)} = I_1^0 E^{(1)} . \tag{17}$$

The interpolated $E^{(0)}$ is then used to obtain a better approximation on Λ^0:

$$\Phi^{(0)}(new) \leftarrow \Phi^{(0)}(old) + E^{(0)} . \tag{18}$$

The steps described here constitute the definition of a *two-level* process. The real MG process is generated by observing that the solution on the coarse grid Λ^1 may be accelerated in turn by an analogous procedure, involving an even coarser lattice Λ^2, etc. . In this way, the familiar V- and W-cycles can be defined. In some cases, it is for theoretical or practical reasons not advantageous to carry through the coarsening as far as possible. We call this procedure a $\Delta = n$ cut-off cycle, where n denotes the distance (expressed in number of levels) between the finest and the coarsest grid used. A cut-off $\Delta = 1$ cycle corresponds exactly to the two-level process discussed earlier. On all levels, ν_1 Kaczmarz relaxation sweeps are carried out before a restriction operation; similarly after a coarse- grid correction (interpolation) ν_2 Kaczmarz sweeps are done on the finer lattice. For gauge theories, the restriction and interpolation operators, as well as the coarse grid fermion matrix, must be selected with care; standard PTMG is such a choice.

2.3 Restriction, interpolation and coarse-grid M

A special property of the staggered form of M is that the fermion fields ψ, and consequently also the solutions Φ, located at the lattice points of the different grids $\Lambda^{(i)}$ are not all equivalent. One must distinguish *pseudoflavors*: in two dimensions there are 4 pseudoflavors and in four dimensions 16. As an example, we show in Fig. 1, how the solutions Φ_a, Φ_b, Φ_c, and Φ_d, corresponding to the four pseudoflavors a, b, c, and d in two dimensions are distributed over the lattice. Because the equation to be satisfied by the solution Φ (Eq. 10) depends upon the pseudoflavor, the fields $\Phi_a ... \Phi_d$ are treated by PTMG as independent fields, i.e., they are restricted and interpolated separately.

In gauge theories, the physical content is gauge invariant: the physical properties of a configuration do not change under a gauge transformation defined in Section 1.1. All configurations differing only by a gauge transformation are therefore physically equivalent. Assuming that the link variable $U_1(i+\frac{1}{2}, j)$ connects pseudoflavor a to b in Figure 1, it does not make sense to compare the specific value of this link variable with e.g. the link variable $U_1(i+\frac{1}{2}, j+2)$, although this

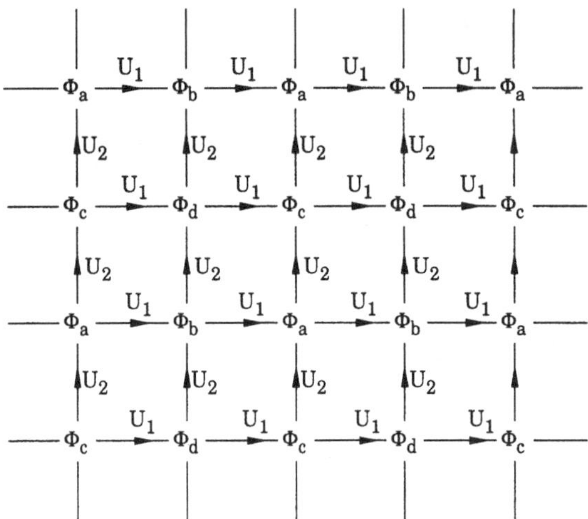

FIGURE 1. Staggered fermions in two dimensions: geometrical distribution of the four pseudoflavors of the field Φ over the lattice.

link connects the same pseudoflavors. Even if these U's are equal, a random gauge transformation, will destroy any relation between them. If, on the other hand, $U_1(i+\frac{1}{2},j) = U_2(i,j+\frac{1}{2})U_2(i,j+\frac{3}{2})U_1(i+\frac{1}{2},j+2)U_2^\dagger(i+1,j+\frac{3}{2})U_2^\dagger(i+1,j+\frac{1}{2})$, this relationship remains invariant under any gauge transformation. In the language of gauge theories, we say that $U_1(i+\frac{1}{2},j+2)$ has been *parallel transported* to the location of $U_1(i+\frac{1}{2},j)$. Consequently, only averaging over parallel-transported quantities should be allowed as part of the definition of the coarse-grid M; the same holds for the definition of the restriction operator I_i^{i+1} for the Φ's. Summarizing, parallel-transporting the relevant quantities removes completely the unphysical *gauge disorder* from the MG process. More details on the precise definitions of restriction, interpolation and coarse grid M may be found in [13, 14, 15].

2.4 Results and outlook

The first goal of the authors of PTMG was to develop an efficient inversion algorithm, which outperforms the commonly used algorithms in all or part of the physically relevant region of the parameters β and m_q. Starting with the massive Schwinger model, this goal was also reached for $SU(2)$ and $SU(3)$ lattice gauge theories in two dimensions.

As a typical example, I present data of a numerical experiment for $SU(3)$ lattice gauge theory on a 128×128 lattice, with quark mass $m_q = 0.01$ [15]. M is inverted, or to be more precise Eq. 10 is solved, for 20 statistically independent gauge configurations, produced by the quenched $SU(3)$ Monte Carlo algorithm, based on a heat-bath algorithm for three different $SU(2)$ subgroups [17], with $\beta = 400$ (gauge field correlation length $\xi = 10$). W-cycles, with on each level two

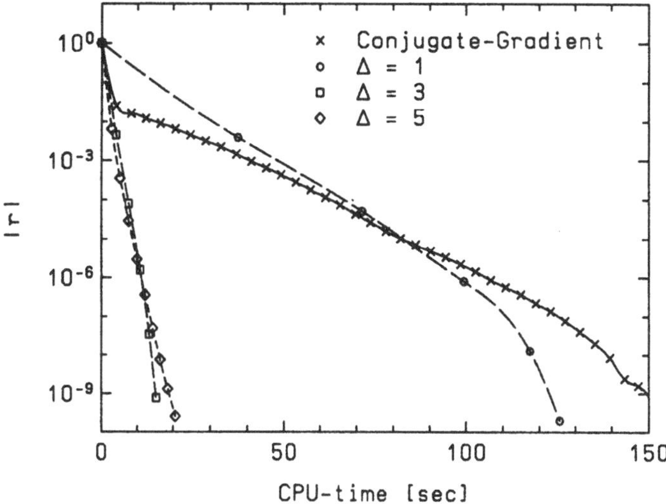

FIGURE 2. Comparison of convergence: PTMG ($\gamma = 2$, $\nu_1 = \nu_2 = 2$, several cycle-depths Δ) vs. Conjugate-Gradient algorithm for the inversion of the $SU(3)$ staggered fermion matrix in two dimensions with $m_q = 0.01$. The norm of the residual is plotted vs. the CPU time in seconds (one processor of a CRAY-YMP). The data shown are averages over 20 quenched $SU(3)$ configurations for $L = 128$ with $\beta = 400$ ($\xi = 10$). The data points for the CG represent 50 CG sweeps, those for the multigrid algorithm two W-cycles.

relaxation sweeps before each coarsening and two sweeps after each correction, are used: $\gamma = 2$ and $\nu_1 = \nu_2 = 2$. The effect of the cycle depth Δ, defined in section 2.2, was also investigated. The numerical experiment consists of measuring for each configuration the norm $|r|$ of the residual after every cycle as well as the amount of CPU time used (one processor of a CRAY-YMP). The results are summarized in Fig. 2, proving at least for this particular experiment the superiority of PTMG over conjugate gradient in realistic computer time.

Although these results are very encouraging, one should not forget that the real goal should be fast solvers for lattice gauge theories in four dimensions, not two. There are good reasons to believe that for correlation lengths ξ around 10, i.e., for configurations with relatively little physical disorder, and with lattice sizes of the order of 128^4, PTMG will also beat conjugate gradient. Especially with full QCD, however, simulations of this size will remain too big a computational task for many years to go, even with strongly increased computer power. Tests with somewhat smaller systems have not been carried out yet because of memory requirements: an efficient implementation of PTMG in four dimensions requires approximately ten times the amount of memory needed for one gauge configuration.

3 Ground-state projection multigrid

It became clear in the previous section that smoothness is not an obvious concept in gauge theories, because of the presence of gauge disorder and its interplay with real physical disorder. Based on the expected connection between smoothness and low-energy eigenstates, several groups of investigators, with varying degrees of sophistication and success, chose their MG interpolation operators in such a way that they project on the ground-state (smoothest) solution of the finer lattice in some sense [18, 19, 20, 21, 22]. In this section I can only give a brief outline of one particular way, how this idea can be realized [18, 19]. For more details and a complete description of the results I must refer to the literature.

3.1 Physical smoothness

Instead of solving

$$(D + m_q)\,\mathrm{X} = f\,, \tag{19}$$

where D is the anti-hermitian matrix defined in Eq. 2, the equivalent system

$$(-D^2 + m_q^2)\,\Phi = f \tag{20}$$

is solved. If Φ is a solution of Eq. 20, then $\mathrm{X} = (D + m_q)^{\dagger}\Phi$ is the corresponding solution of Eq. 19. The main reason for solving Eq. 20 instead of Eq. 19 is that an appropriate choice of interpolation and restriction operator, e.g. by the *variational* method, leads to an algorithm that can never diverge.

In the framework of solving Eq. 20, smoothness of an approximation or solution Φ is given a quantitive meaning by means of the functional

$$s[\Phi] = \frac{\Phi^{\dagger}(-D^2 + m_q^2)\Phi}{\Phi^{\dagger}\Phi}\,. \tag{21}$$

If $s[\Phi] < s[\Psi]$, then Φ is said to be smoother than Ψ. As a consequence, the smoothest field Φ_0 is the eigenstate of the operator $(-D^2 + m_q^2)$ corresponding to the lowest eigenvalue λ_0 of this operator. It is important to note that by this definition smoothness depends on the problem to be solved, because D, is a function of the gauge configuration (Eq. 2). The interpolation operator I_i^{i-1}, as well as the restriction operator I_{i-1}^{i}, must now be defined in such a way that the coarse grid corrections can "take care of" the smooth components on the fine grid.

3.2 Local ground-state projection kernels

Instead of treating a realistic but complicated problem, I will present explicitly, how these kernels may be constructed for a one-dimensional case: inverting a one-dimensional version of the covariant Laplacian plus mass term. I want to stress that this one-dimensional model is used only to show how to construct the kernels

– it does not have any meaning as a *bona fide* lattice gauge theory. Let us define the one-dimensional covariant "Laplacian" in the following way:

$$\Delta_{ij}^{(0)} = \frac{1}{h^2}[-2\delta_{i,j} + U(i+\frac{1}{2})\delta_{j,i+1} + U^\dagger(i-\frac{1}{2})\delta_{j,i-1}] \,. \tag{22}$$

We take the coarsening factor for the MG to be 3, i.e., the number of point on the grid Λ^k will be $\frac{1}{3}$ the number of points on the next-finer grid Λ^{k-1} (in one dimension). The restriction from Λ^{k-1} to Λ^k, is carried out by means of the operator I_{k-1}^k in the following way:

$$\Phi^{(k)}(i^{(k)}) = \sum_{i^{(k-1)}\in\Lambda^{k-1}} I_{k-1}^k(i^{(k)}, i^{(k-1)})\, \Phi^{(k-1)}(i^{(k-1)}) \,; \tag{23}$$

the interpolation operator is taken to be the hermitian conjugate of the the restriction operator: $I_k^{k-1} = {I_{k-1}^k}^\dagger$. We use non-overlapping restriction operators, setting $I_{k-1}^k(i^{(k)}, i^{(k-1)}) = 0$ if $i^{(k-1)} \notin \{3i^{(k)}, 3i^{(k)}+1, 3i^{(k)}+2\}$, i.e., each point $i^{(k)} \in \Lambda^k$ gets contributions from three points on Λ^{k-1}. The interpolation operator is now selected as the lowest eigenstate of a cut-off version of the operator to be inverted: $I_1^0(i^{(0)} \in \{0,1,2\}, i^{(1)} = 0)$ is the solution of the following set of equations that corresponds to the lowest eigenvalue λ_0:

$$\begin{aligned}
\frac{1}{h^2}[-2I_1^0(0,0) + U(\tfrac{1}{2})I_1^0(1,0) \qquad\qquad\qquad\qquad] + m_q^2\, I_1^0(0,0) &= \lambda_0\, I_1^0(0,0)\\
\frac{1}{h^2}[-2I_1^0(1,0) + U(\tfrac{3}{2})I_1^0(2,0) + U(\tfrac{1}{2})^\dagger I_1^0(0,0)] + m_q^2\, I_1^0(1,0) &= \lambda_0\, I_1^0(1,0)\\
\frac{1}{h^2}[-2I_1^0(2,0) \qquad\qquad\qquad + U(\tfrac{3}{2})^\dagger I_1^0(1,0)] + m_q^2\, I_1^0(2,0) &= \lambda_0\, I_1^0(2,0)\,.
\end{aligned}$$

Methods similar in spirit but considerably more complicated have been developed for inverting the fermion matrix. The details may be found in the literature cited above.

It is not *a priori* clear that defining the "local" kernels as the solution of an eigenvalue problem with some cut-off version of the operator $(-D^2 + m^2)$ is really the best one can do. This may well be the reason, why the results obtained with this method are not as good as one could hope for. One way to circumvent this problem will be treated in Section 4.

3.3 Results and outlook

Several variants of this method have been investigated, not only for the staggered fermions [18, 20], but also for the Wilson fermions [22]. Most of the work was limited to lattice gauge theories in two dimensions. $SU(2)$ lattice gauge theories with staggered fermions in four dimensions were also investigated [19]. In the latter study, lattices of sizes up to 18^4 were used and it is shown that a break-even with traditional methods may be reached for not too large systems, albeit with much larger memory requirements.

There seems to be a consensus that, although the practical implementations of projective multigrid methods seem to suffer less from Critical Slowing Down than the traditional algorithms, this problem has not been completely eliminated for the realistic case with non-trivial link variables U. It is completely eliminated, however, if "ideal" kernels are used [23], long-range kernels first proposed and used in the framework of Renormalization Group studies. This result is of theoretical interest only, because the complexity of these kernels makes them useless for practical simulations.

Summarizing, it can be said, that, in spite of some encouraging results, no real practical alternative has been found yet to replace the traditional inversion algorithms in four-dimensional QCD simulations, at least not for the system sizes that will be used during the next few years.

4 Iteratively Smoothing Unigrid (ISU)

As the name says, this method is not a MG algorithm in the usual sense, but rather a unigrid method. The smoothness concept behind this method is the same as for the ground-state projective method (Section 3.1). This method is very recent and not many results have been published yet: a brief outline and the first promising results can be found in [24], more details are contained in [25].

4.1 Interpolation kernels $\mathcal{A}^{[0j]}(x,z)$

The problem solved is the same as in the previous section: Eq. 20. As in all unigrid methods, no coarse grid representations of the system and of the operator $(-D^2 + m^2)$ are computed. Coarse grid corrections are made directly on the variables of the original grid Λ^0 by means of interpolation kernels $\mathcal{A}^{[0j]}$, where j refers to the coarse grid Λ^j.

An important difference with the practical projective multigrid algorithms is the fact that the range of the interpolating kernels, especially the ones responsible for very coarse corrections (high value of j), strongly overlap. In Fig. 3, the different grids and the range of the interpolation kernels are presented for the one-dimensional toy-problem of Section 3.2.

The interpolation kernel for the corrections from level Λ^j is denoted by $\mathcal{A}^{[0j]}(x,z)$, where $x \in \Lambda^j$ and $z \in \Lambda^0$. The way these interpolation kernels are selected is very similar in spirit to the method for the local ground-state projection kernels in Section 3.2. First of all, the condition $\mathcal{A}^{[0j]}(x,z) = 0$ is imposed, if z lies outside the range of x. Then, the remainder of the elements is determined by finding the eigenstate corresponding to the lowest eigenvalue $\lambda_0(x)$ of

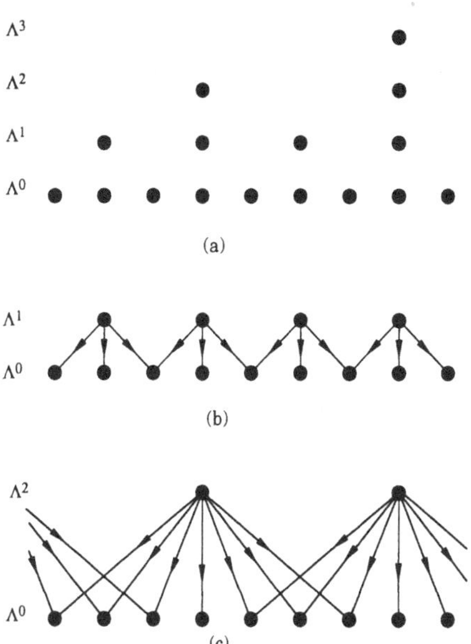

FIGURE 3. Grids and range of interpolation kernels for one-dimensional ISU: (a) schematic picture of Λ^0 through Λ^3, (b) range of interpolation kernels $\mathcal{A}^{[01]}$, (c) range of interpolation kernels $\mathcal{A}^{[02]}$.

the eigenvalue problem:

$$[-\Delta + m^2]\mathcal{A}^{[0j]}(x, z) = \lambda_0(x)\mathcal{A}^{[0j]}(x, z), \tag{24}$$

where $[-\Delta + m^2]$, acting on the coordinates $z \in \Lambda^0$, is the operator to be inverted.

An important question to be asked about this algorithm is the amount of computational work involved. The projection kernels $\mathcal{A}^{[01]}$, i.e., the kernels used for the corrections from Λ^1, are computed directly by inverse iteration. One then computes the remainder of the kernels $\mathcal{A}^{[0j]}$, with $j > 1$, iteratively by means of a MG scheme. As a consequence, the amount of computational work to build the kernels is estimated to be proportional to $V_\Lambda \times \frac{N(N-1)}{2}$, where V_Λ stands for the volume of the lattice and N is the number of grids. After the kernels have been found, the computational work to find the solution Φ is proportional to $V_\Lambda \times N$.

4.2 Results and outlook

In a careful study, the authors of the method have shown that it eliminates CSD completely for the inversion of the covariant Laplacian plus mass term in two-dimensional $SU(2)$. In fact, they consider the operator $(-\Delta - \epsilon_0 + \delta m^2)$. In this expression, Δ stands for the covariant Laplacian in two dimensions, the two-dimensional generalization of Eq. 22. The parameters ϵ_0 and δm^2 are introduced to study the behavior of the inversion algorithm near criticality (lowest eigenvalue

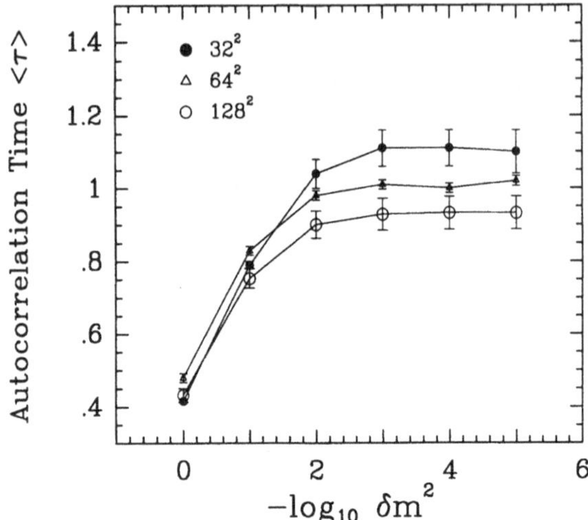

FIGURE 4. Convergence of the inversion of the operator $(-\Delta - \epsilon_0 + \delta m^2)$ by means of ISU: correlation time τ vs. δm^2 for three different lattice sizes. The data are averaged over 40 quenched two-dimensional $SU(2)$ configurations generated with heat-bath algorithm for $\beta = 1$. (unpublished data, courtesy of M. Bäker, G. Mack, M. Speh [24])

of the operator to be inverted near zero). Because Δ is a function of the link variables U, also its spectrum and in particular its lowest eigenvalue depend upon the specific gauge configuration. The authors therefore subtract the lowest eigenvalue ϵ_0 of the operator $-\Delta$ and then add by hand the parameter δm^2 to have complete control over the approach to criticality. Using two-dimensional quenched $SU(2)$ configurations ($\beta = 1.0$), they compute for several values of δm^2 and for different lattice sizes the correlation time τ. This quantity is defined as the decay constant of the exponential decay of the norm of the residual as a function of the number of Unigrid-sweeps: $\tau = 1$ means that one unigrid-sweep reduces $|r|$ by a factor e. From the data, collected in Fig. 4, it follows that CSD has been eliminated completely, actually the method seems to become even more effective for larger lattices. These results are very encouraging and the method will be tried out for the inversion of the fermion matrix.

Although CSD has been eliminated completely, this method might only become competitive with the traditional algorithms in practical QCD simulations for very large lattices, due to the large computational work required for building the kernels.

5 Gauge-Potential Multigrid

Although the physics of a configuration can be completely expressed in gauge invariant terms, this does not necessarily mean that algorithms used e.g. to invert the fermion matrix must be gauge covariant. Just because the physics hidden in a particular configuration does not change under a gauge transformation, it is perfectly acceptable to subject the configurations to a gauge transformation and then work with the transformed configurations. Removing the gauge disorder in the configurations by a gauge transformation and then using traditional MG techniques on this gauge-fixed configuration are key ingredients in a recently proposed MG method to invert the fermion matrix [26, 27].

5.1 The numerical problem

This investigation is carried out for $U(1)$ lattice gauge theory with *reduced staggered fermions* [28]. This fermion formulation is similar to the usual staggered one introduced in Section 1.1, but the number of lattice locations, where the fermion field ψ is defined, is reduced by a factor 2. Just as in the case of the usual staggered fermions, one should distinguish *pseudoflavors* both for the fermion field ψ and for the field Φ (the solution of the linear system to be solved). In two dimensions there are now two pseudoflavors instead of four: a and d; in four dimensions the number of pseudoflavors is eight. Φ_a and Φ_d are distributed over the two-dimensional lattice as in the usual staggered case: Fig. 1; the fields Φ_b and Φ_c, however, are missing in the reduced staggered case.

The reason for distinguishing pseudoflavors is that the equations they must satisfy are different. Let us assume that (in two dimensions) the pseudoflavor a fields are located at the lattice points $(1+2n, 0+2m)$ and correspondingly the pseudoflavor d fields at the points $(0+2k, 1+2l)$ where n, m, k, l are integer numbers. In this case the numerical problem to be solved can be described by giving the two generic equations:

$$
\begin{aligned}
&\frac{1}{2h}[U_1(2+\frac{1}{2},2)\Phi_a(3,2) - U_1^\dagger(2-\frac{1}{2},2)\Phi_a(1,2)] \\
&+\frac{1}{2h}[U_2(2,2+\frac{1}{2})\Phi_d(2,3) - U_2^\dagger(2,2-\frac{1}{2})\Phi_d(2,1)] \\
&+\frac{m_q}{2}[U_2(2,2+\frac{1}{2})\Phi_d(2,3) + U_2^\dagger(2,2-\frac{1}{2})\Phi_d(2,1)] \quad = \quad f_b(2,2)\,, \\
&-\frac{1}{2h}[U_2(3,3+\frac{1}{2})\Phi_a(3,4) - U_2^\dagger(3,3-\frac{1}{2})\Phi_a(3,2)] \\
&+\frac{1}{2h}[U_1(3+\frac{1}{2},3)\Phi_d(4,3) - U_1^\dagger(3-\frac{1}{2},2)\Phi_d(2,3)] \\
&+\frac{m_q}{2}[U_2(3+\frac{1}{2},3)\Phi_a(3,4) + U_2^\dagger(3-\frac{1}{2},3)\Phi_a(3,2)] \quad = \quad f_c(3,3)\,. \qquad (25)
\end{aligned}
$$

5.2 Gauge-potential representation and gauge-fixing

In Section 1.1, lattice gauge theories were defined using two types of variables: the link variables U_μ and the fermion fields ψ. There is a completely equivalent description, where gauge potentials A_μ take the place of the link variables U_μ. If

N_G is the number of generators of the gauge group G ($N_G = 8$ for QCD) and τ^a are the generators of the fundamental representation of G, then the relation between link variables U_μ and the gauge potentials A_μ is given by

$$U_\mu = \exp[ih \sum_{a=1}^{N_G} A_\mu^a \tau^a] . \tag{26}$$

This relation becomes considerably simpler for small gauge potentials ($|hA_\mu^a| \ll 1$): $U_\mu \approx 1 + ih\sum_{a=1}^{N_G} A_\mu^a \tau^a$. For the gauge group $U(1)$, this reduces even further to $U_\mu \approx 1 + ihA_\mu$. Using this A-field discretization for restrictions and interpolations as part of a multi-grid approach may have theoretical advantages [26].

In Section 2.3 the concept of parallel-transport was introduced to remove gauge-disorder and still keep a gauge-covariant formulation. Another well-known method to avoid the problem of gauge-disorder consists of removing, as well as possible, the unphysical degrees of freedom by gauge-fixing in an appropriate gauge. The configurations are subjected to gauge transformations selected in such a way that the A-fields of the gauge-transformed configurations are as smooth as possible, where the word smooth now has its traditional meaning. For their work with $U(1)$ lattice gauge theories in two dimensions, the authors use the *Landau-gauge*, i.e., they impose the condition div $A_\mu = 0$, or in discretized form

$$A_1(i+\frac{1}{2},j) - A_1(i-\frac{1}{2},j) + A_2(i,j+\frac{1}{2}) - A_2(i,j-\frac{1}{2}) = 0 . \tag{27}$$

After this gauge-transformation, the gauge-disorder of the discretized fields A and Φ has been removed and the full range of MG-techniques can now be used to tackle the physical disorder.

5.3 Results and outlook

In a series of controlled experiments for $U(1)$ lattice gauge theories in two dimensions, the effects of the details of the MG-procedure are investigated. Different ways to discretize the equations - central and backward-forward - are tried out; in the latter case a generalized Kaczmarz relaxation scheme, allowing for the simultaneous relaxation of two equations, is found useful. The importance of polynomial acceleration, removing slowly convergent (or even divergent) modes by linearly combining iterants, is investigated as a function of quark-mass m_q and β. The effects of different ways of building the averages for the restriction operator and the corresponding interpolation operators are investigated. Until now the numerical experiments have been carried out on configurations generated by a gaussian approximation [27]. Only configurations with zero global topological charge have been considered and a method has been proposed to handle configurations with non-zero charge [26].

The preliminary results are very encouraging [27]. Although the efficiency decreases for increasing physical disorder (decreasing values of β) and smaller values of the quark-mass m_q, full MG-efficiency is reached for probably the whole physically relevant range of the parameters. Polynomial acceleration seems to be necessary for small quark masses, especially if they are combined with small values of β (large physical disorder).

Although these experiments are very interesting by themselves, the main interest of the lattice gauge theory community is QCD in four dimensions. The generalization of some of the ingredients of this approach to such four-dimensional non-abelian gauge-theories is nontrivial. The urgent need for faster algorithms, however, more than justifies the effort.

6 Concluding remarks

In this survey of attempts to build fast multiscale algorithms for inverting the fermion matrix in QCD simulations, I was forced to be incomplete. I was only able to briefly sketch some approaches and present a couple of results. One multigrid method, based on the Migdal-Kadanoff renormalization group transformations, I had to skip completely [29] and no attention could be paid to more theoretical investigations of the problem [30]. Still I hope that I succeeded in conveying an impression of the very active search for multiscale algorithms in an important field of research.

MC simulations are and will remain an important tool, in some cases even the only tool, to obtain nonperturbative information about physical models. These models describe a wide range of important physical phenomena: elementary particles and their interactions, solids and their phase-transitions, etc. . In these phenomena and in their simulation, many different physical length scales play an essential role. Hence, a multiscale approach may reduce considerably the overall computational work, required to obtain the relevant physical information. For a discussion of ideas and also some results in applying a multiscale approach to many different aspects of MC simulations, I refer to the contribution of A. Brandt at this conference.

In the MC simulations of full QCD, the real computational bottleneck, at this moment, is the inversion of the fermion matrix. Although multiscale approaches may be needed for many more aspects of the simulation (efficient gauge-fixing, efficient updates, etc.), I expect the first contribution to come in the form of a fast multiscale inversion algorithm. If such a solver is found for QCD in four dimensions, the quality of the physical predictions will improve dramatically. Also the need for a multiscale approach to the other aspects of the MC simulations will then become apparent.

References

[1] There are many textbooks describing different aspects of the basic interactions in nature. A good phenomenological introduction to this field can be found in D.H. Perkins, *Introduction to high energy physics* (Addison-Wesley Publishing Company, 1987). A more theoretical approach is taken by T.-P. Cheng and L.-F. Li, *Gauge theory of elementary particle physics* (Oxford University Press, 1984) as well as by C. Itzykson and J.-B. Zuber, *Quantum field theory* (McGraw-Hill International Book Company, 1980).

[2] Perhaps the best didactic introduction to the theory of gravitation and the general theory of relativity is C.W. Misner, K.S. Thorne and J.A. Wheeler, *Gravitation* (W.H. Freeman and Company, 1973).

[3] Several introductions to this field and review articles have been written. The basics are described e.g. by M. Creutz, *Quarks, gluons and lattices* (Cambridge University Press, 1983). Contributions written by some of the leading scientists in this fields have been collected in M. Creutz (editor), *Quantum fields on the computer* (World Scientific, 1992).

[4] L. Susskind, Phys. Rev. D16 (1977) 3031.

[5] K. Wilson, in *New Phenomena in Subnuclear Physics*, edited by A. Zichichi (Plenum Press, 1977).

[6] J.M. Hammersley and D.C. Handscomb, *Monte Carlo Methods* (Chapman and Hall, 1964) Chapter 9.

[7] A.D. Sokal, *Monte Carlo methods in Statistical Mechanics: Foundations and New Algorithms*, Cours de Troisieme Cycle de la Physique en Suisse Romande (Lausanne, June 1989).

[8] A review treating progress and problems of the cluster approach, as well as giving references to the literature: A.D. Sokal, Nucl. Phys. B (Proc.Suppl.) 20 (1991) 55; clusters are also covered by J.J. Binney, N.J. Dowrick, A.J. Fisher, M.E.J. Newman, *The Theory of Critical Phenomena* (Oxford University Press, 1992).

[9] A recent article comparing progress for several models and also giving references to the literature is M. Grabenstein and K. Pinn, J. of Stat. Phys 71 (1993) 607.

[10] S. Duane, A.D. Kennedy, B.J. Pendleton, and D. Roweth, Phys. Lett. B195 (1987) 216.

[11] Results and status of the larg-scale MC simulations in lattice gauge theory are presented in several contributions to *Lattice 92*, Proceedings of the International Symposium on Lattice Field Theory, Amsterdam, The Netherlands,

15-19 September 1992, edited by J. Smit and P. van Baal, Nucl. Phys. B (Proc. Suppl.) 30 (1993). The QCD spectroscopy simulations were reviewed in the plenary talk by A. Ukawa, p. 3.

[12] R. Ben-Av, A. Brandt and S. Solomon, Nucl. Phys. B329 (1990) 193.

[13] M. Harmatz, P.G. Lauwers, R. Ben-Av, A. Brandt, E. Katznelson, S. Solomon, and K. Wolowesky, Nucl.Phys. B (Proc.Suppl.) 20 (1991) 102 and Phys.Lett. B253 (1991) 185;
R. Ben-Av, M. Harmatz, P.G. Lauwers and S. Solomon, Arbeitspapier der GMD Nr. 674, August 1992, accepted for publication in Nucl. Phys. B (1993);
M. Harmatz, P.G. Lauwers, S. Solomon and T. Wittlich, Nucl. Phys. B (Proc. Suppl.) 30 (1993) 192.

[14] S. Solomon, and P.G. Lauwers, in *Workshop on Fermion Algorithms*, eds. H.J. Herrmann and F. Karsch (World Scientific, Singapore, 1991) 149;
P.G. Lauwers, R. Ben-Av, and S. Solomon, Nucl. Phys. B374 (1992) 249.

[15] P.G. Lauwers and T. Wittlich, Int. J. Mod. Phys. C 4 (1993) 609 and Nucl. Phys. B (Proc. Suppl.) 30 (1993) 261.

[16] S. Kaczmarz, Bull. Acad. Polon. Sci. Lett. A. 35 (1937) 355.

[17] N. Cabibbo and E. Marinari, Phys. Lett. 119B (1982) 387.

[18] G. Mack, in *Nonperturbative quantum field theory*, eds. G. 't Hooft et al.(Plenum,1989);
T. Kalkreuter, G. Mack, and M. Speh, in *Fermion Algorithms*,eds. H.J. Herrmann and F.Karsch (World Scientific, 1991) 121;
M. Bäker, T. Kalkreuter, G. Mack, and M. Speh, DESY-preprint DESY-92-126, published in *Proceedings of the 4th International Conference on Physics Computing '92*, edited by R.A. de Groot and J. Nadrchal (World Scientific, 1993);
T. Kalkreuter, Nucl. Phys. B376 (1992) 637 and Phys. Lett. B276 (1992) 485.

[19] G. Mack, T. Kalkreuter, G. Palma, and M. Speh, in *Computational Methods in Field Theory*, Lecture Notes in Physics 409, eds. H. Gausterer and C.B. Lang (Springer Verlag, 1992);
T. Kalkreuter, Nucl. Phys. B (Proc.Suppl.) 30 (1993) 257, Int. J. Mod. Phys. C3 (1992) 1323, and DESY-92-158.

[20] A. Hulsebos, J. Smit, and J.C. Vink, Nucl. Phys. B (Proc.Suppl.) 9 (1989) 512, Nucl. Phys. B331 (1990) 531, Nucl. Phys. B (Proc.Suppl.) 20 (1991) 94, Nucl. Phys. B368 (1992) 379 and in *Fermion Algorithms*,eds. H.J. Herrmann and F.Karsch (World Scientific, 1991) 161.

[21] R.C. Brower, E. Myers, C. Rebbi, K.J.M. Moriarty, in *Multigrid Methods*, S.F. McCormick ed., (Marcel Dekker, New York, 1988);
R.C. Brower, C. Rebbi and E. Vicari, Phys. Rev. D 43 (1991) 43.

[22] R.C. Brower, R.G. Edwards and C. Rebbi, Nucl.Phys.B366 (1991) 689.

[23] T. Kalkreuter, DESY-93-046, to be published in Phys. Rev. D.

[24] M. Bäker, G. Mack, and M. Speh, Nucl. Phys. B (Proc.Suppl.) 30 (1993) 269.

[25] M. Bäker, *Ein Mehrgitterverfahren für ungeordnete Systeme*, Diplomarbeit, Universität Hamburg (1993).

[26] A. Brandt, Nucl.Phys.B (Proc.Suppl.) 26 (1992) 137.

[27] M. Rozantsev, Master's Thesis, Weizmann Institute (in preparation).

[28] C. van den Doel and J. Smit, Nucl. Phys. B228 (1983) 122.

[29] V. Vyas, in *Fermion Algorithms*,eds. H.J. Herrmann and F.Karsch (World Scientific, 1991) 169; Wuppertal Un. Preprints WUB-91-10 and WUB-92-30.

[30] A.D. Sokal, *Some Comments on Multigrid Methods for Computing Propagators*, available from hep-lat/9307020

6

Adaptive Multigrid on Distributed Memory Computers

Hubert Ritzdorf and Klaus Stüben[1]

ABSTRACT [2] A general software package has been developed for solving systems of partial differential equations with adaptive multigrid methods (MLAT) on distributed memory computers. The package supports the dynamic mapping of refinement levels. The general strategy is described and results are reported on compute-intensive problems as well as on some simple problems representing worst-case situations from a parallel efficiency point of view. Inherent limitations of the parallel efficiency will be discussed.

1 Introduction

Generally, adaptive grids are a result of the computation, and dynamically mapping the work load to the processors and achieving load balance are tasks which have to be performed *at run time*. Careful strategies must be employed in order not to destroy the parallel efficiency through communication overhead. In this paper, we present results obtained for the *multi-level adaptive technique* (MLAT [1],[2]) on 2D block-structured, boundary-fitted grids which are widely used in aerodynamic applications. Such grids permit the numerical solution of partial differential equations on geometrically complex domains while keeping regular the local data structure (within each process). We will discuss communication and mapping aspects, the way in which local refinement areas are generated and distributed to the available processors.

MLAT is known to provide very fast solvers on sequential computers and the question is how far this is applicable to parallel machines. While dynamic mapping is required by any adaptive algorithm, an additional problem occurs in MLAT, namely, the problem that each cycle requires substantial global communication (data re-distribution before switching refinement levels). It turns out that the latter has no severe consequences in connection with "complex problems". However,

---

[1]Gesellschaft für Mathematik und Datenverarbeitung mbH,
Postfach 1316, 53731 Sankt Augustin, Germany

[2]This work was supported by the Federal Ministry of Research and Technology (BMFT) under contract no. ITR 9006 (PARANUSS project).

for applications with low arithmetic per grid point, the corresponding overhead may seriously limit the achievable parallel efficiency. Although still acceptable for environments with "ideal" interconnection networks, this limitation is fatal for bus-connected systems like workstation clusters.

Results will, in particular, be presented for the steady-state Euler equations. The emphasis is laid on the adaptive refinement of the shock position. High parallel efficiencies are obtained even for relatively small problems. As worst-case examples, we consider scalar problems with singularities induced by the shape of the domain (re-entrant corners). Based on some simple analysis, we will point out the crucial aspects.

2 Adaptive multigrid (MLAT)

MLAT is essentially an FMG-like process (full multigrid, [1], [2]) which initially works merely on a hierarchy of "global" grids, Ω_ℓ^h ($\ell = 0, 1, 2, ..., \ell_c$), where Ω_0^h denotes the finest global grid given by the user. Only at run time, controlled by certain criteria, local refinement levels $\Omega_{-\ell}^h$ ($\ell = 1, 2, \ldots, \ell_f$), extending over increasingly smaller subdomains of the original domain, will be detected and successively added to the grid hierarchy. If no more local refinement levels are detected, mere multigrid cycling is continued until a reasonable convergence criterion is satisfied.

Let the number of points on the global and the locally refined levels be denoted by N_ℓ and $N_{-\ell}$, respectively. We will consider only *standard coarsening.* In particular, the global grids are nested and the number of points on Ω_ℓ^h is (approximately) $N_\ell = N_0/4^\ell$; the refinement grids are *locally* nested, and the coarse-level restriction of grid $\Omega_{-\ell}^h$ is denoted by

$$\bar{\Omega}_{-\ell+1}^h := \Omega_{-\ell}^h \cap \Omega_{-\ell+1}^h \,. \tag{1}$$

Throughout this paper, we tacitly assume that $N_{-\ell}$ decreases for increasing ℓ such that $N_{mg} = \mathcal{O}(N_0)$ where N_{mg} denotes the total number of grid points involved in the multigrid process.

Analogous to non-adaptive cycles, adaptive ones are defined recursively by means of two-grid methods. Besides the fact that, in the adaptive context, FAS (full approximation scheme [1], [2], [14]) is employed, the only essential difference is that – on any refined grid $\Omega_{-\ell}^h$ – the corresponding two-grid method uses grids $\Omega_{-\ell}^h$ and $\bar{\Omega}_{-\ell+1}^h$ (rather than $\Omega_{-\ell+1}^h$). Along the artificial inner boundaries of $\Omega_{-\ell}^h$, boundary values are usually interpolated from the current approximation on grid $\Omega_{-\ell+1}^h$. The concrete type of interpolation used may be crucial for the speed of convergence as well as for the global discretization error. In many cases, sufficiently accurate standard interpolation may be used (e.g., in case of Poisson-like equations). In other cases, e.g. compressible fluid flow problems, more care has to be taken (cf. Section 3).

Due to the above assumption, the computational work per V-cycle is $\mathcal{O}(N_0)$. Since this is no longer true for different cycle types like F- or W-cycles, such cycles should be avoided in the adaptive context. Moreover, applying FMG in the above fashion, even if it is based on V-cycles, does not yield an $\mathcal{O}(N_0)$-method. Instead, one should employ FMG in a more sophisticated way (e.g. λ-FMG [1], [2], [12]). If robustness requires, for instance, the use of F-cycles rather than V-cycles, one should not use F-cycles directly; more efficiently, for instance, the particular recursive structure of an F-cycle can be combined with the λ-FMG process resulting again in an overall computational work of $\mathcal{O}(N_0)$.

In this paper, we do not consider optimal FMG-implementations, but rather focus on the most crucial aspects from a parallel point of view, in particular those which are specific for the adaptive situation. That is, we focus on the parallel realization of the refinement process itself as well as on the parallel efficiency of plain multigrid cycles applied to the full sequence of grids. (Unless explicitly stated otherwise, we have V-cycles in mind, see above.) Clearly, a more in-depth consideration of the parallel performance has to take the total FMG process into account.

2.1 Parallelization aspects

Since the different multigrid levels are treated sequentially, the only reasonable parallelization strategy is to map each level to as many processors as possible. Formally, this is analogous to the standard way of parallelizing non-adaptive cycles. The well-known deficiencies – decreasing ratio arithmetic/communication and, eventually, less points than processors – just not only apply to the global coarse levels but similarly to the locally refined ones.

There are, however, some new aspects which will be discussed below for the case of *block-structured grids*, i.e., grids which are composed of subgrids each of which is *logically* rectangular (for a very simple example, see Figure 1). Such grids are widely used, for instance, in aerodynamic applications. They build a compromise between geometrical flexibility and simplicity of the data structure.

The block-structure provides a natural basis for the parallelization: Each block is mapped to a different process. (As usually, overlap regions of a certain width have to be introduced in order to allow for an efficient communication.) The minimum number of blocks required to describe a concrete geometry is merely defined by the requirement that the final grid should be "reasonable". Usually, this number is much lower than the number of available processors, P. Consequently, on a parallel machine, large blocks are subdivided further in order to obtain good *load balancing* which typically means that each block should contain the same number of grid points. For instance, the grid in Figure 1 has originally been created as a single-block grid (by a biharmonic grid generator) and was then subdivided into 16 equally sized blocks for use on a 16-processor machine. Generally, good load balancing can be obtained only approximately, which is certainly a tribute we have to pay for, say, the advantages of block-structured grids as opposed to unstructured

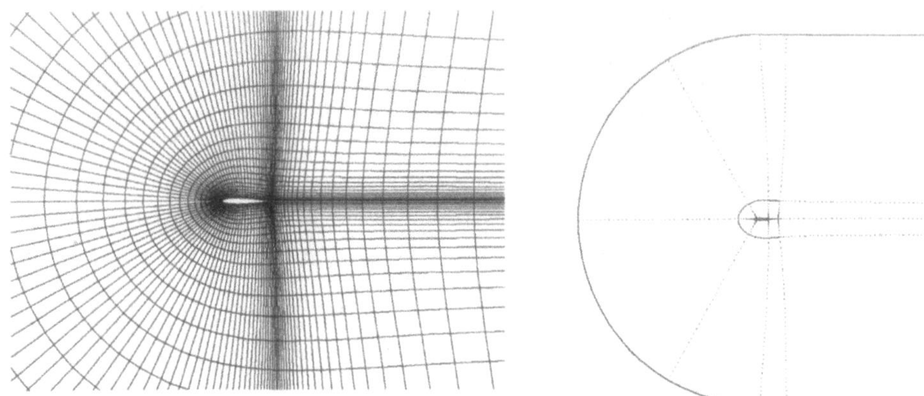

FIGURE 1. Block-structured grid around the NACA0012 airfoil (here: 16 blocks)

grids.

Introducing and mapping new locally refined grids

During the refinement phase, the grid data of each new refinement level is initially distributed across only some of the processes. That is, the parallel FMG process cannot continue in a load balanced way (with the new level being the finest one), until that new level has been re-mapped to all processors. Note that this mapping does not affect the mapping of previous refinement levels.

Generally, obtaining *optimal* load balancing at each stage of the FMG process is too complicated and costly. What is required is an algorithm which *rapidly* re-maps distributed locally refined block-structures to *reasonably* load balanced ones. Omitting complex details as well as some technical restrictions which require certain natural modifications, the essential steps of such an algorithm are simple and outlined in the following. We assume that grid $\Omega^h_{-\ell+1}$ already exists and that the next refined grid, $\Omega^h_{-\ell}$, has to be created and mapped.

1. Each process checks for refinement areas independently of the others. Since we are considering only block-structured grids, each process has to embedd its local refinement area(s) into logically rectangular subgrid(s). If no process detects refinement areas, the refinement process is finished.

2. If refinement areas have been detected, communication is required to analyze the resulting block-structure and to set up the corresponding data-structure. At this point, local "process blocks" should be joined to larger "superblocks" whenever possible in order to obtain a final block-structure with as few blocks as possible. (This gives the maximum freedom for a load-balanced mapping.) The optimal number of grid points each processor should work on, $N^{(P)} = N_{-\ell}/P$, is computed and broadcast.

3. All blocks containing less than $N^{(P)}$ points are distributed immediately (each block to a different process). Small blocks should share the same processor such that the total number of points treated by a processor is as close to $N^{(P)}$ as possible.

4. Since the previous step will not give optimal load balancing with respect to the small blocks, the optimal number of grid points *for the remaining processors* is re-computed: $N^{(P)} = \widetilde{N}_{-\ell}/\widetilde{P}$ where $\widetilde{N}_{-\ell}$ and $\widetilde{P}$ denote the remaining number of grid points and processors, respectively. If there are blocks left containing less than $N^{(P)}$ points, go back to the previous step. Otherwise, proceed to the next step.

5. Blocks containing more than $N^{(P)}$ points have to be subdivided. To be more precise, if a block contains n points, it is subdivided into m subblocks, m being the largest integer $\leq n/N^{(P)}$. Each of the subblocks should approximately contain the same number of points. Note that all blocks can perform their subdivision in parallel.

6. The total number of subblocks created in the previous step cannot exceed $\widetilde{P}$, but it may be smaller, in which case there are free processors (at most equal to the number of blocks which initially had to be subdivided). If this is true, blocks which currently contain the largest subblock(s) are re-subdivided with the number of subblocks increased by one. This is applied to as many blocks as required to make all processors busy.

The result of this procedure is a new block-structure with each block mapped to a different process. Note that the main goal is to minimize the size of the largest block. In fact, this is the most crucial point in trying to obtain approximate load balancing, much more important than trying to get all the small blocks perfectly load balanced. For reasons of high parallel efficiency, one might want to impose additional constraints like, e.g., preserving nearest neighbor relations or minimizing the cost for data re-distribution within cycling (see Section 2.1). Apart from the fact that the underlying goals are conflicting, it is very hard to realize "optimal" algorithms for general block-structured grids. Since it is not clear a priori, whether or not such more sophisticated algorithms would really pay in practice, we have not yet invested much work in this direction.

We want to emphasize that each of the essential mapping steps can be performed in parallel and all communication can be arranged to be either nearest neighbor or along embedded trees. Thus, the communication overhead is merely $\mathcal{O}(log(P))$. In practice, the total work required for re-mapping is negligible if compared to the rest of the work. An examplary sequence of three successive, block-structured refinement areas, obtained when solving the Euler equations on the grid as depicted in Figure 1, is shown in Figure 2 (for more details, see Section 3).

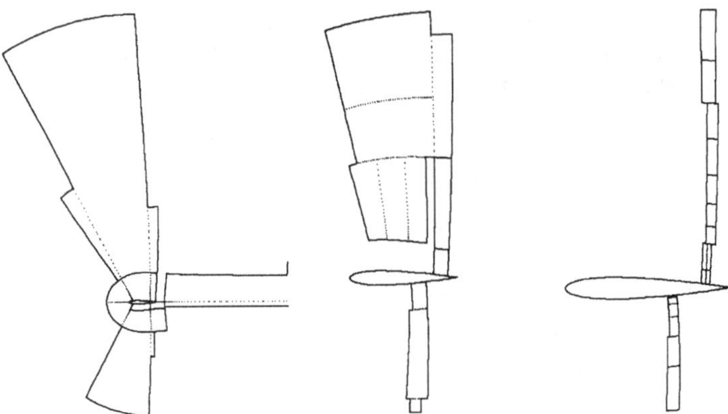

FIGURE 2. Hierarchy of 3 block-structured refinement areas.

Data re-distribution within each multigrid cycle

Due to the above mapping, data has to be re-distributed whenever locally refined levels are switched during a multigrid cycle. (This can only be avoided by using much more complicated mapping strategies, see Section 4.3.) This re-distribution should be done such that all arithmetic required in the grid transfer (corrections, residuals) can be performed in a load-balanced way.

To be more precise, let us assume that we have just finished relaxation on grid $\Omega^h_{-\ell+1}$ and that we want to transfer corrections from that level to the next finer one, $\Omega^h_{-\ell}$. At this point, the relevant correction data is contained on the coarse-level subgrid $\bar{\Omega}^h_{-\ell+1}$. Since this subgrid is distributed only over some of the processors, one should *first* distribute the data to the processes of grid $\Omega^h_{-\ell}$ and *only then* perform the actual interpolation and correction. Similarly, during the fine-to-coarse transfer, all necessary computations (evaluation of residuals, application of the full weighting operator, etc.) should be done on the *fine* level; only the data which is really relevant for the coarser level should then be re-distributed. Note that now, according to the definition of FAS, two types of grid functions have to be re-distributed, namely, residuals and current approximations.

In addition to load balancing, this way of re-distributing data has another obvious advantage: the amount of data to be re-distributed is the smallest possible.

Treatment of the critical levels

Both for the coarsest global grids and the finest local grids, the number of grid points may finally become smaller than the number of processors. We call the corresponding levels the *critical* ones.

Concerning the critical *fine* levels, we reduce the number of processors gradually from level to level. To be more specific, the decision on the number of processors which will stay active on grid $\Omega^h_{-\ell}$ is based on the number $\bar{N}_{-\ell+1}$ of points contained in the coarse-level subgrid $\bar{\Omega}^h_{-\ell+1}$: the number of active processors on grid $\Omega^h_{-\ell}$ is $P_{-\ell} = \min\{P, \bar{N}_{-\ell+1}\}$. Consequently, on each of the critical fine levels, each processor is acting on a fixed number of grid points (4 on the average). In addition to the usual communication overhead (inter-level as well as intra-level), this introduces arithmetic overhead in the parallel cycle. Generally, the total overhead caused by the critical levels depends merely on P. For instance, it is $\mathcal{O}(log(P))$ if $N_{-\ell}$ decreases geometrically[3].

Clearly, in order to keep the overhead caused by the critical *coarse* levels of the same order, one has to proceed similarly on these levels. For a small number of processors, however, one might as well skip these levels totally, and simply resolve the new coarsest-grid equations sufficiently well (e.g. by additional relaxation steps).

2.2 AVAILABLE SOFTWARE

The communication tasks required on block-structured grids are independent of the actual application. A comfortable and flexible library of highlevel FORTRAN routines has been developed (COMLIB [5]), which perform all communication required on such grids. This includes both local and global communication on single grid levels (provision of overlap areas, overlap update, computation of global quantities, etc.), analyzing, mapping and load balancing of new refinement levels, global data re-distribution as well as all inter-level communication.

On the one hand side, one may regard the COMLIB as a user interface to a parallel machine, freeing the user from the need to use any parallel language construct. In addition, and much more important, the use of the provided routines drastically simplifies the development of parallel application programs. Each process "sees" merely a single logically rectangular grid; the complex grid structure as a whole is solely managed by the COMLIB and never visible to the programmer. Thus, basically, the programmer's work is reduced to what he would have to do on standard sequential computers and for single-block grids. Finally, the COMLIB itself is based on a portable message passing programming model which has been implemented on a wide variety of different architectures, thus giving portability among all these machines (PARMACS [6]).

Remark: In the current library release, two of the steps described in Section 2.1 are not yet realized. Firstly, the distributed local refinement blocks are not "joined" as mentioned in Step 2. Secondly, only one grid block is treated by each processor (cf. Step 3). (The creation of more than one process per processor is not supported on all new architectures.)

The results presented in the following have been obtained by the general program package L_iSS [11], [8] which has been developed for the parallel multigrid

[3] We here think of "real" parallel systems. Of course, for workstation clusters this is not true.

solution of large classes of systems of partial differential equations on distributed memory computers. L_iSS is based on the COMLIB and can thus handle general block-structured grids.

3 Results for the Euler equations

In the following, we consider the steady-state Euler equations

$$\frac{\partial f}{\partial x} + \frac{\partial g}{\partial y} = 0 \quad \text{where} \quad f = \begin{bmatrix} \rho u \\ \rho u^2 + p \\ \rho uv \\ (E+p)u \end{bmatrix}, \quad g = \begin{bmatrix} \rho v \\ \rho uv \\ \rho v^2 + p \\ (E+p)v \end{bmatrix}$$

with ρ, u, v, E and p denoting the density, the cartesian velocity components, the total energy and the pressure, respectively. In addition, we assume the state equation $p = (\gamma - 1)(E - \frac{1}{2}\rho(u^2 + v^2))$.

As test examples, we consider flows around the NACA0012 airfoil with $M_\infty =$ 0.85, angle of attack 1.0° (Example 1) and $M_\infty = 0.8$, angle of attack 1.25° (Example 2). In the first example, we have a strong shock at the lower surface, in Example 2 only a very weak one (see Figure 3).

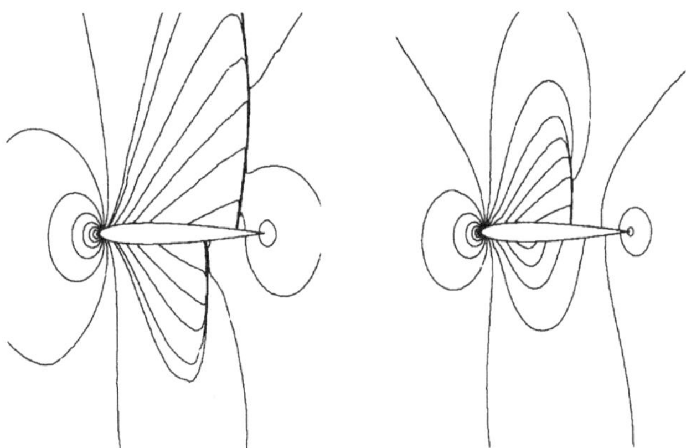

FIGURE 3. Pressure distribution for Examples 1 and 2

Following [4], we use a finite-volume discretization based on Osher's flux-difference splitter. In contrast to [4], however, we apply it to a vertex-centered distribution of unknowns. The computational grid, Ω_0^h, and its corresponding subdivision into 16 blocks (for the use on 16 processors) are shown in Figure 1.

Values at points along inner boundaries of refinement areas are not interpolated from coarse-level values, but rather discretized conservatively by applying Osher's scheme to special control volumes (cf. Figure 4a). This turned out to be important; using non-conservative formulas instead (e.g. cubic interpolation) may cause not only a deterioration of the accuracy, but also a considerably worse multigrid convergence (cf. [3]). Note that the unknowns along inner boundaries are incorporated into the multigrid process in just the same way as all the other unknowns.

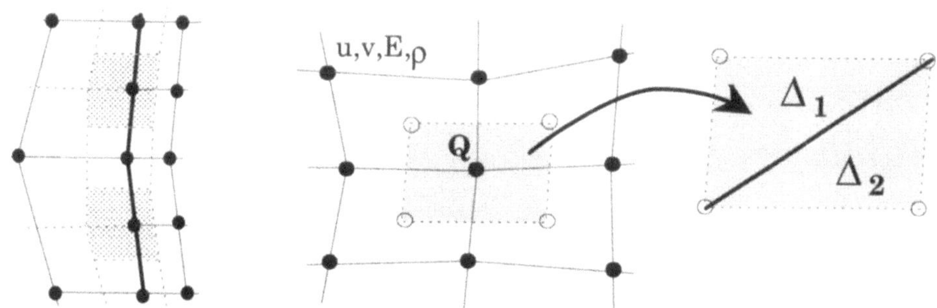

FIGURE 4. a) Discretization at inner boundary, b) Refinement criterion

For the self-adaptive grid refinement, a heuristic criterion, based on the *finite-element residual*, turned out to be well suited (cf. [13]). It not only detects critical areas but also yields a natural stopping criterion for the refinement process (in contrast to most criteria used in practice, e.g. those based on gradients). The essential idea is as follows. For each point of the current level, Q, its corresponding control volume is subdivided into two triangles, Δ_1 and Δ_2 (cf. Figure 4b). For each triangle, we compute linear functions approximating u, v, ρ and E (based on their current nodal values), and – by inserting these functions into the Euler equations – corresponding *residual vectors* r_1^h and r_2^h. With r_{ij}^h denoting the j–th component of r_i^h, we define the control quantity

$$r^h(Q) = \sum_{i=1}^{2} \sum_{j=1}^{4} \int_{\Delta_i} | r_{ij}^h | \, dx \, .$$

Figure 5 shows contour lines for the finite element residual r^h in case of Example 1, plotted on grid Ω_0^h.

Given some tolerance, ε, points with $r^h(Q) \geq \varepsilon$ will be marked for refinement. After all points of the current level have been processed this way, marked points will be embedded into blocks (Step 1 in Section 2.1) and the refinement and mapping algorithm outlined in Section 2.1 yields the next refinement level. This process is applied recursively to add more levels.

For Example 1 and $\varepsilon = 10^{-3}$, we obtain 3 levels of block-structured refinement areas (depicted in Figure 2). The corresponding composite grid is shown in Figure 6. Figure 7 compares the pressure distribution, computed on the finest global grid, Ω_0^h, and the locally refined one, respectively.

FIGURE 5. FE-residual contour lines for NACA0012 (Example 1)

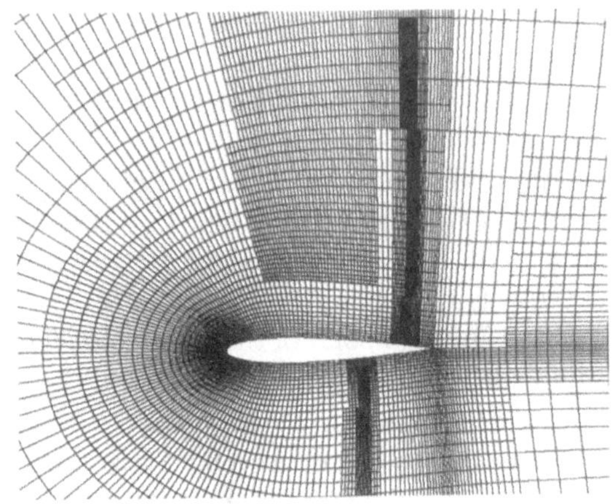

FIGURE 6. Composite grid (Example 1)

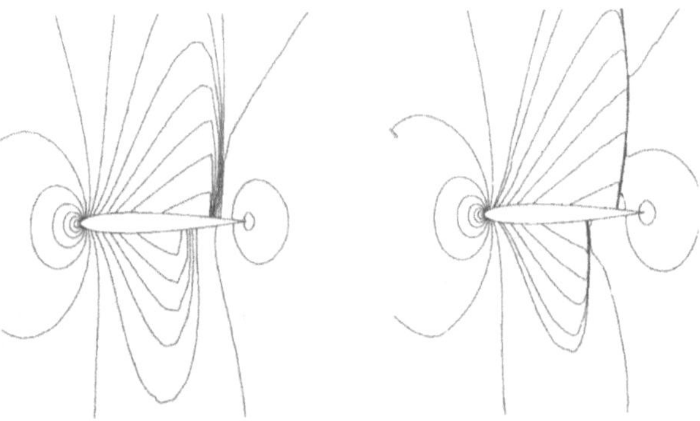

FIGURE 7. Pressure distribution without/with local refinements (Example 1)

Generally, the parallel efficiency, $E(N_0, P)$, is defined by

$$E(N_0,P) = \frac{1}{P}\frac{T(N_0,1)}{T(N_0,P)}; \quad \mathcal{E}(N_0,P) = \frac{1}{P}\frac{\sum_{i=1}^{P} a_i}{\max_i (a_i + c_i)} \tag{2}$$

where $T(N_0, p)$ denotes the computing time required (by the same program) on p processors. Since, usually, storage limitations do not permit the solution of the complete problem on a single processor, we measure parallel efficiency in terms of $\mathcal{E}(N_0, P)$ instead, see (2). Here, a_i and c_i denote the (wall-clock) time for the arithmetic and the total communication time (including idle times), respectively, on processor i. We have $E(N_0, P) \approx \mathcal{E}(N_0, P)$ if all processes are synchronized when the parallel application and the time measurements start, if each process (block) is mapped to a different processor, if the parallel algorithm does not involve substantial additional arithmetic, and if the floating point performance of the nodes does not depend too sensitively on the grid size (such that we can assume $T(N_0, 1) \approx \sum_{i=1}^{P} a_i$). Below, all this is approximately true.

Table 1 shows convergence factors (ρ), number of grid points on the composite grid (N_{cg}) and parallel efficiencies $\mathcal{E}$ per cycle[4] measured on the Intel iPSC/860 for $P = 16$. The first row contains results for the finest global grid Ω_0^h (i.e. no refinements), the other rows refer to an increasing number of refinement levels (3 and 4 in case of Example 1 and 2, respectively).

	Example 1			Example 2		
finest level	ρ	N_{cg}	$\mathcal{E}$/cyc	ρ	N_{cg}	$\mathcal{E}$/cyc
0	0.33	3200	71.4%	0.31	3200	70.8%
-1	0.43	8446	↓	0.31	11224	↓
-2	0.56	12123	↓	0.37	21507	↓
-3	0.5	13866	67.3%	0.40	26925	↓
-4	—	—	—	0.40	28771	69.4%

TABLE 1. Numerical results measured on the iPSC/860 (P=16)

Although the finest global grid is relatively coarse (3200 points, i.e. only 200 points per processor), the parallel efficiency measured for the final adaptive cycles is rather high, namely, 67.3% and 69.4% for the two examples. Clearly, the best we can expect, is the efficiency of the corresponding cycles without refinements, i.e., 71.4% and 70.8%, respectively. That is, the effective loss in parallel efficiency due to the introduction of refinement levels is very small. It is essentially caused by the increased number of grids with a deteriorated arithmetic/communication ratio and by non-optimal load-balancing of the refinement levels. Due to the high arithmetic

[4]For reasons of robustness, we used F-cycles rather than V-cycles (cf. the corresponding remarks in Section 2).

work per grid point (many 100 floating point operations), the communication work required for global data re-distribution has by far the lowest influence (cf. also the next section).

This indicates that, for compute-intensive problems, we do not have to expect a severe performance degradation for adaptive multigrid cycles on parallel machines if compared to their non-adaptive counterparts.

4 Worst-case considerations

In this section, we first consider scalar problems with *boundary-induced singularities.* For many such problems, reasonable refinement strategies are known *a priori.* Concrete measurements are performed for the Poisson equation. Clearly, this particular problem can be solved very efficiently by different approaches without using local refinements. We here regard it merely as a worst-case problem for parallel machines (low arithmetic per grid point, many refinement levels). Afterwards, we will consider the limits of the parallel efficiency.

4.1 Boundary-induced singularities

It is well-known that a corner at the domain boundary typically causes a singularity in the solution of elliptic boundary value problems near that corner, the strength of which depends on the size of the inner angle ϕ ($\pi/2 < \phi \leq 2\pi$) (cf. Figure 8). Using second order differencing on a uniform grid Ω_0^h of mesh size h_0, generally results in a global discretization error (measured in the maximum norm) of $\mathcal{O}(h_0^\eta)$ with some $\eta < 2$. For instance, for the Poisson equation with Dirichlet boundary conditions, we have (essentially) $\eta = \pi/\phi$.

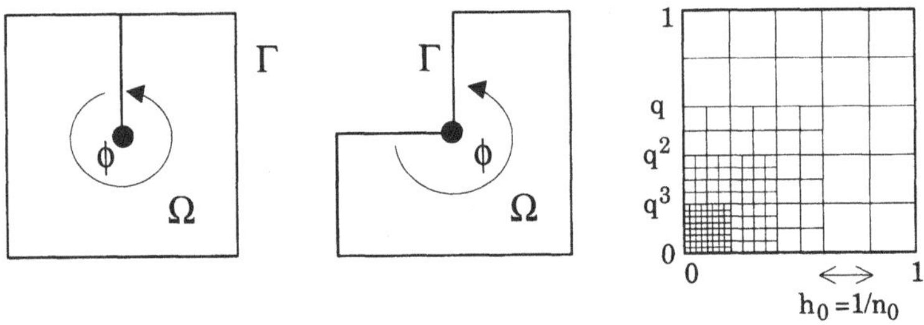

FIGURE 8. Exemplary domains and simplified computational grid

One way to obtain $\mathcal{O}(h_0^2)$ accuracy is to refine the grid locally towards the corner [1], [7], [12]. The "optimal meshsize" at the distance r from the corner, $\mathcal{H}(r)$, typically is $\mathcal{H}(r) = h_0(r/R)^{1-\eta/2}$ (R denotes the "radius" of the domain, measured from the singular point). Note that the finest mesh size to be used near the corner, $h_\star$, is defined by $\mathcal{H}(h_\star) = h_\star$ which obviously satisfies $h_\star^\eta = \mathcal{O}(h_0^2)$ (i.e.,

$h_\star$ is the mesh size which, if used globally, would give a global error of just the required order).

Since we employ only mesh sizes $h_{-\ell} = h_0/2^\ell$, we may select $h_{-\ell}$ for all points with $r_\ell \le r < r_{\ell+1}$ where r_ℓ is defined by $\mathcal{H}(r_\ell) = h_{-\ell}$, i.e., $r_\ell = q^\ell R$ with $q = 0.5^{2/(2-\eta)}$ ($< 1/2$). Assuming that the singular corner is located in the origin, a reasonable hierarchy of grids $\Omega^h_{-\ell}$ ($\ell = 1, 2, \ldots, \ell_f$) is recursively obtained by discretizing

$$\Omega_{-\ell} = ([-r_\ell, r_\ell] \times [-r_\ell, r_\ell]) \cap \Omega \quad \text{where} \quad r_\ell = q^\ell R, \quad q = 0.5^{2/(2-\eta)} \tag{3}$$

with respect to $h_{-\ell}$. (We slightly enlarge the size of $\Omega_{-\ell}$ such that its interior boundary coincides with grid lines of the previous grid, $\Omega^h_{-\ell+1}$.) The sequence of grids is terminated for $\ell_f = max\,\{\ell : r_\ell > h_{-\ell}\} \approx log_{2q}(h_0/R)$.

To be more specific, let us consider the Poisson equation with Dirichlet boundary conditions on the first domain in Figure 8 (corresponding to the case $\eta = 1/2$ and $q = 0.5^{4/3} \approx 0.4$). For simplicity, we formally restrict the computations to the unit square ($R = 1$), assuming that the refinement strategy (3) is done in one corner (see Figure 8). Denoting $h_0 = 1/n_0$, the number of refined levels becomes

$$\ell_f \approx log_{2q}(h_0) = \frac{2-\eta}{\eta} log_2(n_0) = 3\, log_2(n_0) \tag{4}$$

which is three times larger than the number of global multigrid levels.

Figure 9 shows parallel V-cycle efficiencies (3 smoothing steps per level), measured on the iPSC/860 for different numbers of processors and different values of n_0. For each grid level, the processor mapping is *boxwise* (cf. Figure 10).

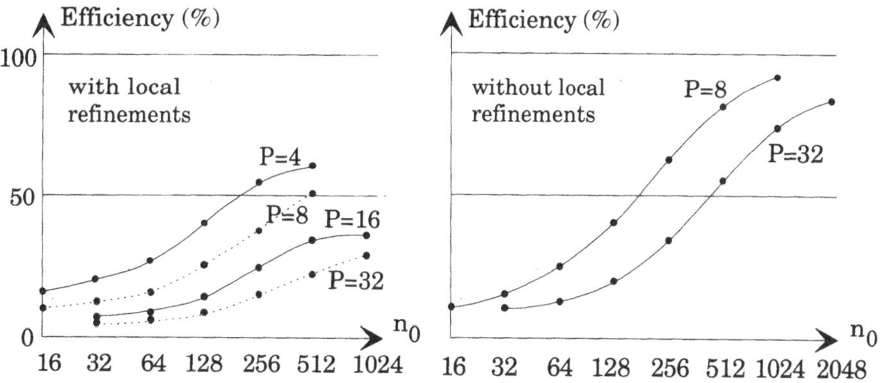

FIGURE 9. Parallel V-cycle efficiency on iPSC/860 (Poisson equation, $\eta = 1/2$)

For the same n_0, the total *intra-level* communication (and also the additional arithmetic required on the critical levels due to a reduced number of processors) is of the same order for cycles with and without local refinements. This is because the number of grid points decreases geometrically both towards the coarsest global as well as towards the finest refinement grid (with factors approximately 1/4 and $4q^2$, respectively). The corresponding overhead for adaptive cycles is just higher by a factor depending only on η.

For small n_0, this overhead dominates the additional overhead caused by inter-level communication (global data re-distribution) in adaptive cycles, and, consequently, cycles with and without local refinements behave similarly with respect to their parallel efficiency (cf. Figure 9). For increasing n_0 (and fixed P), however, the increase of the intra-level communication is of lower order (compared to the increase of arithmetic), while the inter-level communication overhead grows at the same rate. (To be more precise, the essential cost is due to the data *transmission*; the corresponding startup cost is also of lower order). This results in a saturation of the parallel efficiency of adaptive cycles as clearly seen in the figure.

4.2 Maximal efficiency

We have seen in Figure 9 that the maximally achievable V-cycle efficiency, $E_\infty^{(P)}$ (the limit of $E(N_0, P)$ for fixed P and $N_0 \longrightarrow \infty$), may be rather low for problems with low arithmetic per point. In order to discuss this aspect somewhat further, we want to roughly estimate $E_\infty^{(P)}$. This is fairly simple, since, for increasing N_0, the only communication overhead left is merely due to the data-transmission in the re-distribution steps. Compared to the arithmetic, all the other overhead (including all overhead caused by the critical levels) is of lower order.

Since we are only interested in the most crucial aspects, we make some idealizing assumptions. First, we assume an "ideal" network of processors. That is, the time required for transmitting i double precision numbers from one processor to *any* other one is given by $t(i) = \alpha + \beta i$ (α = startup time, β = time for transmitting one number). In addition, any different pairs of processors can perform their communication fully in parallel. Second, we assume that all levels are load-balanced and re-mapping has been done such that each processor keeps as many grid points for itself as possible. Finally, we assume arithmetic and communication to be fully synchronized.

Let us first consider interpolation from grid $\Omega^h_{-\ell+1}$ to grid $\Omega^h_{-\ell}$ for the particular refinement process considered in the previous section. All data relevant for interpolation is distributed across those processors whose geometrical area overlaps with $\bar{\Omega}^h_{-\ell+1}$ (shaded region in Figure 10). Assuming that $\bar{\Omega}^h_{-\ell+1}$ extends over the area of at least one processor[5], according to our idealizing assumptions, the amount of data to be sent during re-distribution (worst case among the processors), is effectively

$$L_{-\ell} = (N_{-\ell+1} - N_{-\ell}/4)/P \quad (\ell \geq 1). \tag{5}$$

By summing up, and ignoring terms which are not important for our asymptotic consideration, the total amount of data to be sent in the coarse-to-fine re-

[5]This means that the processor grid satisfies $P = p_1 \times p_2$ with $p_i \geq 1/q$.

distribution steps of one V-cycle is seen to be effectively

$$L = \sum_{\ell \geq 1} L_{-\ell} = \frac{1}{P}\left(N_0 + \frac{3}{4}\sum_{\ell \geq 1} N_{-\ell}\right) = \frac{3}{4P}\left(\frac{4}{3}N_0 + \sum_{\ell \geq 1} N_{-\ell}\right) = \frac{3}{4}\frac{N_{mg}}{P}. \quad (6)$$

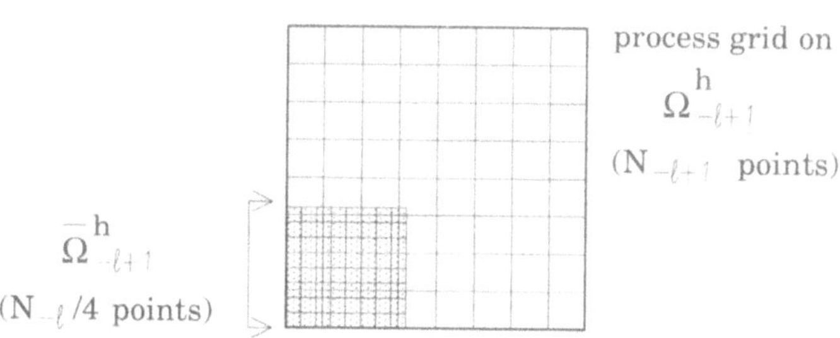

FIGURE 10. Data re-distribution involving grids $\Omega^h_{-\ell+1}$ and $\Omega^h_{-\ell}$

Although derived for a particular sequence of refined grids, it is clear that (6) is also relevant in general: Assuming *any* sequence of locally refined grids, then (5) is still true

if there is at least one processor whose geometrical area on $\Omega^h_{-\ell+1}$ *lies completely inside the next refinement grid,* $\bar{\Omega}^h_{-\ell+1}$. $(\star)$

(This corresponds to the requirement in Footnote 5.) If this happens to be true for each level, we obtain the same approximation for L as above. In this sense, (6) characterizes the worst-case for a general sequence of refined grids.

Thus, assuming now more generally that we are solving any elliptic PDE system of m unknowns by adaptive V-cycles and that the (FAS) fine-to-coarse data re-mapping costs twice as much as the coarse-to-fine one, the worst-case total re-distribution cost per cycle is approximately $3m\beta L$. Assuming finally the sequential work per cycle to be approximately σN_{mg} (with σ denoting the cost for a two-grid-cycle per grid point, not counting the cost for solving the corresponding coarse-grid equations), we obtain the following worst-case approximation for the maximum parallel efficiency per cycle:

$$E_{\infty}^{(P)} = \frac{1}{P}\frac{\sigma N_{mg}}{\sigma N_{mg}/P + 3m\beta L} = \frac{\sigma}{\sigma + \frac{9}{4}m\beta}. \quad (7)$$

That is, this approximation is valid if $(\star)$ is true for each ℓ. Otherwise, it is too pessimistic. Note that (7) is independent of P.

For the case shown in Figure 9 (Poisson equation), we have $m = 1$ and approximately $\sigma = 25$ (in terms of number of floating point operations). Since the iPSC/860 has as a realistic node performance of about 5 MFlops and a bandwidth of 2.7 MBytes/sec, we can assume $\beta \approx 15$. From (7) we obtain $E_\infty^{(P)} \approx 0.43$ if ($\star$) is true for each ℓ, i.e., if the processor grid consists of at least 3 processors in either direction (cf. Footnote 5 for $q = 0.4$). This is in fairly good agreement with the result shown in Figure 9 for $P = 16$. For $P = 8$ and $P = 4$, the requirement ($\star$) is never true and thus (7) is too pessimistic.

In this concrete situation, the limitation of the efficiency is not too severe. However, the current trend in hardware development shows a very strong increase in node performance while, at the same time, the communication bandwidth increases only slowly. For instance, IBM (with its SP1) aims at a peak performance of over 500 MFlops per node in the near future. On such machines, the maximally achievable efficiency will become much worse than above. Assuming a fictitious machine with 128 MFlops and a bandwidth of 4 MBytes/sec (the approximate peak values of the CM5), we obtain $\beta \approx 256$ resulting in only $E_\infty^{(P)} \approx 0.04$.

In any case, however, since the maximally achievable efficiency does not depend on P, asymptotically, the scalability properties of adaptive cycles are similar to those of non-adaptive ones. That is, the speedup $S(N_0, P)$ behaves as $\mathcal{O}(P)/log(P)$ if $P \longrightarrow \infty$ and the grid size per processor, N_0/P, is kept fixed. It is just the constant which is (possibly much) smaller for adaptive cycles.

4.3 Conclusions

Clearly, the larger σ, the less severe (7) becomes. This is particularly the case for *systems* of equations where we typically have $\sigma = \mathcal{O}(m^2)$. In fact, for the problem treated in Section 3 (Euler equations, many 100 floating point operations per grid point), the data re-distribution had no essential negative influence on the parallel performance, at least not within the range of machines and grid sizes which we were able to test.

However, one should keep in mind that (7) is based on the idealized assumptions stated at the beginning of Section 4.2. In particular, the *independency of P* is only true for ideal networks for which the re-distribution work can fully be shared between all processors involved. In a real parallel system, how good this idealization is (can be) met, will strongly depend on the concrete network as well as on the number of processors. Generally, we have to expect that the parallel efficiency will not only be limited, but that the limit will also get worse for increasing P.

As an extreme case of a more severe situation, consider a *single-bus connected system* (e.g., workstation cluster) such that different processors cannot be assumed to send data in parallel any more. For the particular model problem considered in Section 4.1, for instance, (5) has to be replaced by $L_{-\ell} = q^2\,(N_{-\ell+1} - N_{-\ell}/4)$ resulting in $L = \frac{3}{4}q^2 N_{mg}$. The corresponding maximum efficiency,

$$E_\infty^{(P)} = \frac{\sigma}{\sigma + \frac{9}{4}\,q^2\,\beta\,P}\,, \tag{8}$$

now quickly tends to 0 if P increases; adaptive multigrid cycles are not scalable at all on such systems. In fact, the maximally achievable speedup, $S_\infty^{(P)} = P\,E_\infty^{(P)}$, is not only bounded from above but is, if $\frac{9}{4}\, q^2 \beta \geq \sigma$, even smaller than 1, independently of P! Incidentally, for clusters of high speed workstations (connected by Ethernet), β will typically be of the order of many hundreds. That is, unless σ is very large, we cannot expect any significant speedup on such systems, even not if N_0 is arbitrarily large.

Let us return to systems with a real parallel network. Even if the influence of data re-distribution then is not substantial for sufficiently complex problems, one might want to reduce this overhead as far as possible. Clearly, this can only be achieved by more sophisticated mapping strategies than the one sketched in Section 2.1. The general goal should be to avoid ($\star$) as far as possible and to map refinement grids such that as many processors as possible can share the final re-distribution work. Since the location of new refinement areas is not known in advance, this means that a complicated multi-stage re-mapping strategy is required.

In case there is some information on the location of refinement grids, there may be straightforward ways to proceed. For instance, for point singularities as considered in Section 4.1, one might use stripwise rather than boxwise mapping (cf. Figure 11a). In our model case, instead of (5), we then obtain $L_{-\ell} = (q\, N_{-\ell+1} - N_{-\ell}/4)/P$. By summing up and observing that $N_{-\ell}/4 \approx q^2\, N_{-\ell+1}$, we obtain $L = \frac{3}{4} \frac{q}{1+q} N_{mg}/P$ which gives

$$E_\infty^{(P)} = \frac{\sigma}{\sigma + \frac{9}{4} \frac{q}{1+q}\, \beta}\,. \tag{9}$$

Compared to (7), the communication term is not only significantly smaller (by at least 66%) but also decreases for decreasing q.

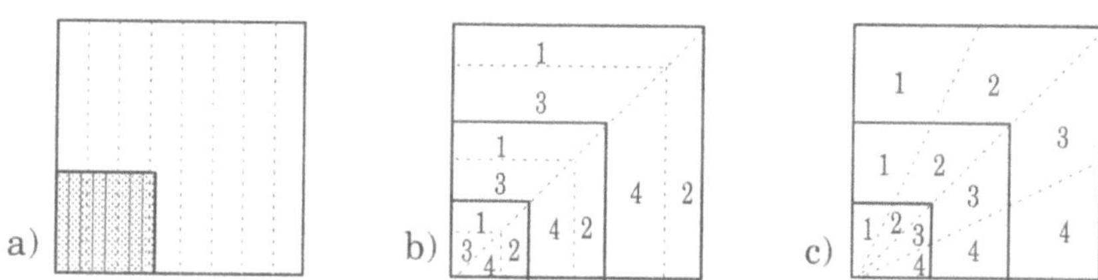

FIGURE 11. Different mapping strategies

If we assign all processors to both $\Omega^h_{-\ell}$ *and* $\bar{\Omega}^h_{-\ell+1}$ for each ℓ, re-distribution may even be totally avoided. Two possibilities are sketched in Figure 11b-c for the case $P = 4$. The first one [10] implies that each processor obtains as many different subregions as there are refined levels, a disadvantage which is avoided by the second mapping. Note that such strategies, in particular the one in Figure 11b, might be used as a basis for a mapping strategy also in general situations. The implementation of the multigrid algorithm itself may then however become quite cumbersome. To our knowledge, such strategies have not yet been tested in practice.

Finally, we would like to briefly comment on competitive asynchroneous multi-level strategies. Due to their additional *vertical parallelism*, different processors can be assigned to different levels which, up to a certain extent, should remove the typical efficiency degradation of standard parallel multigrid towards smaller grids in the hierarchy. AFAC (*asynchroneous fast adaptive composite grid method* [9]) has been developed (and is meaningful only) for adaptive grids. It allows the finest global grid and all refined grids to be treated *simultaneously*. There are, however, two severe drawbacks of this approach.

First, from a convergence point of view, two AFAC cycles roughly correspond to one MLAT cycle. Consequently, for complex problems (such that the parallel efficiency of MLAT is around 50% or higher), AFAC cannot compete with MLAT. Second, AFAC requires global data re-distribution in essentially the same way as MLAT, except that this re-distribution is performed "outside" each cycle. However, just because vertical parallelism is exploited, less processors will be available per level and, consequently, less processors can share the work for data re-distribution. In fact, a closer view shows that the essential communication term of AFAC (corresponding to the one of MLAT in (7)), is no longer independent of P, but rather grows like $\mathcal{O}(log_2(P))$. That is, unless P is relatively small, the maximally achievable parallel efficiency will be considerably lower for AFAC than for MLAT.

Summarizing, AFAC might be superior to MLAT only in special situations. For instance, for applications with low arithmetic per point, solved on machines with relatively small P and high startup costs. Note furthermore that re-distribution-free mappings (as indicated in Figure 11 for MLAT) do not exist for AFAC.

References

[1] Brandt, A.: *Multi-Level Adaptive Solutions to Boundary-Value Problems*, Math. Comp. 31, pp. 333–390, 1977.

[2] Brandt, A.: *Multigrid techniques: 1984 guide with applications to fluid dynamics*, GMD-Studie 85, St. Augustin, 1984.

[3] Canu, J.; Ritzdorf, H.: *Adaptive, block-structured multigrid on local memory machines*, Proceedings of 9th GAMM Seminar held at Kiel 1993, to appear.

[4] Hemker, P.W.; Koren, B.: *A Non-Linear Multigrid Method for the Steady Euler Equations*, Notes on Numerical Fluid Dynamics 26, pp. 175–196, Vieweg, Braunschweig, 1989.

[5] Hempel, R.; Ritzdorf, H.: *The GMD communications subroutine library for grid-oriented problems*, Arbeitspapiere der GMD 589, St. Augustin, 1991.

[6] Hempel, R.; Hoppe, H.C.; Supalov, A.: *PARMACS 6.0 Library Interface Specification*, GMD-report, 1993 (available through GMD).

[7] Kaspar, W.; Remke, R.: *Die numerische Behandlung der Poisson Gleichung auf einem Gebiet mit einspringenden Ecken*, Computing 22, pp. 141–151, 1979.

[8] Lonsdale, G.; Ritzdorf, H.; Stüben, K.: *The L_iSS package*, Arbeitspapiere der GMD 745, St. Augustin, 1993.

[9] McCormick, S.: *Multilevel adaptive methods for partial differential equations*, SIAM, Frontiers in Applied Mathematics, Vol. 6, Philadelphia, 1989.

[10] Mierendorff, H.: *Parallelization of multigrid methods with local refinements for a class of nonshared memory systems*, in: "Multigrid Methods: Theory, Applications, and Supercomputing" (McCormick, S., ed.), pp. 449–465, Marcel Dekker, 1988.

[11] Ritzdorf, H.; Schüller, A.; Steckel, B.; Stüben, K.: *L_iSS - An environment for the parallel multigrid solution of partial differential equations on general 2D domains*, Parallel Computing, to appear 1993.

[12] Ritzdorf, H.: *Lokal verfeinerte Mehrgitter-Methoden für Gebiete mit einspringenden Ecken*, Diplomarbeit, Universität Bonn, 1984.

[13] Sonar, Th.: *Strong and weak norm refinement indicators based on the finite element residual for compressible flow*, Impact of Computing in Science and Engineering, to appear 1993.

[14] Stüben, K.; Trottenberg, U.: *Multigrid methods: Fundamental algorithms, model problem analysis and applications*, in "Multigrid methods" (Hackbusch, W.; Trottenberg, U., eds.), Lecture Notes in Mathematics Vol. 960, Springer, Berlin, 1982.

7

Multicomputer–Multigrid Solution of Parabolic Partial Differential Equations

Stefan Vandewalle[1] **and Graham Horton**[2]

ABSTRACT [3] We discuss the numerical computation of approximations to the solution of parabolic partial differential equations by using multigrid methods on parallel computer systems. The paper focuses on algorithms that operate on the whole of the space-time grid, treating the time-dimension as just another spatial dimension. Three different algorithms that have appeared earlier in the literature are recalled; their theoretical convergence properties are analyzed by Fourier mode analysis, and their parallel complexities are investigated.

1 Introduction

The time-accurate numerical solution of parabolic partial differential equations (PDEs) is a time-consuming computational procedure in many scientific and engineering disciplines. The application of efficient numerical algorithms and the use of advanced parallel computer architectures are therefore of great importance in order to lower the required computation time.

Traditionally, time-dependent PDEs are solved as a sequence of boundary value problems defined on successive time-levels. The great potential of multigrid as a rapid solver for these boundary value problems was realized from the early days of multi-level algorithms research, see, e.g., [2, 3, 24, 18]. Over the years various improvements to the basic algorithm have been suggested: *modified nested iteration* techniques based on the similarity of time-dependent PDEs to parameter-

[1]Senior Research Assistant of the National Fund for Scientific Research (N.F.W.O), Belgium. Katholieke Universiteit Leuven, Department of Computer Science, Celestijnenlaan 200A, B-3001 Leuven, Belgium.

[2]Lehrstuhl für Rechnerstrukturen (IMMD 3), Universität Erlangen – Nürnberg, Martensstr. 3, D-91058 Erlangen, Germany.

[3]The following text presents research results of the Belgian Incentive Program "Information Technology" - Computer Science of the future, initiated by the Belgian State - Prime Minister's Service - Science Policy office. The scientific responsibility is assumed by its authors.

dependent continuation problems ([9, 11]), *modified multigrid cycle types* to solve for an incremental solution instead of the full PDE solution ([4, 8]), *τ-extrapolation* and *frozen-τ* techniques which allow time-stepping to proceed on coarse spatial grids with possibly large time-steps while retaining fine grid accuracy ([3, 18, 7, 8]), and *double discretization* methods to circumvent certain stability problems ([7, 8]).

The parallel implementation of the multigrid algorithm for boundary value problems has been the subject of numerous studies by many authors. A comprehensive overview of some this work is given in [20]. A comparison of the parallel performance of various time-stepping methods for parabolic problems, including explicit, implicit and line-implicit methods, is presented in [32]. By these and other studies it was made clear that the essentially sequential nature of the time-stepping procedure imposes serious limitations on the obtainable parallel performance. No matter how many time-steps are to be computed, the obtainable degree of parallelism is restricted by the parallelism in the multigrid solver used to compute the solution in one single time-step. This is especially disappointing as the number of time-steps is often many times larger than the size of the spatial mesh.

This observation has led to the development of algorithms that operate on more than one time-level simultaneously; that is to say, on grids extending in space and in time, further called *space-time grids*. One such algorithm is the *parallel time-stepping* method ([33]). It is closely related to the class of *windowed relaxation* methods ([23]). The parallel time-stepping method is the obvious extension of any standard iterative technique to multiple time-levels: while the solution is being computed on the first time-level by applying the iterative method to a starting approximation, the approximations to the solutions on subsequent time-levels are being updated by the same iterative method, or, possibly, by an other one. The overlap of computations on different time-levels enables the use of many processors, especially when slowly convergent iterative solvers are used. With rapidly converging iterative solvers like multigrid, however, only few processors can be used effectively, and the method loses most of its advantages. A second method, *multigrid waveform relaxation*, originated by combining the multigrid idea with the waveform relaxation method, an iterative solver commonly used in electrical engineering practice for solving large nonlinear systems of ordinary differential equations ([19, 30, 27, 28, 29, 31, 17]). *Parabolic multigrid* is a method that extends the elliptic multigrid idea to the set of equations obtained after discretizing a parabolic problem in space and time ([10, 5]). Its time-parallel variant was recently the subject of much further study ([1, 6, 13, 15, 14]). The latter two methods were analyzed and compared in [26]. Finally, a fourth method is the *space-time multigrid* method, which was developed only recently by addressing some of the convergence problems that arose within the time-parallel multigrid method ([16]).

The latter three multigrid methods are presented in §2. They turn out to be closely related, as multigrid methods defined on grids extending in space and in time. Their theoretical convergence characteristics are studied in §3 by means of a two-grid Fourier mode analysis. We derive their parallel complexities in §4, and compare the results with the complexity of standard time-stepping, and that of an 'optimal' direct solver. We end in §5 with some concluding remarks.

2 Multigrid methods on space-time grids

We shall concentrate on the problem of numerically computing the solution to a model problem, in case the d-dimensional heat equation,

$$u_t - \Delta u = f(x,t) \qquad x \in \Omega = (0,1)^d,\ 0 < t \leq T\ , \tag{1}$$

subject to the usual initial and boundary conditions

$$\begin{aligned} u(x,0) &= g(x)\ , \quad x \in \Omega\ , & (2)\\ u(x,t) &= h(x,t)\ , \quad x \in \partial\Omega\ ,\ 0 < t \leq T\ . & (3) \end{aligned}$$

For notational simplicity, we consider the one-dimensional problem, discretized in space using central differences on a regular grid with grid spacing Δx, and discretized in time with the backward Euler method, on a set of time-levels with constant time-increment Δt. This discretization leads to a large linear system of equations in the unknowns $u_{i,j}$ with $i = 1, \ldots, 1/\Delta x - 1$ and $j = 1, \ldots, T/\Delta t$, that approximate the PDE solution at the grid points (x_i, t_j) with $x_i = i \cdot \Delta x$ and $t_j = j \cdot \Delta t$. The grid will be denoted further by Ω_h, with h standing for the pair $(\Delta x, \Delta t)$ characterizing the size of the grid. The equations on Ω_h are of the form

$$-\frac{1}{(\Delta x)^2} u_{i-1,j} + \left(\frac{2}{(\Delta x)^2} + \frac{1}{\Delta t}\right) u_{i,j} - \frac{1}{(\Delta x)^2} u_{i+1,j} - \frac{1}{\Delta t} u_{i,j-1} = f(x_i, t_j)\ , \tag{4}$$

or, with the parameter λ_h defined as $\Delta t / \Delta x^2$,

$$-\lambda_h\, u_{i-1,j} + (2\lambda_h + 1)\, u_{i,j} - \lambda_h\, u_{i+1,j} - u_{i,j-1} = \Delta t\, f(x_i, t_j)\ . \tag{5}$$

Note that parameter λ_h can be considered as a measure of the degree of anisotropy of the discrete operator. In the case of a very large λ_h, the set of equations is essentially decoupled in time, and corresponds to a set of (almost) independent discrete boundary value problems, one per time-level. In the case of very small λ_h, the set of equations is (almost) decoupled in space, and corresponds to a set of first order linear recurrences, one per spatial grid point.

The principal components of any multigrid method are the coarsening strategy, the discretization on each grid level, and the smoothing and intergrid transfer operators. Each of the three algorithms discussed below uses the 'natural' discretization corresponding to (5) on each grid level. (Of course, the value of λ_h may differ from one grid level to the next.) Hence, for a given fine grid, the three methods solve the same set of equations. They differ only in the choice of coarsening strategy and multigrid operators.

The *multigrid waveform relaxation* method employs a semi-coarsening strategy, with coarsening only in the spatial dimension. The standard smoother is a zebra Gauss-Seidel method (i.e., red/black line-relaxation), with lines parallel to the time-axis. It can be shown that this smoother is robust w.r.t. λ_h ([26]). Note that the time-line solver is particularly simple as it only involves the forward evaluation of first order recurrence relations. The intergrid transfer operators are the standard ones used in combination with semi-coarsening. In the sequel, we shall use the linear prolongation (I_H^h) and full weighting (I_h^H) formulae, with stencils whose non-zero values extend in the spatial dimension only,

$$I_H^h : \frac{1}{2}\left]\begin{array}{ccc} 0 & 0 & 0 \\ 1 & 2 & 1 \\ 0 & 0 & 0 \end{array}\right[\quad \text{and} \quad I_h^H : \frac{1}{4}\left[\begin{array}{ccc} 0 & 0 & 0 \\ 1 & 2 & 1 \\ 0 & 0 & 0 \end{array}\right] . \tag{6}$$

The method easily extends to higher dimensional problems, and PDEs different from our model problem (1). The method was first presented in [19] for linear problems, and in [30] for nonlinear ones. A large number of examples, illustrating typical multigrid convergence rates are given in [27, 28, 25]. Its application for solving time-periodic parabolic problems is analyzed in [29]. Its implementation on small-scale and medium-scale multiprocessors is discussed in the above references. Large-scale and massively parallel implementation are documented in [31] and in [17], respectively.

The *parabolic multigrid* or *time parallel multigrid* method differs from the multigrid waveform relaxation method only in the choice of smoothing operator. It applies a standard spatial smoother replicated on each time-level. Non-smoothed old values are used whenever values at grid points on previous time-levels are needed. The red/black smoother, for example, consists of one point-wise relaxation step on all red grid points at all time-levels concurrently, followed by a similar operation on all black points. Note that colouring is only w.r.t. the spatial dimension. Further inspection reveals an interesting relation to the waveform relaxation method (for the system of equations (5)). While the waveform method solves each system of equations in one time-line exactly, the parabolic multigrid method solves them approximately, by doing one Jacobi relaxation step.

The parabolic multigrid method was described first in [10], and a theoretical analysis for a one-dimensional model problem followed in [5]. The parallel implementation of the method, its application to the Navier-Stokes equations, and the use of different time-parallel smoothers is discussed in [1, 6, 13, 14, 21]. Its combination with extrapolation techniques is the subject of [15].

The most recent method studied in this paper is the *space-time multigrid* method. It is based on a semi-coarsening strategy, with coarsening in space or in time depending on the current value of λ_h. If λ_h is larger than a certain threshold value λ_{crit}, coarsening is in the spatial dimension, and restriction and prolongation operators are the standard ones given in (6). If λ_h is smaller than λ_{crit}, coarsening is in the time-dimension. In that case special intergrid operators are used, whose

TABLE 1. Averaged convergence factor of space-time multigrid V(2,1)-cycle on rectangular grids for the one-dimensional model problem with backward Euler discretization ($n_s = 1/\Delta x$, $n_t = T/\Delta t$, $\lambda_h = \Delta t/(\Delta x)^2$).

$n_s \times n_t$	32×32	32×64	32×128	32×256	32×512	32×1024
$\lambda_h = 1/64$	0.023	0.031	0.048	0.061	0.074	0.086
$\lambda_h = 1/4$	0.049	0.077	0.11	0.15	0.14	0.14
$\lambda_h = 1/2$	0.081	0.13	0.14	0.13	0.13	0.13
$\lambda_h = 1$	0.095	0.11	0.12	0.13	0.13	0.12
$\lambda_h = 2$	0.092	0.10	0.10	0.097	0.09	0.082
$\lambda_h = 4$	0.083	0.091	0.093	0.087	0.081	0.081
$\lambda_h = 64$	0.020	0.019	0.021	0.022	0.023	0.023

stencils are given by

$$I_H^h \ : \ \left]\begin{matrix} 0 & 1 & 0 \\ 0 & 1 & 0 \\ 0 & 0 & 0 \end{matrix}\right[\quad \text{and} \quad I_h^H \ : \ \frac{1}{2}\begin{bmatrix} 0 & 0 & 0 \\ 0 & 1 & 0 \\ 0 & 1 & 0 \end{bmatrix} . \tag{7}$$

The method is further based on a point-wise red/black smoother, with standard colouring of the entire space-time grid.

As explained in [16], by this choice of operators and smoother the space-time multigrid method approaches an exact solver in the limiting cases of λ_h going to ∞ and λ_h going to 0. In [16], the method is also discussed in combination with different time-discretization methods (Crank-Nicolson and second order backward differentiation), and numerical results are provided for the *two*-dimensional model problem. Some numerical results of a computational experiment are presented in Table 1. They illustrate the very good convergence of the method, for different values of λ_h and for different mesh sizes. In this table we concentrated on the most interesting case for practical purposes, where a fixed-size spatial problem is integrated over various large time intervals.

3 Two-grid Fourier mode analysis

In this section we shall analyze the two-grid variants of the three multigrid methods presented in the previous section. We consider the one-dimensional model problem, discretized with the backward Euler method on a space-time grid Ω_h with $(n_s + 1) \times (n_t + 1)$ grid points; i.e., $n_s = 1/\Delta x$ and $n_t = T/\Delta t$. The two-grid method makes use of an additional grid, Ω_H, derived from Ω_h by doubling the mesh size in the space dimension ($H = (2\Delta x, \Delta t)$) or the time dimension ($H = (\Delta x, 2\Delta t)$) .

By a two-grid cycle the error e^{old} of an approximation to the solution on Ω_h is transformed into a new error e^{new}, with $e^{new} = M_h^H e^{old}$, where M_h^H is the so-called two-grid iteration matrix. This matrix is given by

$$M_h^H = S_h^{\nu_2} \left(I_h - I_H^h L_H^{-1} I_h^H L_h\right) S_h^{\nu_1} , \tag{8}$$

where S_h is the smoothing operator on Ω_h; ν_1 and ν_2 are the numbers of pre- and post-smoothing iterations; I_h, I_H^h, I_h^H, are the identity, prolongation, and restriction operators. L_H and L_h are discretized differential operators on Ω_H and Ω_h. It can be shown that the entries of M_h^H depend on λ_h and λ_H only, and not of the particular values of the discretization parameters h and H.

The properties of the two-grid iteration matrix are often determined, or approximately calculated, in the frequency domain, by a so-called *exponential Fourier mode analysis* ([2]). This analysis can be regarded as an analysis for special model problems, namely those with periodic boundary conditions. This analysis shows that multiplication with matrix M_h^H leaves certain linear spaces of exponential Fourier modes invariant. More precisely, it can be shown that M_h^H is equivalent to a block-diagonal matrix, whose diagonal blocks are matrices of rank at most 4. The general expression for the diagonal blocks is called the *Fourier mode symbol* of the two-grid operator. This symbol is easily found to be

$$\hat{M}_h^H(\theta) = \hat{S}_h^{\nu_2}(\theta) \left(I_h - \hat{I}_H^h(\theta)\, \hat{L}_H^{-1}(\theta)\, \hat{I}_h^H(\theta)\, \hat{L}_h(\theta) \right) \hat{S}_h^{\nu_1}(\theta) , \tag{9}$$

where $\hat{S}_h(\theta)$, $\hat{I}_h^H(\theta)$, $\hat{I}_H^h(\theta)$, $\hat{L}_h(\theta)$, and $\hat{L}_H(\theta)$ denote the symbols of the smoothing operator, restriction operator, prolongation operator, fine grid PDE operator, and coarse grid PDE operator. Precise formulae of these symbols and a further discussion can be found in [16, 26].

The convergence of the two-grid cycle is characterized by the *Fourier mode convergence factor*,

$$\rho = \max\{\kappa(\hat{M}_h^H(\theta)) : \theta \in \Theta_{\tilde{s}}\} . \tag{10}$$

where $\kappa(\cdot)$ denotes the spectral radius operator, and where the set of frequencies $\Theta_{\tilde{s}}$ is given by

$$\Theta_{\tilde{s}} = \{(\theta_s, \theta_t) : \theta_\alpha = 2\pi k_\alpha / n_\alpha,\ k_\alpha = -n_\alpha/4,\ -n_\alpha/4+1, \ldots, n_\alpha/4-1\}. \tag{11}$$

(We assumed that n_s and n_t are multiples of 4.) The value of ρ usually shows very good agreement with actual convergence factors obtained on Ω_h. Its calculation is straightforward, by numerically computing $\kappa(\hat{M}_h^H(\theta)))$ and by optimizing this over the discrete set $\Theta_{\tilde{s}}$. We have calculated two-grid Fourier mode convergence factors for the multigrid waveform relaxation method, the parabolic multigrid method, and the space-time multigrid method. The results are graphically depicted as functions of the parameter λ_h in Figure 1.

The waveform relaxation picture clearly illustrates the robustness of the method across the entire range of λ_h values. The parabolic two-grid method performs satisfactorily for large values of λ_h, i.e., when the problems on each time-level are more or less decoupled. In that case parabolic multigrid is equivalent to a standard elliptic multigrid method for a problem extending in space only. The method

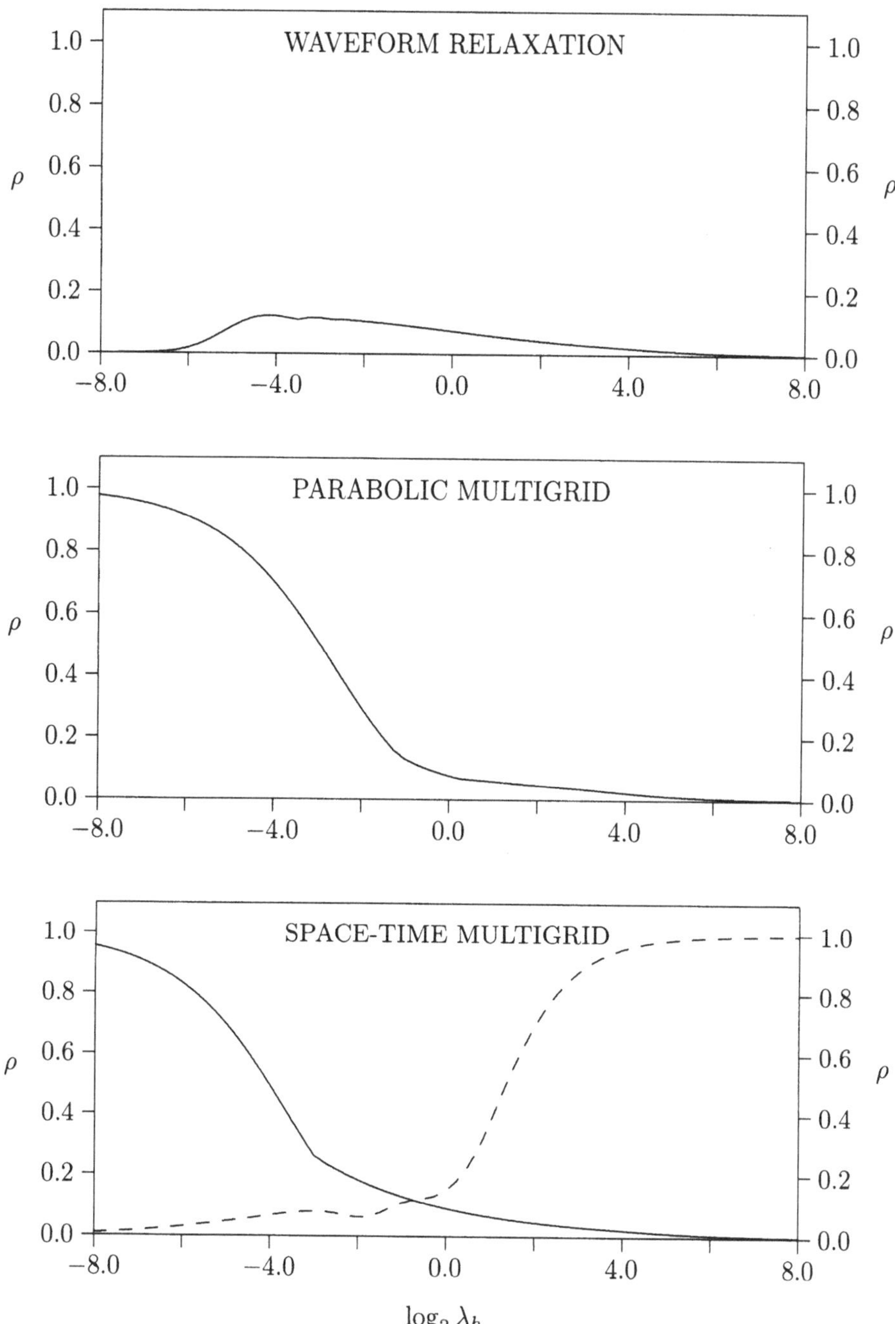

FIGURE 1. Two-grid Fourier mode convergence factor for backward Euler discretization (two-grid cycle with 2 pre- and 1 post-smoothing steps, $n_s = n_t = 128$).

fails, however, completely for small values of λ_h. Finally, two curves are drawn in the space-time multigrid picture. The solid line corresponds to the *space*-coarsening strategy, while the dashed line corresponds to the *time*-coarsening strategy. The intersection of both curves determines the value of λ_{crit}. As such for any λ_h it follows that the ρ of the space-time two-grid method is always on the lower of the two curves. Hence, the method is robust for any λ_h.

4 Parallel complexity

Parallel complexity is a theoretical measure of an algorithm based on the assumption that an unlimited number of processors is available for its execution. It describes the asymptotic dependence of the parallel computation time of the algorithm on the size of the input. Any communication requirements are disregarded.

Consider the d-dimensional model problem; let n_s denote the spatial sidelength of the space-time grid and n_t the sidelength in the time direction. We shall derive the parallel complexities of the *nested iteration* variants of the multigrid methods described in §2. Standard multigrid arguments show these algorithms to achieve discretization accuracy, given that the convergence rate of a V-cycle is independent of the grid size, and a fixed number of V-cycles is used per grid level.

The parallel complexity of a *standard time-stepping* method is given by

$$\mathcal{O}(\ n_t \log^2(n_s)\ )\ . \tag{12}$$

This is easily seen by observing that the n_t time-steps are executed sequentially, and by noting that the parallel complexity of the modified nested iteration algorithm on a grid with sidelength n_s is given by $\mathcal{O}(\log^2(n_s))$, see, e.g., [20].

As derived in [17], the nested iteration *multigrid waveform relaxation* algorithm with a fixed number of V-cycles per grid level has a complexity equal to

$$\mathcal{O}(\ \log(n_t)\log^2(n_s)\ )\ . \tag{13}$$

The complexity of evaluating the recurrence relations in the smoother is $\mathcal{O}(\log(n_t))$, if a parallel cyclic reduction or recursive doubling method is used. This is at the same time the total complexity of the operations at a single grid level, as the other operations are $\mathcal{O}(\ 1\ )$ computations. Summing the total number of grid levels visited in the nested iteration method on a grid hierarchy with $\mathcal{O}(\ \log(n_s)\ )$ grid levels, we arrive at (13).

The *parabolic multigrid* method is identical to the previous method, except for its smoother, which is a point-wise algorithm. The complexity of the latter is $\mathcal{O}(\ 1\ )$. This immediately leads to the following expression for the parallel complexity of the nested iteration parabolic multigrid algorithm:

$$\mathcal{O}(\ \log^2(n_s)\ )\ . \tag{14}$$

Finally, from [16] we recall the complexity of the nested iteration *space-time multigrid* algorithm,

$$\mathcal{O}(\ (\log(n_s) + \log(n_t))^2\ ) \ . \tag{15}$$

The correctness of this formula is easily realized by considering the point-wise nature of all involved operators, and the fact that the total number of grid levels is given by $\mathcal{O}(\log(n_s) + \log(n_t))$. Note that the complexity of the method is $\mathcal{O}(\log^2(n))$, where n is the total number of variables. This agrees with the classical formula for the parallel complexity of the multigrid method for elliptic problems.

In [34] an information theoretic lower bound is derived for the cost of solving linear PDEs. By considering how many data values are required to calculate a single solution value, it is shown that the parallel complexity must a least grow as $\mathcal{O}(\log(n))$, *independent of the algorithm used.* For problem (1) it is easy to come up with an algorithm that achieves this optimal complexity, see, e.g., [22, §3.3.3] and [12, §5.6.2].

Spatial discretization and incorporation of the boundary conditions, transforms (1) into a system of ordinary differential equations (ODEs),

$$\dot{U} - LU = F(t) \ , \quad U(0) = U_0 \ . \tag{16}$$

L is the discrete d-dimensional Laplace operator. Let Q be an orthogonal matrix that diagonalizes L, i.e., $Q^T L Q = \Lambda$, with $\Lambda = \operatorname{diag}(\lambda_1, \lambda_2, \ldots)$. (Its columns are the eigenvectors of L.) For spatial discretization with standard finite differences, matrices Q and Λ are well-known in terms of sine-functions. Setting $U(t) = Q\tilde{U}(t)$, $F(t) = Q\tilde{F}(t)$ and $U_0 = Q\tilde{U}_0$, system (16) can be rewritten as a system of ODEs whose equations are decoupled,

$$\dot{\tilde{U}} - \Lambda\tilde{U} = \tilde{F}(t) \ , \quad \tilde{U}(0) = \tilde{U}_0 \ . \tag{17}$$

The following three-step algorithm results. *Step 1*: Compute $\tilde{F}_j = Q^T F(t_j)$ for every time level t_j, and compute $\tilde{U}_0 = Q^T U_0$. *Step 2*: Discretize and solve system (17) for the values $\tilde{U}_j \approx Q^T U(t_j)$, using the $\tilde{F}_j$ and $\tilde{U}_0$ values. *Step 3*: Compute $U_j = Q\tilde{U}_j$ for every time level t_j.

Steps 1 and 3 require $\mathcal{O}(\log(n_s))$ parallel steps, since the computations on each time-level can proceed concurrently, and since multiplication by Q or Q^T can be performed by means of the d-dimensional Fast Fourier Transform (FFT). Step 2 requires $\mathcal{O}(\log(n_t))$ parallel computations, when parallel cyclic reduction (CR) is used to calculate the linear recurrences that arise in the ODE solver. Hence, the parallel complexity of the FFT/CR algorithm is given by

$$\mathcal{O}(\log(n_s) + \log(n_t)) \ . \tag{18}$$

5 Concluding remarks

We have discussed various ways of extending standard time-stepping multigrid to methods that solve on several time-steps simultaneously. The potential for parallelism for this class of problems is thereby greatly increased. We considered multigrid waveform relaxation, space-time multigrid and the parabolic multigrid method.

Two-grid Fourier analysis shows both principal methods to be robust w.r.t. the space-time grid aspect ratio λ_h, both achieving a fast two-grid convergence rate. The parabolic multigrid method is, however, limited to the large λ_h case. The latter scheme has a parallel complexity which is independent of the number of time levels n_t, whereas that of the former two methods is polylogarithmic in the size of the time dimension. All three methods compare favourably in this respect to the standard time-stepping scheme, whose parallel complexity remains linear in n_t, a severe restriction on the achievable parallelism.

The direct method outlined in §4 has a lower parallel complexity than any of the multigrid methods described in §2. Its application, however, is restricted to simple linear problems on rectangular grids. (See [22, p. 66] for conditions on the applicability of an algorithm similar to FFT/CR.) Waveform relaxation and parabolic multigrid have been shown to be applicable to a much wider class of linear and nonlinear problems. Moreover, they can be extended immediately to non-rectangular domains. Although current experience with the space-time multigrid method is limited to the model problem, we expect it to be applicable also to more difficult problems.

Further work will include the implementation of each method on a massively parallel computer and the investigation of non-linear problems. In addition we intend to gain experience with the methods on MIMD machines and demonstrate the improvements in efficiency obtainable when time-parallelism is introduced into an otherwise standard multigrid scheme.

References

[1] P. Bastian, J. Burmeister, and G. Horton. Implementation of a parallel multigrid method for parabolic partial differential equations. In W. Hackbusch, editor, *Parallel Algorithms for PDEs (Proceedings of the 6th GAMM Seminar Kiel, January 19-21, 1990)*, pages 18–27, Wiesbaden, 1990. Vieweg Verlag.

[2] A. Brandt. Multi-level adaptive solutions to boundary-value problems. *Math. Comp.*, 31:333–390, 1977.

[3] A. Brandt. Multi-level adaptive finite-element methods I: Variational problems. In J. Frehse, D. Pallaschke, and U. Trottenberg, editors, *Special Topics of Applied Mathematics*, pages 91–128, Amsterdam, 1980. North-Holland.

[4] A. Brandt and J. Greenwald. Parabolic multigrid revisited. In W. Hackbusch and U. Trottenberg, editors, *Multigrid methods III (Proceedings of the third European Multigrid Conference, Bonn, 1990)*, pages 143–154, number 98 in ISNM, Basel, 1991. Birkhaüser Verlag.

[5] J. Burmeister. *Paralleles Lösen diskreter parabolischer Probleme mit Mehrgittertechniken.* Diplomarbeit, Universität Kiel, 1985.

[6] J. Burmeister and G. Horton. Time-parallel multigrid solution of the Navier-Stokes equations. In W. Hackbusch and U. Trottenberg, editors, *Multigrid methods III (Proceedings of the third European Multigrid Conference, Bonn, 1990)*, pages 155–166, number 98 in ISNM, Basel, 1991. Birkhaüser Verlag.

[7] E. Gendler. *Multigrid methods for time-dependent parabolic equations.* Master's thesis, The Weizmann Institute of Science, Rehovot, Israel, August 1986.

[8] J. Greenwald. *Multigrid Techniques for Parabolic Problems.* Ph.D.-thesis, The Weizmann Institute of Science, June 1992.

[9] W. Hackbusch. Multigrid solution of continuation problems. In R. Ansorge, Th. Meis, and W. Törnig, editors, *Iterative Solution of Nonlinear Systems of Equations*, number 953 in Lecture Notes in Mathematics, pages 20–45, Berlin, 1982. Springer-Verlag.

[10] W. Hackbusch. Parabolic multi-grid methods. In R. Glowinski and J.-L. Lions, editors, *Computing Methods in Applied Sciences and Engineering VI*, pages 189–197, Amsterdam, 1984. North Holland.

[11] W. Hackbusch. *Multi-Grid Methods and Applications.* Springer Verlag, Berlin, 1985.

[12] R. Hockney and C. Jesshope. *Parallel Computers : Architecture, Programming, and Algorithms.* Adam Hilger Ltd., Bristol, UK, 1981.

[13] G. Horton. Time-parallel multigrid solution of the Navier-Stokes equations. In C. Brebbia, editor, *Applications of Supercomputers in Engineering.* Elsevier, August 1991.

[14] G. Horton. The time-parallel multigrid method. *Communic. in Appl.-Num. Meth.*, 8:585–595, 1992.

[15] G. Horton and R. Knirsch. A time-parallel multigrid-extrapolation method for parabolic partial differential equations. *Parallel Computing*, 18:21–29, 1992.

[16] G. Horton and S. Vandewalle. A space-time multigrid method for parabolic P.D.E.s. Technical Report IMMD 3, 6/93, Universität Erlangen-Nürnberg, Martensstrasse 3, D-91058 Erlangen, Germany, July 1993.

[17] G. Horton, S. Vandewalle, and P. Worley. An algorithm with polylog parallel complexity for solving parabolic partial differential eqations. Technical Report IMMD 3, 8/93, Universität Erlangen-Nürnberg, Martensstrasse 3, D-91058 Erlangen, Germany, July 1993.

[18] T. Kroll. *Multigrid Solution of Parabolic Problems.* Master's thesis, Institut für Angewandte Mathematik, Universität Bonn, 1981.

[19] C. Lubich and A. Ostermann. Multigrid dynamic iteration for parabolic equations. *BIT*, 27:216–234, 1987.

[20] O. McBryan, P. Frederickson, J. Linden, A. Schüller, K. Solchenbach, K. Stüben, C. Thole, and U. Trottenberg. Multigrid methods on parallel computers — a survey of recent developments. *IMPACT of Computing in Science and Engineering*, 3:1–75, 1991.

[21] C.W. Oosterlee and P. Wesseling. Multigrid schemes for time-dependent incompressible Navier-Stokes equations. Report no. 92-102, Delft University of Technology, Faculty of Technical Mathematics and Informatics, 1992.

[22] M. Pickering. *An Introduction to Fast Fourier Transform Methods for Partial Differential Equations, with Applications.* John Wiley and Sons Inc., New York, 1986.

[23] J. Saltz and V. Naik. Towards developing robust algorithms for solving partial differential equations on MIMD machines. *Parallel Computing*, 6:19–44, 1988.

[24] K. Solchenbach. *Einsatz schneller elliptischer Löser zur Lösung nichtlinearer parabolischer Anfangsrandwertaufgaben.* Diplomarbeit, GMD, Bonn, 1980.

[25] S. Vandewalle. *Parallel Multigrid Waveform Relaxation for Parabolic Problems.* B.G. Teubner Verlag, Stuttgart, 1993.

[26] S. Vandewalle and G. Horton. Fourier mode analysis of the multigrid waveform relaxation and time-parallel multigrid methods. Technical Report IMMD 3, 7/93, Universität Erlangen-Nürnberg, Martensstrasse 3, D-91058 Erlangen, Germany, July 1993.

[27] S. Vandewalle and R. Piessens. Numerical experiments with nonlinear multigrid waveform relaxation on a parallel processor. *Applied Numerical Mathematics*, 8(2):149–161, 1991.

[28] S. Vandewalle and R. Piessens. Efficient parallel algorithms for solving initial-boundary value and time-periodic parabolic partial differential equations. *SIAM J. Sci. Stat. Comput.*, 13(6):1330–1346, November 1992.

[29] S. Vandewalle and R. Piessens. On dynamic iteration methods for solving time-periodic differential equations. *SIAM J. Num. Anal.*, 30(1):286–303, February 1993.

[30] S. Vandewalle and D. Roose. The parallel waveform relaxation multigrid method. In G. Rodrigue, editor, *Parallel Processing for Scientific Computing*, pages 152–156, Philadelphia, 1989. Proceedings of the Third SIAM Conference

on Parallel Processing for Scientific Computing, Los Angeles, December 1-4, 1987, SIAM.

[31] S. Vandewalle and E. Van de Velde. Space-time concurrent multigrid waveform relaxation. Technical Report CRPC-93-2, Center for Research on Parallel Computation, California Institute of Technology, April 1993.

[32] S. Vandewalle, R. Van Driessche, and R. Piessens. The parallel performance of standard parabolic marching schemes. *Int. J. High Speed Computing*, 3(1):1–29, 1991.

[33] D. Womble. A time-stepping algorithm for parallel computers. *SIAM J. Sci. Stat. Comput.*, 11(5):824–837, September 1990.

[34] P. Worley. Limits on parallelism in the numerical solution of linear PDEs. *SIAM J. Sci. Stat. Comput.*, 12(1):1–35, January 1991.

8

Multilevel Solution of Integral and Integro-differential Equations in Contact Mechanics and Lubrication

C.H. Venner[1] and A.A. Lubrecht[2]

1 Introduction

Since their introduction multilevel techniques have influenced many fields in science. Wherever large scale computations are needed, e.g. fundamental research, applied research and design and development of technical equipment, they have reduced computational cost and/or created the ability to solve increasingly complex and extensive problems. In this paper we present an example from the field of tribology, i.e. the science and technology of interacting surfaces in relative motion, or, to use a more popular definition, the science of friction, lubrication and wear. In particular we will describe the essential steps leading to an efficient multilevel solver for the simulation of lubricated concentrated contacts. However, the techniques described in this paper are neither restricted to this problem, nor are they restricted to the field of contact mechanics and lubrication. In fact, they can be of interest for any problem described by Fredholm integral and integro-differential equations.

This paper is organized in the following way. First section 2 introduces lubricated concentrated contacts and provides some background of the research. Subsequently, section 3 presents the equations describing the two characteristic modes of operation of such a contact generally considered in theoretical studies. In section 4 the discrete equations are given for both cases. In section 5 the multilevel solution of the equations is addressed. First it is explained how to obtain stable relaxation schemes that efficiently smooth the error. Subsequently several aspects related to the coarse grid correction cycle are discussed. With this coarse grid correction cycle, even for extreme operating conditions, a solution can be obtained with an error smaller than the discretization error, employing a 2-FMG algorithm.

---

[1]University of Twente, Enschede, The Netherlands

[2]University of Twente, Enschede, The Netherlands & SKF Engineering Research Centre B.V., Nieuwegein, the Netherlands

However, when taking a close look at the algorithm it becomes obvious that the efficiency, although greatly improved compared to single grid solvers, is still far removed from the $O(h^{-d})$ (d being the dimension of the domain) efficiency usually obtained for elliptic problems. This is due to a multi-summation (Fredholm integral) appearing in the equations, the evaluation of which requires $O(h^{-2d})$ operations. In section 6 a multilevel algorithm "Multilevel Multi-Integration," is explained that enables an evaluation in $O(h^{-d})$ operations while maintaining accuracy, as is demonstrated. Merging this algorithm with the techniques explained in section 5 yields a multilevel solver with overall complexity $O(h^{-d})$. The paper is concluded with a characteristic calculational result for a concentrated contact and an outline of directions for future research.

2 Concentrated Contacts

Lubricated concentrated contacts are common in technical equipment and everyday life. An example is the contact between the rolling element (ball) and the inner or outer raceway in a rolling element (ball) bearing. In general these contacts are lubricated with oil in order to separate the opposing surfaces by a thin oil film, transferring the load from one surface to the other. In that case, friction will be small (minimum power-loss) and wear of the surfaces will be nearly absent. The shape and the thickness of this lubricant film generally depend on the surface velocities, the contact load, and the geometry of the surfaces. In the case of concentrated contacts, two additional effects have to be considered. The pressure in these contacts can range up to 2.0 GPa, and consequently both the elastic deformation of the surfaces (even for steel components) and the pressure-dependence of lubricant properties, e.g. the viscosity, must be included in the analysis.

The purpose of a numerical simulation of these contacts is threefold. Firstly to reveal the mechanisms determining the film formation and the pressure in the contact, both globally, (on the scale of the entire contact) as well as locally, i.e. what is the effect of particular local surface features, e.g. surface roughness. Secondly, for design purposes, to predict the thickness and shape of the lubricant film given the operating conditions. Finally, even an ideally lubricated contact eventually breaks down due to (sub)surface fatigue, i.e. due to the stresses in the material caused by the pressure at the surface. To obtain optimal service life, insight into the mechanisms initiating this failure is essential.

Figure 1 shows the model of a concentrated contact generally used in theoretical studies: two elastic bodies of paraboloidal shape in relative motion subjected to a certain contact load, and the equivalent reduced configuration, i.e. the contact between a single paraboloid and a flat surface. Displayed in figure 1 is a so-called "point contact". A special case often studied separately is the "line contact" which is the simplification to infinitely wide bodies ($R^1_{x_2} = R^2_{x_2} = \infty$, and F is replaced by a load per unit width).

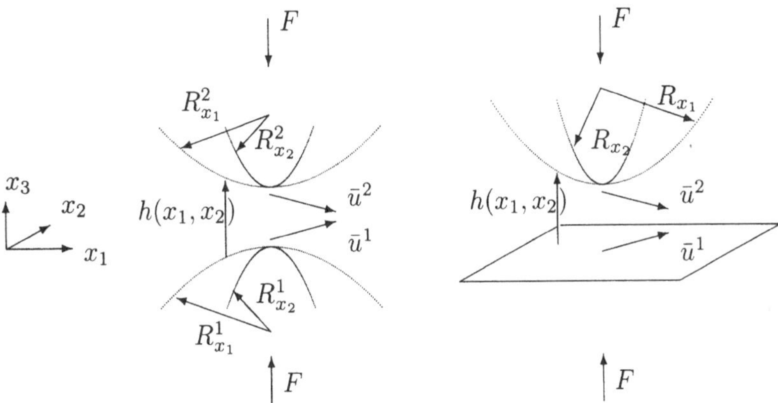

FIGURE 1. *The EHL point contact and the reduced geometry used in the theoretical analysis;* R_{x_i} *= reduced radius of curvature in* x_i *direction:* $1/R_{x_i} = 1/R^1_{x_i} + 1/R^2_{x_i}$. $h =$ *film thickness*

In theoretical studies two extreme modes of operation are distinguished: "dry contact" and "fully lubricated contact". In the first case the bodies are simply pressed together by the given load which causes them to deform elastically yielding a specific pressure distribution at the surface and stresses in the material. In the second case they are fully separated by a thin lubricant film and the pressure on the surfaces, (and the stresses in the material) is not only determined by elastic effects but also by the flow of lubricant through the gap. Of these problems, the lubricated contact is the more complex to solve numerically and, to understand some of the problems involved, it is advantageous to study the dry contact problem as a prelude. This approach is followed throughout this paper. However, the dry contact is also of interest in its own right, because, due to the very thin film in practical contacts, it already allows the prediction of contact stresses and the associated subsurface stress field to a good approximation.

3 Equations

In order to reduce the number of parameters, it is common practice to introduce dimensionless variables. This also applies to the equations presented here. The dimensionless variables used are based on the Hertzian theory of elasticity [4, 7, 5] and are denoted by uppercase characters, e.g. X_1, H. Their exact definition is not essential to the present paper and can be found in Lubrecht [8] and Venner [13].

3.1 Dry contact

Let $X \in \Omega \subseteq \mathbb{R}^d$, (d=1,2) then the (dimensionless) gap between the two surfaces, can be given as:

$$H(X) = H_0 + G(X) + W(X) \tag{1}$$

where $G(X)$ is a known function, containing the undeformed surface geometry e.g. the paraboloids in the case of perfectly smooth surfaces, H_0 is a constant discussed below, and $W(X)$ denotes the elastic deformation. Approximating both elements by elastic half-spaces, e.g. see Johnson [5], and Love [7]:

$$W(X) = \int_\Omega K(X,Y)P(Y)\,dY, \tag{2}$$

The kernel $K(X,Y)$ depends only on the dimension d of the problem and the distance $|X-Y|$. For example for $d=2$; $K = 1/|X-Y|$. This multi-integral plays an important role with respect to numerical solution of the problem as it determines the type of relaxation, see section 5.1. Furthermore, if no special measures are taken, the evaluation of its discrete counterpart will be very time consuming. This aspect is treated in detail in section 6.

The (integration) constant H_0 is determined by a socalled global constraint:

$$\int_\Omega P(X)\,dX = c \tag{3}$$

where c is a constant depending only on the dimension d of the domain Ω. In physical terms (3) imposes the force balance between the integral over the pressure and the externally applied contact load.

In the case of a dry contact the two elements are simply pressed together by the external load and, in its simplest form, it can now be described as: solve $P(X)$, and the integration constant H_0 from:

$$H(X) = 0 \quad X \in \Omega \quad P = 0 \text{ on } \partial\Omega \tag{4}$$

and the global constraint (3). However, this formulation is not suited for numerical solution, as the domain Ω is not known a priori. Furthermore, several physical constraints must be included, i.e. the surfaces cannot penetrate each other and the pressure cannot drop below the ambient pressure ($P = 0$). To incorporate these conditions and deal with the unkown domain, equation (4) is extended into the domain Ω' ($\Omega \subset \Omega'$) and written as a complementarity equation. As a result, the problem to be solved is: for a given geometry $G(X)$, and the kernel as given above, determine $P(X)$, i.e. solve the Fredholm integral equation of the first kind (5), subject to $P(X) \geq 0$ and the global constraint (3).

$$H_0 + G(X) + \int_{\Omega'} K(X,Y)P(Y)\,dY = 0 \quad X \in \Omega' \quad P = 0 \text{ on } \partial\Omega' \tag{5}$$

3.2 Lubricated contact

In the dry contact problem the pressure in the contact is determined by elasticity only. This is no longer true if lubrication is taken into account in which case the lubricant flow in the gap also plays an important role. The lubricated contact problem is described by two equations. The first equation is the so-called Reynolds equation [12]. This equation is the basic equation of lubrication and it relates the pressure in the lubricant to the geometry of the domain and the surface velocities:

$$\nabla \cdot \left[\frac{\bar{\rho} H^3}{\bar{\eta}} \nabla P - \bar{\lambda} \bar{\rho} H \right] - \lambda_1 \frac{\partial(\bar{\rho} H)}{\partial T} = 0 \quad X \in \Omega \tag{6}$$

For details the reader is referred to [6, 8, 12, 13] $\bar{\lambda}$ is a vector containing the dimensionless surface velocities λ_1, and λ_2. Furthermore, $\bar{\eta}$ is the dimensionless lubricant viscosity (relative to the viscosity at ambient pressure) and $\bar{\rho}$ denotes the dimensionless lubricant density (also relative to the density at ambient pressure). Both viscosity and density depend on the pressure and are obtained from empirical equations. For details the reader is referred to [8, 13]. The dependance of the density on the pressure is relatively weak, however, the dependance of the viscosity on the pressure is very strong, i.e. it increases roughly exponentially with increasing pressure.

For the description of the essential elements of the numerical solver we restrict ourselves to the steady state situation with the velocities of both surfaces aligned in the X_1 direction. Furthermore, usually liquid lubricants are applied, i.e. the pressure cannot drop below the vapour pressure. As a result the problem must be solved as a complementarity problem and the domain Ω is not known. The problem is extended into a domain Ω' adding the equation $P \geq 0$. On the boundary of the domain (Ω') $P = 0$ is assumed (ambient pressure defined as $P = 0$). As a result the equation for the pressure reads:

$$\nabla \cdot (\epsilon \nabla P) - \frac{\partial(\bar{\rho} H)}{\partial X_1} = 0 \tag{7}$$

with $P \geq 0$ and ϵ is defined by:

$$\epsilon = \frac{\bar{\rho} H^3}{\bar{\eta} \lambda}$$

The second equation is the equation for the film thickness, i.e. equation (1):

$$H(X) = H_0 + G(X) + \int_{\Omega} K(X,Y) P(Y)\, dY, \quad X \in \Omega' \tag{8}$$

with $K = \ln|X - Y|$ if $d = 1$ and $K = 1/|X - Y|$ if $d = 2$. Finally, the global condition should be satisfied:

$$\int_{\Omega'} P(X)\, dX = c \tag{9}$$

Summarizing, given a specific contact case (defined by $G(X)$, λ, and two parameters appearing in the viscosity pressure equation), the lubricated contact problem consists of solving $P(X)$ and $H(X)$ from (7), (8) and (9).

With respect to the numerical solution of this system of equations it is important to note that, as a result of the roughly exponential viscosity pressure dependance, ϵ in (7) for realistic conditions varies many orders of magnitude over the domain. In the outer region it is very large, due to large H and small $\bar{\eta}$ whereas in the central region it almost vanishes due to the large $\bar{\eta}$ and small H. Consequently, in the outer region the problem behaves as a partial differential problem whereas in the central region the integral aspects dominate. Furthermore, if $d = 2$ there is an additional complication. With decreasing ϵ, equation (7) becomes strongly "anisotropic", as the coupling in X_2 direction weakens and eventually nearly vanishes. In the limit only the weak indirect coupling via the multi-integral remains. This behaviour has an important consequence for the relaxation to be used in a multilevel solver.

4 Discretization

Let Ω' be given by $\{X \in I\!R^1 | X_a \leq X \leq X_b\}$ if $d = 1$ and by $\{X = (X_1, X_2) \in I\!R^2 | X_a \leq X_1 \leq X_b \wedge -Y_a \leq X_2 \leq Y_a\}$ if $d = 2$. This domain is covered with a uniform grid with mesh size h. Let i denote the gridpoint $i = (i_1, ... i_d)$, then the elastic deformation integral can discretized as:

$$W^h(X_i^h) = W_i^h \stackrel{\text{def}}{=} \int_\Omega K(X_i^h, Y)\hat{P}^h(Y)\, dY = h^d \sum_j K_{i,j}^{hh} P_j^h, \qquad (10)$$

where $\hat{P}^h$ is a piecewise polynomial function of degree $2s - 1$ and $\hat{P}^h(Y_j^h) = P_j^h$, the coefficients $K_{i,j}^{hh}$ are calculated such that equation (10) holds. The factor h^d is introduced to ensure that both K and P are of comparable size on grids with different mesh sizes. The discretization error made in this process will be of the order h^{2s}. For all results that will be presented in this paper $s = 1$.

4.1 Dry contact

Using (10) equation (5) at any gridpoint i is discretized as:

$$H_0 + G(X_i^h) + W_i^h = 0 \quad X_i \in \Omega' \quad P_i^h = 0 \text{ on } \partial\Omega' \qquad (11)$$

subject to $P_i^h \geq 0$ and the global condition reads:

$$h^d \sum_i P_i^h - c = 0 \qquad (12)$$

4.2 LUBRICATED CONTACT

Using a second order accurate central discretization for the first term of equation (7) and a first order upstream discretization of the second term leads to the following equation to be satisfied at each non boundary site i, $(X_a + i_1h, -Y_a + i_2h)$ $(d = 2)$, subject to the cavitation condition: $P_i^h \geq 0$.

$$\begin{aligned} h^{-2}\big(&\epsilon_{(i_1-1/2,i_2)}(P^h_{(i_1-1,i_2)} - P^h_{(i_1,i_2)}) + \epsilon_{(i_1+1/2,i_2)}(P^h_{(i_1+1,i_2)} - P^h_{(i_1,i_2)}) \\ &+\epsilon_{(i_1,i_2-1/2)}(P^h_{(i_1,i_2-1)} - P^h_{(i_1,i_2)}) + \epsilon_{(i_1,i_2+1/2)}(P^h_{(i_1,i_2+1)} - P^h_{(i_1,i_2)})\big) \\ &- h^{-1}(\bar{\rho}_{(i_1,i_2)}H^h_{(i_1,i_2)} - \bar{\rho}_{(i_1-1,i_2)}H^h_{(i_1-1,i_2)}) = 0 \end{aligned} \tag{13}$$

where $\epsilon_{(i_1\pm1/2,i_2)}$ and $\epsilon_{(i_1,i_2\pm1/2)}$ denote the value of ϵ at the *intermediate* locations and are approximated using for example:

$$\epsilon_{(i_1-1/2,i_2)} \equiv (\epsilon_{(i_1,i_2)} + \epsilon_{(i_1-1,i_2)})/2$$

with:

$$\epsilon_{(i_1,i_2)} = \frac{\bar{\rho}(P^h_{(i_1,i_2)})(H^h_{(i_1,i_2)})^3}{\bar{\eta}(P^h_{(i_1,i_2)})\lambda}$$

The discretized film thickness equation reads:

$$H_i = H_0 + G(X_i^h) + h^d \sum_j K^{hh}_{i,j} P^h_j \tag{14}$$

and, as for the dry contact problem, the discretized global condition reads:

$$h^d \sum_i P_i^h - c = 0 \tag{15}$$

5 Multilevel Solution

In this section emphasis is on *solving* the discrete equations employing the usual multigrid processes, i.e. relaxation to smooth the error, a coarse grid correction cycle (using the F.A.S. because of the non-linearity), to accelerate convergence, all embedded in the well known FMG structure. First the subject of relaxation is addressed, where, again the dry contact problem is used as a prelude to the more involved lubricated contact problem. Subsequently different aspects characteristic for the present problems and essential for an efficient coarse grid correction cycle are discussed. The techniques explained here (added to the usual multigrid techniques) result in solution of the problems to discretization error in a 2-FMG algorithm, also for extreme conditions, where generally W cycles are needed to obtain optimal cycle convergence.

5.1 Relaxation for the dry contact

The objective is to solve the Fredholm integral equation (11). Let $\tilde{P}^h$ denote an approximation to the solution and $\tilde{W}^h$ the associated integrals. A straightforward approach to improve this approximation is to scan the entire grid, at each location applying changes to $\tilde{P}^h$ to satisfy (11) (with $\tilde{W}^h$). After all points have been visited the new approximation to P^h is used to update the discrete integrals.

Such a one point relaxation, either as a collective displacement scheme (Jacobi) or as a simultaneous displacement scheme (Gauss-Seidel) is generally an effective error smoother for partial differential equations. However, it may be less suitable for integral equations. For example, when applied to the present equations it is unstable because of too large an accumulation of changes of $\tilde{P}^h$ in the integrals during one relaxation, (even when integrals are updated while relaxing) resulting in amplification of smooth error components. For the type of kernels appearing here, a local behaviour of the relaxation (and stability) can be obtained by means of distributive relaxation. At each point changes to the current solution are applied also at a number of neighbouring points, with certain pre-set distribution weights such that the changed values satisfy a weighted sum (i.e. some pre-set linear combination) of several (sometimes just one) neighbouring discrete equations. The relaxation is called first order distributive if it will not change the sum $\sum P_i^h$. More generally a distribution order r leaves $\sum Q(x_i^h)P_i^h$ unchanged for any polynomial Q of degree less than r. For a 2 dimensional problem, on a uniform grid, $r = 2$ distributive relaxation has the following stencil of changes:

$$\frac{1}{4}\delta_{(i_1,i_2)} \begin{bmatrix} 0 & -1 & 0 \\ -1 & 4 & -1 \\ 0 & -1 & 0 \end{bmatrix} \tag{16}$$

For the present kernels, the influence of such distributed changes applied at point i on the discrete integrals at points j decays fast with increasing $|i-j|$, i.e. proportional to the second derivative (general r^{th} derivative) of the kernel. Hence, the effect of changes at a given point is limited only to discrete integrals at locations in its immediate vicinity and the relaxation is effectively local. This can be shown with a local mode analysis. Disregarding the influence of the complementarity condition and the global condition, such an analysis for the present problem ($s = 1$) yields an asymptotic smoothing rate ($\bar{\mu}$) of 0.40 for $d = 1$, and 0.45 for $d = 2$ relaxing (11) (for P) as a simultaneous displacement scheme (see Venner [13]), i.e. (re-)evaluating the multi-summation only after all sites i have been visited. Smaller values can be obtained if the integrals are locally updated. In that case the result depends on the order in which the points are visited, e.g. lexicographic, red-black, etc. This option was discarded from an integral evaluation point of view, as local updates are computationally expensive whereas evaluating them all at once can be done fast as will be explained in section 6. Furthermore, $\bar{\mu} = 0.45$ already enables solution to the discretization error in a 1-FMG or 2-FMG algorithm, with just a few pre/post relaxations, which is satisfactory from a practical point of view.

5.2 RELAXATION FOR THE LUBRICATED CONTACT

In designing a relaxation for the problem we focus on solving P^h from the discrete approximation to (7), (i.e. equation (13) if $d = 2$), alternated with recomputing or updating H^h using (14). This particular choice is again induced by the possibility to evaluate all integrals at once in a fast manner. As mentioned in section 3.2, the main problem when numerically solving the lubricated contact problem is that ϵ in (7) varies many orders of magnitude over the domain. Obviously for a multilevel solver it is essential that the relaxation scheme effectively smooths the error over the entire domain, i.e. for both small and large ϵ. As a first step in designing such a relaxation process, a linearized model, characteristic for the local behaviour of the full problem, was studied:

$$\epsilon\Delta P - \frac{\partial H}{\partial X_1} = 0 \tag{17}$$

with H given by (8) with a fixed H_0, ϵ a given constant and $P = 0$ on the boundary. With the usual 5 point discretization for ΔP if $d = 2$, or 3-point if $d = 1$, a first order upstream discretization for $\partial H/\partial X_1$, and using (14), a local mode analysis can be performed. This analysis shows that the smoothing rate of a given relaxation will depend on ϵ/h^2. Obviously for large values of ϵ/h^2 this analysis yields asymptotic smoothing rates as obtained for the discrete Poisson problem (see ([1])), e.g. $\bar{\mu} = 0.5$ ($d = 2$) and $\bar{\mu} = 1/\sqrt{5}$, ($d = 1$) for lexicographic Gauss-Seidel relaxation. However, for small values of ϵ/h^2 such a relaxation becomes unstable. This instability is caused by the accumulation of changes in the integrals, which, via the film thickness, affect equation (17). As in the case of the dry contact problem, this accumulation can be limited by distributive relaxation. In fact, from equation (17), one might expect smoothing rates for the limiting case $\epsilon = 0$, to be as good as for the dry contact problem, already with a distribution that is one order lower. This is indeed true for $d = 1$ where a first order distributive relaxation has $\bar{\mu} = 0.40$.

However, such a distributive relaxation has rather poor smoothing behaviour (when compared to the schemes mentioned above) for large ϵ/h^2. Therefore, the key to an efficient multilevel solver for (17) is a combination, i.e. to apply a different relaxation depending on the value of ϵ/h^2 on the grid. Strictly speaking there is an optimal value of the switch limit between the two relaxations. However, as we are not after asymptotic convergence already a crude criterion serves well. The final step to an efficient relaxation for the full problem is then to realize that relaxation is a local process. Therefore, in the complete problem, the local ratio of ϵ/h^2 can be used as a criterion for the type of changes to be applied. For example, using simple one point Gauss-Seidel changes in regions of large ϵ/h^2, and Jacobi first order distributive changes in regions of small ϵ/h^2, yields a stable and efficient relaxation for the one-dimensional problem. A detailed explanation can be found in [13].

The above reasoning also applies to the case $d = 2$. A simple Gauss-Seidel relaxation either pointwise or line-wise is an effective smoother for large ϵ/h^2, whereas for small ϵ/h^2 distributed relaxation is needed. However, there is the additional complication of the vanishing coupling in X_2 direction. Therefore, for small ϵ/h^2, a distributive line relaxation should be used, i.e. for each line of constant X_2 solve all changes (to be applied distributively) simultaneously from equation (17) and (14). Subsequently, an effective smoother for the problem regardless of ϵ is again best obtained by a combination of relaxations, e.g. by applying either simple Gauss-Seidel line relaxation or distributive line relaxation depending on the value of ϵ/h^2. This approach can then be extended to a hybrid line relaxation scheme for the full problem, i.e. in regions of large ϵ/h^2 changes as prescribed by Gauss-Seidel line relaxation are solved and in regions of small ϵ/h^2 changes as prescribed by the distributive line relaxation are solved. For further details and analysis the reader is referred to [13].

5.3 Global condition

Equation (3) links the integral over P to the global constant H_0. This relation is treated similarly to global equations in multilevel solvers of differential equations, see [1]. The residual of the discrete global condition, i.e. (12), is calculated on every level, transferred to coarser grids, and the equation is only treated (relaxed) on the coarsest grid. As equation (3) does not directly link the H_0 to the pressure integral, (the constant does not appear in it) equation (12) is relaxed by changing H_0 in the direction driving the residual of this equation to zero.

5.4 Complementarity condition

The complementarity condition ($P \geq 0$) introduces a free boundary. As a result distributed changes may be computed which after application result in a violation of the complementarity condition. They are then forced to comply, thereby introducing long range disturbances, since the distribution is altered. For 'simple' boundaries between the domains Ω and Ω' the convergence is not adversely affected. When the boundary becomes complex, convergence can be degraded. In a similar way the boundary of Ω' requires special attention. The simplest option is to not apply the neighbouring changes for i near the boundary. Alternatively one can modify the distribution. The complementarity condition also requires special attention in the coarse grid correction cycle. The transfer of residuals to coarser grids should be done carefully, i.e. it should be ensured that the residual of the equation valid at the given location, i.e. (11) or $P_i = 0$, is transferred. Furthermore, in the vicinity of the free boundary injection must be used to avoid mixing information from cavitated and non cavitated points in one coarse grid right hand side. For a detailed discussion on the multigrid treatment of a free boundary the reader is referred to Brandt and Cryer [2]

5.5 Film thickness equation

A special point of attention in the coarse grid correction cycle for the lubricated contact problem is the treatment of equation (14). Given an approximation to P^h, this equation can be solved exactly at any time. Hence, if the integrals are recalculated or updated after relaxing for P^h, equation (14) will have zero residuals at the time of transfer to the coarse grid. However, this by no means implies a zero coarse grid right hand side. To deal with the non linearity the Full Approximation Scheme is used which naturally must apply to *all* equations, i.e. also to equation (14). Further details regarding the treatment of this equation can be found in [8, 13].

6 Multilevel Multi-integration

Implementing the techniques explained in the previous section in a FMG algorithm yields stable solvers that are indeed fast compared to single grid solvers. However, their efficiency is still far removed from the usual $O(h^{-d})$ efficiency obtained for simple elliptic problems as the $O(h^{-2d})$ operations needed for the evaluation of the discretized multi-integral, equation (14) will determine the computing time. In this section we briefly explain how the $O(h^{-d})$ can be restored and describe "multilevel multi-integration", a fast algorithm for evaluation of the multi-integrals. An extensive treatment of the subject can be found in [3].

The time consuming nature of numerical evaluation (solution) of integral equations as (2) has long been recognized, and traditionally far field assumptions have been applied to speed up computation in parts of the domain Ω' where the kernel K is relatively small (thus the name: far field). Here, however, we will make use of the smoothness properties of these kernels, thereby replacing the values of K in some points by interpolations in order to reduce the complexity. When the kernel is sufficiently smooth the work can be reduced to $O(h^{-d})$ operations, using integration (summation) on coarser grids. For potential-type kernels, which exhibit a non-smooth (singular) behaviour for $X = Y$, the complexity can be reduced to $O(h^{-d}\log(h^{-d}))$ operations, given the requirement that the additional error made in this process should be smaller than the error made in discretizing the equation (2).

6.1 Discretization

Recalling the discretized integral:

$$W^h(X_i^h) = W_i^h \stackrel{\text{def}}{=} \int_\Omega K(X_i^h, Y)\hat{P}^h(Y)\, dY = h^d \sum_j K_{i,j}^{hh} P_j^h, \tag{18}$$

For convenience we will introduce only one coarser grid, with mesh size $H = 2h$ and $N \simeq n/2^d$ points; the indices on this coarse grid will be denoted by uppercase characters. The two grids will be arranged such that $X^h_{2I} = X^H_I$. Furthermore, it is necessary to define transfer operators between the two grids, such as the coarse-to-fine interpolation operator Π^h_H. The index on which such an operator works is denoted by a dot and the new index appears after the square bracket: $\tilde{K}^{hh}_{i,j} = [\Pi^h_H K^{hH}_{i,\cdot}]_j$. In later sections we will use more than two grids and it will be convenient to refer to them as levels, starting with the coarsest level which will be called level 1.

6.2 Smooth kernel coarse grid integration

In this section, the fine grid integrals are approximated by coarse grid integrals in order to decrease the computational work involved in performing the integration. We will require that the error made in this coarse grid integration process is smaller than the fine grid discretization error. As a first step we will approximate the values of W^h_i where the point with index i also belongs to the coarse grid ($i = 2I$). Whenever the kernel $K(X,Y)$ is smooth with respect to the variable Y, we can approximate K by $\tilde{K}$:

$$\tilde{K}^{hh}_{i,j} = [\Pi^h_H K^{hH}_{i,\cdot}]_j \tag{19}$$

with $K^{hH}_{i,J}$ given by $K^{hH}_{i,J} = K^{hh}_{i,2J}$, hence equation (10) can be approximated by:

$$\begin{aligned} W^h_i \simeq \tilde{W}^h_i &\stackrel{\text{def}}{=} h^d \sum_j \tilde{K}^{hh}_{i,j} P^h_j = h^d \sum_j [\Pi^h_H K^{hH}_{i,\cdot}]_j P^h_j \\ &= h^d \sum_J K^{hH}_{i,J} [(\Pi^h_H)^T P^h_{\cdot}]_J = H^d \sum_J K^{hH}_{i,J} P^H_J \end{aligned} \tag{20}$$

On the coarse grid, some of the values of K are replaced by interpolations. This is implemented by letting the adjoint interpolation operator work on P^h in defining the coarse grid function P^H:

$$P^H_J \stackrel{\text{def}}{=} 2^{-d} [(\Pi^h_H)^T P^h_{\cdot}]_J \tag{21}$$

where $(\Pi^h_H)^T$ is the adjoint operator of Π^h_H and $P^h_{2J} \simeq P^H_J$ when P^h is a smooth function.

As a second step the values of the integrals in the fine grid points that do not belong to the coarse grid ($i = 2I + 1$) are calculated by interpolation from the fine grid points ($i = 2I$) (equation (20)). Again, this can be performed whenever $K(X,Y)$ is also smooth with respect to X.

$$\tilde{W}^h_i \simeq [\Pi^h_H W^H_{\cdot}]_i \tag{22}$$

where

$$W^H_I \stackrel{\text{def}}{=} \tilde{W}^h_{2I} = H^d \sum_J K^{HH}_{I,J} P^H_J \tag{23}$$

and therefore $K^{HH}_{I,J} = K^{hh}_{2I,2J}$.

The problem of calculating equation (10) has thus been reduced to a similar problem (23) on a coarser grid. Since the number of points on this grid will be $N \simeq n/2^d$, the total work of the multi-summation (23) relative to the fine grid work will be much smaller. This process of coarsening is then repeated, using even coarser grids, until a grid with $N' = O(\sqrt{n})$ points is reached. On this grid the integration (summation) is actually performed, since further coarsening would not reduce the overall complexity, because the work involved, for instance, in the fine-to-coarse grid transfer, is already of the order of n operations. Note that the number of levels required to reach a grid with $N' = O(\sqrt{n})$ is inversely proportional to the dimension d.

6.3 Non-smooth kernel coarse grid integration

Until now, the kernel K was assumed to be sufficiently smooth over the entire domain Ω; for a more quantitative description of the smoothness required, the reader is referred to [3]. A number of kernels of practical interest, however, does not fulfil this requirement; for instance, the potential-type kernels of interest in the dry contact problem, $K(X,Y) = \ln|X-Y|$ and $K(X,Y) = |X-Y|^{-1}$ are non-smooth (singular) in the neighbourhood of $X = Y$. Fortunately, their smoothness increases rapidly with increasing distance $|X-Y|$. Since the kernel is smooth in a large portion of the domain we will approach this problem in the same way as outlined above. In order to keep the additional error, made in the coarse grid integration process, below the required level we will do some extra (correction) work in the neighbourhood of the singularity. We will start by deriving an exact expression that will replace equation (20) for the case that the fine grid point i belongs also to the coarse grid ($i = 2I$).

$$\begin{aligned} W_i^h = h^d \sum_j K_{i,j}^{hh} P_j^h &= h^d \sum_j \tilde{K}_{i,j}^{hh} P_j^h + h^d \sum_j (K_{i,j}^{hh} - \tilde{K}_{i,j}^{hh}) P_j^h \\ &= h^d \sum_j [I_H^h K_{i,.}^{hH}]_j P_j^h + h^d \sum_j (K_{i,j}^{hh} - \tilde{K}_{i,j}^{hh}) P_j^h \\ &= W_I^H + h^d \sum_j (K_{i,j}^{hh} - \tilde{K}_{i,j}^{hh}) P_j^h \end{aligned} \tag{24}$$

In this derivation equations (19), (20) and (23) have been used.
Now it can be shown that the correction term $(K_{i,j}^{hh} - \tilde{K}_{i,j}^{hh}) \to 0$ as $|i-j| \to \infty$. Remember that $\tilde{K}$ is obtained by interpolation from K itself and that K becomes smoother with increasing $|i-j|$. To be more precise:

$$(K_{i,j}^{hh} - \tilde{K}_{i,j}^{hh}) = \begin{cases} 0 & \text{for } i = 2I,\ j = 2J; \\ O(h^{2p} K^{(2p)}(\xi)) & \text{for } i = 2I,\ j = 2J+1. \end{cases} \tag{25}$$

where $2p$ is the order of interpolation used (only even orders are considered here) to obtain $\tilde{K}$ (equation (19)) and $K^{(2p)}(\xi)$ is the $2p$'th derivative of K at some intermediate point. Thus, whenever this derivative of K becomes small, the correction term will become small and can be neglected. Clearly this is no longer true for $i \simeq j$ in the case of the singular smooth kernels mentioned above and thus we will have to carry out the corrections in a neighbourhood of $i = j$ ($|j-i| \leq m$ or $i - m \leq j \leq i + m$ if $d = 1$). The precise shape of this neighbourhood in higher dimensions is discussed in [3]. Equation (24) can therefore be simplified to:

$$W_i^h \simeq W_I^H + h^d \sum_{|j-i|\leq m} (K_{i,j}^{hh} - \tilde{K}_{i,j}^{hh}) P_j^h \tag{26}$$

If point i is not in the coarse grid ($i = 2I + 1$), another coarse grid approximation $\hat{K}$ to K is defined:

l	$k = l$	$k = l-1$	$k = l-2$	$k = l-3$	$k = l-4$	$k = l-5$
2	2.3 10^{-1}	2.3 10^{-1}	–	–	–	–
	1.1 10^{-1}	1.0 10^{+0}	–	–	–	–
3	7.7 10^{-2}	7.6 10^{-2}	7.6 10^{-2}	–	–	–
	1.4 10^{+0}	2.6 10^{+0}	*2.2 10^{+0}	–	–	–
4	1.5 10^{-2}	1.5 10^{-2}	1.5 10^{-2}	1.5 10^{-2}	–	–
	1.9 10^{+1}	9.9 10^{+0}	6.3 10^{+0}	6.0 10^{+0}	–	–
5	$\sim$ 4 10^{-3}	4.6 10^{-3}	4.5 10^{-3}	4.4 10^{-3}	4.7 10^{-3}	–
	$\sim$ 3 10^{+2}	5.7 10^{+1}	2.1 10^{+1}	*1.8 10^{+1}	1.7 10^{+1}	–
6	$\sim$ 1 10^{-3}		1.3 10^{-3}	1.0 10^{-3}	9.5 10^{-4}	1.6 10^{-3}
	$\sim$ 5 10^{+3}		1.0 10^{+2}	6.7 10^{+1}	6.4 10^{+1}	6.3 10^{+1}
7	$\sim$ 3 10^{-4}			4.0 10^{-4}	3.4 10^{-4}	3.8 10^{-4}
	$\sim$ 8 10^{+4}			2.8 10^{+2}	*2.5 10^{+2}	2.5 10^{+2}

Table 1: Error and computing time (sec) in multi-integral on level l, while performing the multi-summation on level k.
Two-dimensional problem, s=1, $K = 1/|X - Y|$, employing sixth order transfers (see [3, 13]). $k = l$ is direct summation, $*$ denotes summation level with $\sqrt{n}$ points.

$$\hat{K}_{i,j}^{hh} = [I\!I_H^h K_{.,j}^{Hh}]_i \tag{27}$$

where $\hat{K}_{I,j}^{Hh} = K_{2I,j}^{hh}$. In terms of this new kernel we can derive an expression similar to equation (24) for the integrals in points with index $i = 2I + 1$.

$$\begin{aligned} W_i^h = h^d \sum_j K_{i,j}^{hh} P_j^h &= h^d \sum_j \hat{K}_{i,j}^{hh} P_j^h + h^d \sum_j (K_{i,j}^{hh} - \hat{K}_{i,j}^{hh}) P_j^h \\ &= h^d \sum_j [I\!I_H^h K_{.,j}^{Hh}]_i P_j^h + h^d \sum_j (K_{i,j}^{hh} - \hat{K}_{i,j}^{hh}) P_j^h \\ &\simeq [I\!I_H^h W_.^H]_i + h^d \sum_j (K_{i,j}^{hh} - \hat{K}_{i,j}^{hh}) P_j^h \end{aligned} \tag{28}$$

Far from the singularity the correction terms become small and are neglected. Equation (28) then reduces to:

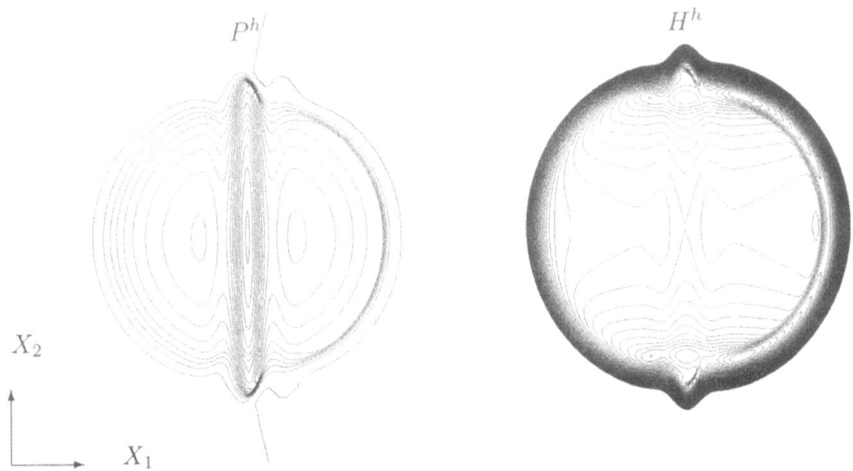

FIGURE 2. *Contour plots of pressure and film thickness in an EHL point contact.*

$$W_i^h \simeq [\Pi_H^h W_.^H]_i + h^d \sum_{|j-i|\leq m} (K_{i,j}^{hh} - \hat{K}_{i,j}^{hh}) P_j^h \tag{29}$$

If $K(X,Y)$ has similar smoothness properties in X and Y, and identical interpolation operators are used in equations (19) and (27), the correction term in equation (28) is similar to (25), but since the interpolation is carried out with respect to the i index, it will be non-zero for all j.

$$(K_{i,j}^{hh} - \hat{K}_{i,j}^{hh}) = O(h^{2p} K^{(2p)}(\xi)) \qquad (\forall j, i = 2I+1) \tag{30}$$

This will result in larger errors (approximately three times as large) in the points $i = 2I+1$, as compared to the points $i = 2I$ (a factor of two comes from equations (29) and (30), a factor of one comes from the approximation in equation (28)). In [3] the optimal values of m and $2p$ are derived in order to minimize the total work. It is shown that for the potential-type kernels mentioned above, a total work proportional to $h^{-d} \log h^{-d}$ can be obtained, as can be seem from table 1, taken from [3].

7 Results

The multilevel evaluation algorithm explained in section 6, can be straightforwardly merged with the techniques explained in section 5, yielding multilevel solvers for the problems considered here with complexity $O(h^{-d} \log h^{-d})$. Note that this is effectively the optimal $O(h^{-d})$ complexity, as $\log(h^{-d})$ increases only slowly with decreasing mesh size. These solvers have subsequently been applied to a wide variety of contact situations. For detailed engineering applications the reader is referred to [8, 9, 10, 11, 13, 14, 15]. In this paper, due to space limitations only one set of results is presented in Figure 2: a contour plot of the pressure and the film thickness for a lubricated stationary contact with a ridge (local surface feature), solved on a grid of $513 * 513$ points.

8 Further Developments

The incentive behind the application and development of these numerical techniques is the requirement to obtain a tribological model that can predict successful operation or failure of highly loaded concentrated contacts. Such a model, which must be transient by nature, should be capable of describing rheological and thermal effects, and it has to accomplish *locally* a very detailed analysis. Current research is directed along two avenues: Improvement of the physical mathematical model, and further algorithmic development. The latter involves extension to transient situations, and to higher order approximations using double discretization. Furthermore, research has started to incorporate local grid refinement techniques, which, due to the multi-integral raises interesting fundamental algorithmic questions to be answered.

ACKNOWLEDGEMENTS

The authors wish to thank Prof. A. Brandt of the Weizmann Institute of Science, Rehovot, Israel, for his encouragement and his support of this work over the last decade. Visits to Prof. Brandt were supported by the faculty of Mechanical Engineering and the "Hogeschoolfonds" of the University of Twente, and by the foundations KiVI, NWO and Stichting Tribologie. Long term financial support for this research was received from the foundation "Niels Stensen" and from the Royal Netherlands Academy of Arts and Sciences. Finally the authors wish to thank Dr. H.H. Wittmeyer, Managing Director of SKF Engineering & Research Centre B.V., for his kind permission to publish this paper.

References

[1] BRANDT, A., 1984, "Multigrid Techniques: 1984 Guide with Applications to Fluid Dynamics", monograph, available as GMD studien **85** from GMD, postfach 1240, Schloss Birlinghofen D-5205, St. Augustin 1, BRD.

[2] BRANDT, A., AND CRYER, C.W., 1983, "Multigrid algorithms for the solution of linear complementarity problems arising from free boundary problems," *SIAM J. Sci. Stat. Comput.*, **4, 4**, pp. 655-684.

[3] BRANDT, A., AND LUBRECHT, A.A., 1990, "Multilevel Multi-Integration and Fast Solution of Integral Equations", Journal of Computational Physics, **90**, pp. 348-370.

[4] HERTZ, H., 1881, "Über die Berührung fester elastischer Körper," *J. für die reine und angew. Math.*, **92**, pp. 156-171.

[5] JOHNSON, K.L., 1985, "Contact mechanics," Cambridge University Press.

[6] LANGLOIS, W.E., 1964, "Slow viscous flow," The Macmillan Company, New York.

[7] LOVE, A.E.H., 1944, "A treatise on the mathematical theory of elasticity," 4th edition, Dover publications, New York.

[8] LUBRECHT, A.A., 1987, "Numerical Solution of the EHL Line and Point Contact Problem Using Multigrid Techniques," Ph.D. Thesis, University of Twente, Enschede, The Netherlands, ISBN 90-9001583-3.

[9] LUBRECHT, A.A., VENNER, C.H., LANE, S., JACOBSON, B., AND IOANNIDES, E., 1990, "Surface Damage - Comparison of Theoretical and Experimental Endurance Lives of Rolling Bearings," *Proceedings of the 1990 Japan International Tribology Conference, Nagoya, Japan*, **1**, pp. 185-190.

[10] LUBRECHT, A.A., DWYER-JOYCE, R.S., AND IOANNIDES, E., 1991, "Analysis of the Influence of Indentations on Contact Life," proceedings of the 18th Leeds-Lyon Symposium on Tribology, pp. 173-181.

[11] LUBRECHT, A.A., AND VENNER, C.H., 1992, "Aspects of Two-Sided Surface Waviness in an EHL Line Contact," presented at the 1992 Leeds-Lyon Conference on Tribology, Leeds, U.K.

[12] REYNOLDS, O., 1886, "On the theory of lubrication and its application to Mr Beauchamp Tower's experiments, including an experimental determination of the viscosity of olive oil," *Phil. Trans. R. Soc.*, **177**, pp. 157-234.

[13] VENNER, C.H., 1991, "Multilevel Solution of the EHL Line and Point Contact Problems," Ph.D. Thesis, University of Twente, Enschede, The Netherlands, ISBN 90-9003974-0.

[14] VENNER, C.H., AND LUBRECHT, A.A., 1992, "Transient Analysis of Surface Features in an EHL line Contact in the Case of Sliding," presented at the 1992 ASME-STLE tribology conference in San Diego, U.S.A.

[15] VENNER, C.H., AND LUBRECHT, A.A., 1993, "Numerical Simulation of a Transverse Ridge in a Circular EHL Contact, under Rolling/Sliding," under preparation.

Part II

Contributed Papers

1

A Multi-Grid Method for Calculation of Turbulence and Combustion

X.S. Bai and L. Fuchs[1]

ABSTRACT The application of the Multi-Grid (MG) method to the calculation of turbulent reacting flows is considered. Turbulence is handled by using the $k-\epsilon$ model. The eddy-dissipation concept based on a reduced global chemical reaction scheme is used for modeling the chemical reactions. For low Reynolds number laminar flows the MG efficiency is best, with the convergence rate in the order of 0.8. For uniformly spaced grids the convergence rate can be better, and for highly skewed grid, slower. The introduction of turbulence and combustion generally slows down the converging process. However, the MG method still demonstrates considerable acceleration over the single grid solver.

1 Introduction

Turbulent reacting flows involve many different processes, such as, advection, diffusion, turbulence, chemical reactions and heat transfer. The different processes are described by a system of partial differential equations (PDE), an ordinary differential equation and some algebraic relations. These equations are highly coupled, corresponding to the interaction between the different physical processes. A key issue in calculating turbulent reacting flows is therefore the numerical efficiency, so that numerical prediction will be appropriate for engineering design.

Efficient solvers for elliptic partial differential equations could be tailored using Multi-Grid (MG) methods [1]. Previously, variants of MG methods have been applied to incompressible laminar flows [2] and isothermal turbulent flows [3,4]. In this paper, the application of MG method to turbulent reacting flows in cylindrical coordinates is discussed.

Turbulent reacting flows are unsteady. They can be calculated by Direct Numerical Simulation (DNS) [5]. However, DNS is applicable only for low or moderate Reynolds number (Re) situations and rather simple geometries. For non-trivial ge-

[1]Department of Mechanics/Applied CFD
Royal Institute of Technology, S-100 44 Stockholm

ometries and flows of practical interest, DNS are not applicable. If one is interested in the (time-) averaged properties rather than the turbulent fluctuations, one can use some simple modeling technique, such as $k-\epsilon$ model to handle turbulence [6]. If one is not interested in intermediary and low concentration species, one may neglect some elementary chemical reactions [7-8]. Turbulence-chemistry interaction can be handled by using simplified models such as the Eddy- Dissipation Concept (EDC) [9]. Previous research [10-13] had shown that such simplified models yield reasonably good results in many cases.

The MG calculation is performed on a modeled annular gas turbine combustion chamber. The basic solver shown in this paper is very efficient for low Reynolds number flow calculations. Extension to the calculation of turbulent reacting flows has shown that variable density field (gas expansion process) and eddy viscosity calculation make the convergence slower compared to the optimal laminar case. The calculation shows, however, that MG method reduces computational time by up to two orders of magnitude when compared with the single grid relaxations, in the calculation of high Reynolds number turbulent reacting flows.

2 Mathematical models

2.1 Governing Equations for Turbulent Reacting Flows

When the mean properties of the flow field are of interest, as in many engineering applications, the Reynolds averaged equations have to be used. For "closing" these equations one has to use a turbulence model, (e.g. the $k-\epsilon$ equations). Let the averaged pressure and density be denoted by p, ρ. Let u, v, w denote the velocity components in axial, radial and azimuthal directions, respectively. The conservation of mass and momentum are as follows.

$$\frac{\partial \rho}{\partial t}+\frac{1}{r}\frac{\partial \rho r v}{\partial r}+\frac{1}{r}\frac{\partial \rho w}{\partial \theta}+\frac{\partial \rho u}{\partial x}=0 \tag{1}$$

$$\frac{\partial \rho u}{\partial t}+\frac{1}{r}\frac{\partial \rho r u v}{\partial r}+\frac{1}{r}\frac{\partial \rho u w}{\partial \theta}+\frac{\partial \rho u u}{\partial x}=-\frac{\partial p}{\partial x}+\frac{1}{r}\frac{\partial r\tau_{rx}}{\partial r}+\frac{1}{r}\frac{\partial \tau_{\theta x}}{\partial \theta}+\frac{\partial \tau_{xx}}{\partial x} \tag{2}$$

$$\frac{\partial \rho v}{\partial t}+\frac{1}{r}\frac{\partial \rho r v v}{\partial r}+\frac{1}{r}\frac{\partial \rho v w}{\partial \theta}+\frac{\partial \rho u v}{\partial x}-\frac{\rho w^2}{r}=-\frac{\partial p}{\partial r}+\frac{1}{r}\frac{\partial r\tau_{rr}}{\partial r}+\frac{1}{r}\frac{\partial \tau_{r\theta}}{\partial \theta}+\frac{\partial \tau_{rx}}{\partial x}-\frac{\tau_{\theta\theta}}{r} \tag{3}$$

$$\frac{\partial \rho w}{\partial t}+\frac{1}{r^2}\frac{\partial \rho r^2 v w}{\partial r}+\frac{1}{r}\frac{\partial \rho w w}{\partial \theta}+\frac{\partial \rho u w}{\partial x}=-\frac{1}{r}\frac{\partial p}{\partial \theta}+\frac{1}{r^2}\frac{\partial r^2\tau_{r\theta}}{\partial r}+\frac{1}{r}\frac{\partial \tau_{\theta\theta}}{\partial \theta}+\frac{\partial \tau_{\theta x}}{\partial x} \tag{4}$$

$\tau_{rr}, \tau_{rx}, ...$ are the components of the stress tensor.

$$\tau_{rx} = \mu_{eff}[\frac{\partial u}{\partial r} + \frac{\partial v}{\partial x}] \quad \tau_{xx} = \mu_{eff}[2\frac{\partial u}{\partial x} - \frac{2}{3}\nabla \cdot \mathbf{v}]$$

$$\tau_{\theta x} = \mu_{eff}[\frac{\partial w}{\partial x} + \frac{1}{r}\frac{\partial u}{\partial \theta}] \quad \tau_{rr} = \mu_{eff}[2\frac{\partial v}{\partial r} - \frac{2}{3}\nabla \cdot \mathbf{v}]$$

$$\tau_{r\theta} = \mu_{eff}[r\frac{\partial w/r}{\partial r} + \frac{1}{r}\frac{\partial v}{\partial \theta}] \quad \tau_{\theta\theta} = \mu_{eff}[2(\frac{1}{r}\frac{\partial w}{\partial \theta} + \frac{v}{r}) - \frac{2}{3}\nabla \cdot \mathbf{v}]$$

where $\mu_{eff} = \mu_L + \mu_t$ and $\mu_t = \rho C_\mu k^2/\epsilon$. μ_L, μ_t are the laminar viscosity and turbulent eddy viscosity, respectively. k is the turbulent kinetic energy and ϵ its dissipation rate. The two-equation $k - \epsilon$ model is given by

$$\frac{\partial \rho k}{\partial t} + \frac{1}{r}\frac{\partial \rho r v k}{\partial r} + \frac{1}{r}\frac{\partial \rho w k}{\partial \theta} + \frac{\partial \rho u k}{\partial x} = \frac{1}{r}\frac{\partial}{\partial r}(r\frac{\mu_{eff}}{P_k}\frac{\partial k}{\partial r}) + \frac{1}{r}\frac{\partial}{\partial \theta}(\frac{\mu_{eff}}{rP_k}\frac{\partial k}{\partial \theta})$$
$$+ \frac{\partial}{\partial x}(\frac{\mu_{eff}}{P_k}\frac{\partial k}{\partial x}) + S_k \tag{5}$$

$$\frac{\partial \rho \epsilon}{\partial t} + \frac{1}{r}\frac{\partial \rho r v \epsilon}{\partial r} + \frac{1}{r}\frac{\partial \rho w \epsilon}{\partial \theta} + \frac{\partial \rho u \epsilon}{\partial x} = \frac{1}{r}\frac{\partial}{\partial r}(r\frac{\mu_{eff}}{P_\epsilon}\frac{\partial \epsilon}{\partial r}) + \frac{1}{r}\frac{\partial}{\partial \theta}(\frac{\mu_{eff}}{rP_\epsilon}\frac{\partial \epsilon}{\partial \theta})$$
$$+ \frac{\partial}{\partial x}(\frac{\mu_{eff}}{P_\epsilon}\frac{\partial \epsilon}{\partial x}) + S_\epsilon \tag{6}$$

where S_k, S_ϵ are the source terms in the k and ϵ equations, respectively.

We define the specific enthalpy h as $h = \int_{T_0}^{T} C_P dT, (T_0 = 25C)$, then:

$$\frac{\partial \rho h}{\partial t} + \frac{1}{r}\frac{\partial \rho r v h}{\partial r} + \frac{1}{r}\frac{\partial \rho w h}{\partial \theta} + \frac{\partial \rho u h}{\partial x} = \frac{1}{r}\frac{\partial}{\partial r}(r\frac{\mu_{eff}}{P_h}\frac{\partial h}{\partial r}) - \frac{\partial p}{\partial t}$$
$$+ \frac{1}{r}\frac{\partial}{\partial \theta}(\frac{\mu_{eff}}{rP_h}\frac{\partial h}{\partial \theta}) + \frac{\partial}{\partial x}(\frac{\mu_{eff}}{P_h}\frac{\partial h}{\partial x}) + H_i^0 R_i + S_r \tag{7}$$

R_i is the reaction rate of species i, H_i^0 is the enthalpy formation of i. S_r is the source term due to thermal radiation (which is neglected in this calculation, so that $S_r = 0$).

For species i the mass fraction m_i - equation is:

$$\frac{\partial \rho m_i}{\partial t} + \frac{1}{r}\frac{\partial \rho r v m_i}{\partial r} + \frac{1}{r}\frac{\partial \rho w m_i}{\partial \theta} + \frac{\partial \rho u m_i}{\partial x} = \frac{1}{r}\frac{\partial}{\partial r}(r\frac{\mu_{eff}}{P_m}\frac{\partial m_i}{\partial r})$$
$$+ \frac{1}{r}\frac{\partial}{\partial \theta}(\frac{\mu_{eff}}{rP_m}\frac{\partial m_i}{\partial \theta}) + \frac{\partial}{\partial x}(\frac{\mu_{eff}}{P_m}\frac{\partial m_i}{\partial x}) - R_i \tag{8}$$

The reaction rate, R_i, is a function of mass fractions, temperature, and influenced by turbulence:

$$R_i = f(m_i, T, k, \epsilon, ...) \tag{9}$$

This relation will be discussed later in the next section.

The system of equations above (1-9) is completed by the equation of state.

At inlet the boundary conditions for u, v, w are given by a certain profile and the inlet mass flux. At outlet they are given by setting the second derivatives of the velocity vector to zero. At walls u, v, w vanish. The boundary conditions for k and ϵ are as follows: at inlet, the turbulent kinetic energy is taken to be proportional (a few percents) to the inlet kinetic energy. At solid walls, we use the so called "wall functions" [6]. At outlet, we force the second streamwise derivatives of k and ϵ to vanish.

The boundary conditions for h can be specified as follows: at inlet they are given by certain values, at outlet the zero second derivative is used. At wall they are given by specifying certain heat flux or the temperature itself. The boundary conditions on m_i at inlet and outlet are treated in a similar way as h. At wall we set the normal derivative of m_i to zero.

In the equations above, the following parameters values are used: $C_\mu = 0.09, P_k = 1.0, P_\epsilon = 1.22, P_h = 0.7, P_m = 0.7$; [13].

2.2 Chemical Reaction Scheme

For a hydrocarbon fuel (such as C_3H_8), one may use the following two-step global reaction scheme [7].

$$C_3H_8 + 3.5(O_2 + 3.76N_2) \rightarrow 3CO + 4H_2O + 13.16N_2$$

$$3CO + 1.5(O_2 + 3.76N_2) \rightarrow 3CO_2 + 5.64N_2$$

For each of these steps, the reaction rate is computed by:

$$R_1' = \min(R_1^A, B_1\rho \min(m_{C_3H_8}, \frac{m_{air}}{r_1})\frac{\epsilon}{k})$$

$$R_2' = \min(R_2^A, B_2\rho \min(m_{CO}, \frac{m_{air}}{r_2})\frac{\epsilon}{k})$$

where R_1^A, R_2^A are reaction rates computed by Arrhenius Law [7]. $r_1 = 10.92, r_2 = 2.451$ are stoichiometric constants of the reaction steps and $B_1 = B_2 = 2.0$ are model constants. The reaction rate for species C_3H_8, CO, O_2 can be computed by:

$$R_{C_3H_8} = R_1' \tag{10}$$

$$R_{CO} = R_2' - r_3R_1' \tag{11}$$

$$R_{O_2} = r_1R_1' + r_2R_2' \tag{12}$$

where $r_3 = 1.909$.

3 Solution Method

3.1 Mass Flux Conserving Transformation

The continuity equation is written in a strong conservation form. This formulation simplifies the handling of the axial singularity *and* also ensures global mass conservation in the MG process in a straightforward manner. Let

$$\rho r u = U \quad \rho r v = V \quad \rho w = W \tag{13}$$

then the continuity equation in conservative form becomes:

$$\frac{\partial \rho r}{\partial t} + \frac{\partial V}{\partial r} + \frac{\partial W}{\partial \theta} + \frac{\partial U}{\partial x} = 0 \tag{14}$$

The momentum equations and the transport equations for a scalar f ($f = k$, ϵ, h, m_i, ...), in terms of the new dependent variables, can be written as:

$$\frac{\partial U}{\partial t} + \frac{\partial UV/(\rho r)}{\partial r} + \frac{\partial WU/(\rho r)}{\partial \theta} + \frac{\partial UU/(\rho r)}{\partial x} = -r\frac{\partial p}{\partial x} + \frac{\partial r\tau_{rx}}{\partial r} + \frac{\partial \tau_{\theta x}}{\partial \theta} + \frac{\partial r\tau_{xx}}{\partial x} \tag{15}$$

$$\frac{\partial \rho f r}{\partial t} + \frac{\partial V f}{\partial r} + \frac{\partial W f}{\partial \theta} + \frac{\partial U f}{\partial x} = \frac{\partial}{\partial r}(r\frac{\mu_{eff}}{P_f}\frac{\partial f}{\partial r})$$

$$+ \frac{\partial}{\partial \theta}(\frac{\mu_{eff}}{rP_f}\frac{\partial f}{\partial \theta}) + \frac{\partial}{\partial x}(r\frac{\mu_{eff}}{P_f}\frac{\partial f}{\partial x}) + rS_f \tag{16}$$

3.2 Discretisation

The discretisation is done on a staggered grid. The components of the velocity vector are defined at the center of the corresponding cell side. Scalars are defined at the cell center. All terms with the possible exception of the convective terms are approximated by central differences. The convective terms in all the equations are approximated by the first order scheme. When only steady state is sought, as in the cases considered here, one may use a quasi-time marching technique for the relaxation of the equation. By analogy to a time dependent term we define a "correction" term:

$$\frac{\partial \phi}{\partial t} \sim \frac{\phi^{n+1} - \phi^n}{\Delta t} \sim \frac{\beta U_*}{\Delta x}\Delta \phi \tag{17}$$

The superscript n represents a pseudo, n-th time step, while $\Delta\phi$ represents the correction during each iteration. U_* is the characteristic velocity, usually taken as the maximum of the inlets velocities. Δx is the characteristic spatial mesh size. $\beta \geq 0$ acts as a relaxation parameter (Usually, the smaller the value of β is, the faster the convergence. However, occasionally when β is too small, the relaxation process may diverge).

3.3 Multi-Grid Solver

The discretized equations result in a system in which all the dependent variables are coupled together. The SIMPLE scheme [14] is used for "decoupling" velocity and pressure. Since we use the mass flux conserving transformation, the coupling between density and velocity is only through source terms. The energy equation and the other transport equations are updated after updating the velocity and pressure ("sequential" relaxation [13]). The relaxation is linewise, in the radial direction, for all the equations.

To accelerate the convergence of the basic line by line solver, a MG method is used. The solution procedure starts on a coarse grid doing several V-cycle MG relaxations in the fully approximate storage (FAS) mode [2,3]. After converging to a certain level, the variables are transferred to a finer grid. This procedure is repeated until the finest grid is reached and the converged solution is obtained. The transfer of scalars to coarser grids is done by volume averaging, whereas the components of the velocity vectors are transferred by area (flux conserving) averaging. Mass flux conserving restriction is a necessary condition for the convergence in the coarse grid. The corrections are interpolated to fine grids by trilinear interpolations.

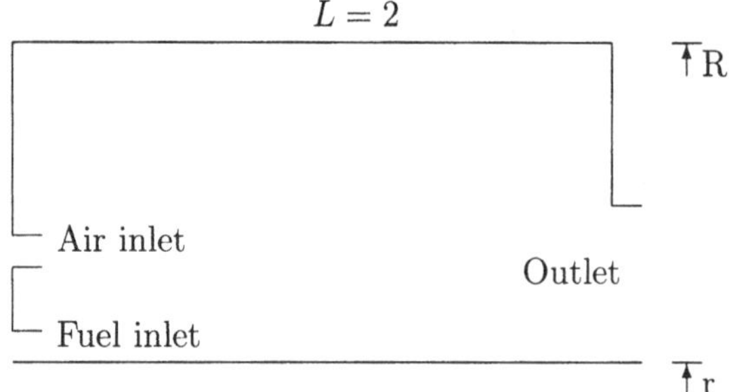

FIGURE 1. A sketch of the model annular gas turbine combustion chamber

4 Numerical Examples

The above solver is tested on a cylinder combustion chamber (a simplified model of annular gas turbine combustor). As shown in Fig.1, the inner radius r can be changed. The outer radius is $R = 1 + r$. The chamber length is $L = 2$. We compute only a sector of the chamber, assuming periodicity (with the sector angle $\alpha = 34.4^o$). Two inlets supply air and fuel (C_3H_8), respectively. The fuel/air equivalence ratio is 0.88. The coarsest grid is 5x3x3, which is refined by 3 levels (Grid 1) and 4 levels (Grid 2).

4.1 Analysis of the Basic Solver

In order to achieve high MG convergence efficiency for turbulent reacting flow calculation, it is necessary to obtain high efficiency for a single convection-diffusion equation calculation. In the following, instead of illustrating a single equation calculation, we show the behaviour for calculation of Navier-Stokes equations for laminar flow.

As depicted in Fig.2, the grid has important influence on the convergence process. When the grid is refined, e.g., with 4 levels (Grid 2), the single grid relaxation becomes less efficient. The convergence rate of the MG scheme, on the other hand, is almost "grid-independent". Another feature is that the MG convergence process is monotone, while single grid relaxations converge in a non-uniform manner.

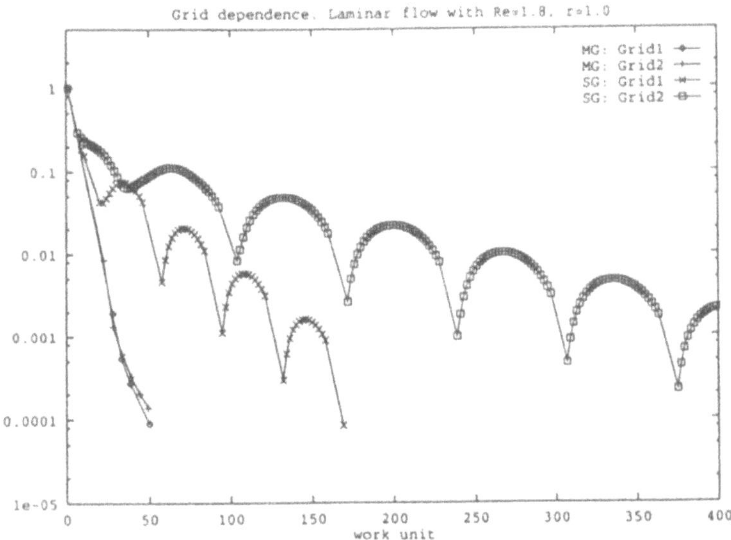

FIGURE 2. Convergence history for Grid 1 and Grid 2

In this calculation $Re = 1.8$. The inner radius is $r = 1.0$ which keeps the mesh spacing ratio close to unity. The mesh spacing ratio (aspect ratio) varies from $h_x : h_r : h_\theta = 1 : 0.83 : 1$ to $2 : 1.66 : 1$. Such near-uniform grid distribution yields a rather fast convergence rate (0.8). The MG efficiency deteriorates when the inner radius r decreases. As seen from Fig.3, when the inner radius is $r = 0.06$, the grid aspect ratio varies in the range from $2 : 1.66 : 1$ to as much as $33.3 : 27.6 : 1$. On this grid both MG and single-grid relaxation become less efficient. The "smoother" has to be modified in order to retain the previously attained MG efficiency.

Next consider larger values of Reynolds numbers ($Re = 1.8$ to $Re = 18800$). In the former case the flow is laminar, but is not in the latter. As long as the convective and diffusive terms are of comparable order of magnitude, and the

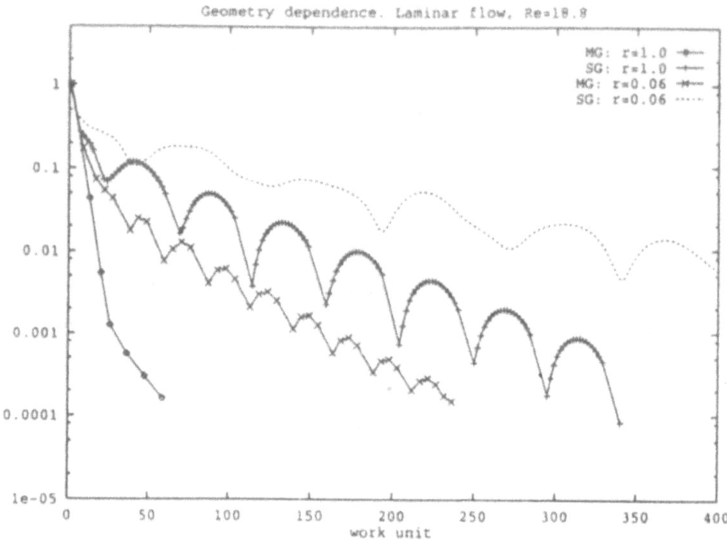

FIGURE 3. Convergence history for $r = 1.0$ and $r = 0.06$ on Grid 1

mesh spacing allows the resolution of all possible scales, one is able to compute the flow with rather good efficiency (Fig. 4). Naturally, for larger values of *Re* one has to use a time-dependent solver, to account for the unsteadiness of the flow. Attempting to solve a steady-state flow, results in a "non-converging solution" (see Fig. 4). Once the local mesh Reynolds number (Re_h) (based on the local speed and local mesh spacing) is large (in fact for $Re_h > 2$), the grid cannot support the smallest scales and hence, one has to add "viscosity" artificially, so that the resulting scales can be supported on the grid. The increase in the effective diffusivity in turbulent flows is an expression of this type of "viscosity". Thus, at higher values of *Re*, one has to introduce a turbulence model. These models has the effect of restoring the (high-frequency) ellipticity of the system, allowing the computation of converged solution by a MG solver.

4.2 Calculation of Turbulence and Combustion

To compute turbulent flows, one may use the $k-\epsilon$ turbulence model. The turbulent transport is modeled by an eddy viscosity, which is considerably larger than the molecular viscosity. Therefore it is possible to obtain a stationary, time-averaged, solution.

For turbulent flow calculations, one has to handle the coupling between k, ϵ and the velocity- and the pressure-fields. The coupling is essentially through the eddy viscosity. The sequential ("segregated") relaxation method described earlier [13] is more stable, though it may be less efficient (in comparison with a local,

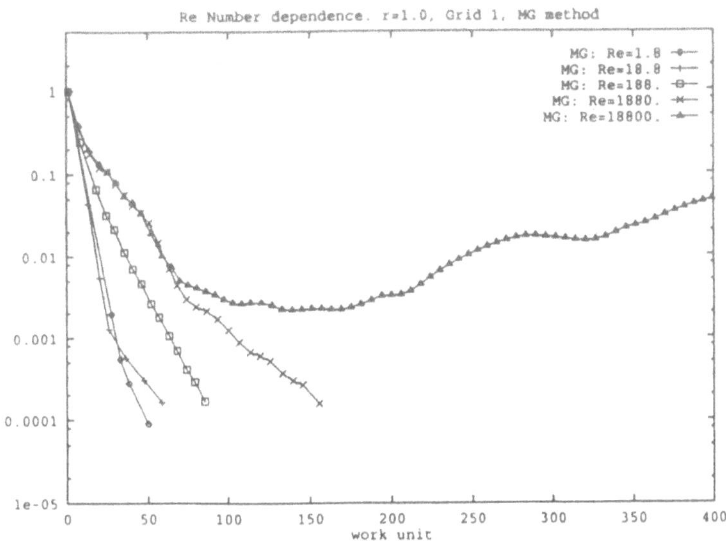

FIGURE 4. Convergence history for different Re

"block", Newton's solution of the system). A calculation based on Grid 1 and $r = 0.06$ is shown in Fig.5. As seen, the MG solver is more efficient and stable.

In the calculation of turbulent combustion, the effect of hot gas expansion (i.e. density variations) introduces further coupling among the dependent variables (velocity, pressure, k, ϵ, μ_t, enthalpy and mass fractions). Fig.6 and Fig.7 depict the convergence behaviour for combustion chambers with $r = 1.0$ and $r = 0.06$, respectively. The calculations are carried out on Grid 1 (with 3 MG levels). In the former case the single grid relaxation is more efficient than in the latter case, due to the relatively uniform grid distribution and better smoothing properties on less skewed grids. The situation is similar also when MG is used. However, in both cases the MG solver improves the convergence rate considerably. When compared with Fig.5, the gas expansion process make the convergence of the $k - \epsilon$ part less efficient. The non-monotonicity of the convergence can be seen from these figures. A possible way of improving the total convergence rate would call for a coupled relaxation of the $k - \epsilon$ equations, as these two dependent variables are linked in a highly non-linear manner.

5 Concluding Remarks

The Multi-Grid method has been applied to the calculation of turbulent reacting flows. The mathematical equations are the Reynolds averaged Navier-Stokes equations together with two-equation $k - \epsilon$ model and a two-step hydrocarbon

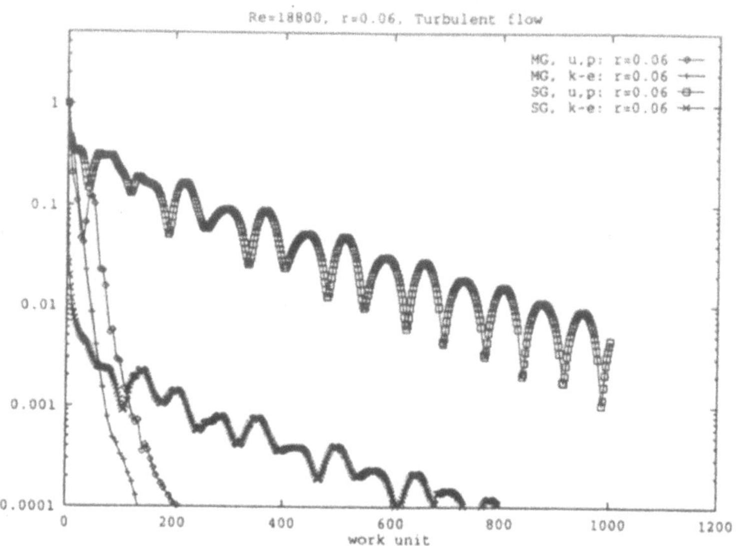

FIGURE 5. Convergence history of *turbulent* flow calculations

oxidation mechanisms. The calculation is done on a modeled annular gas turbine combustion chamber. The basic MG scheme and single grid relaxation scheme are studied for simple cases (low Reynolds number flow) and further extended to complex turbulent reacting flows. The results have shown that:

(1). The basic MG solver is nearly independent of grid resolution, while single grid relaxation is very slow for fine grids. The mesh aspect ratio has considerable influence on the MG efficiency.

(2). The MG method is efficient compared with single grid relaxation for calculation of turbulent reacting flows. However, the present segregated coupling strategy is non-optimal. The introduction of eddy viscosity and gas expansion slows down the convergence process. Also, the non-monotone convergence of k and ϵ indicates that the system has to be handled in a different way. One such possibility would be the coupled relaxation of these two variables.

Acknowledgement

This work has been supported by NUTEK (the Swedish National Board for Industrial and Technical Development).

References

[1] Brandt, A.: "Multi-Level Adaptive Solution to Boundary- Value Problem", Mathematics of Computation, Vol. 31, No. 138, pp. 333-390, 1977.

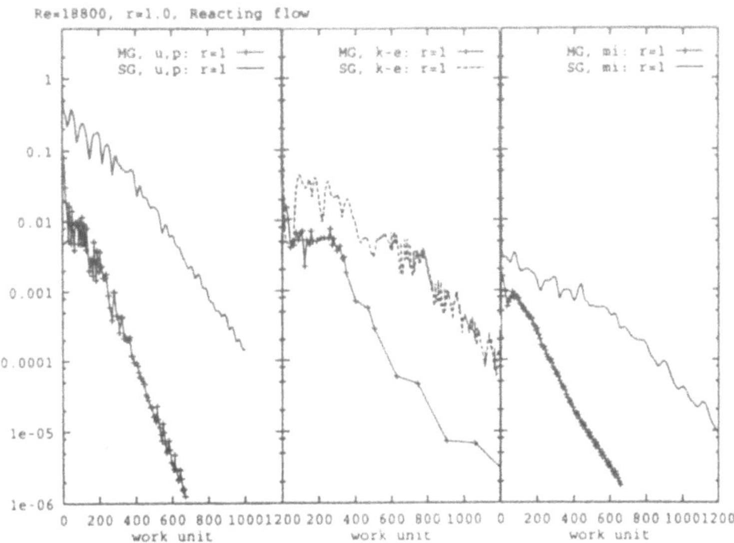

FIGURE 6. Convergence history for *turbulent reacting* flows at $r = 1.0$

[2] Fuchs, L. and Zhao, H.S.: "Solution of Three-Dimensional Viscous Incompressible Flows by a Multi-Grid Method", Int. J. for Numer. Methods in Fluid, Vol.4, pp.539-555, 1984.

[3] Bai, X.S. and Fuchs, L.:"A Fast Multi-Grid Method for 3-D Turbulent Incompressible Flows", Int. J. of Numer. Method for Heat and Fluid Flows, Vol. 2 , 1992.

[4] Lien,F.S. and Leschziner, A.: "Multi-Grid Convergence Acceleration for Complex Flow including Turbulence", Int. series of Numerical Mathematics, vol.98, (Multigrid Methods III, Hackbusch and Trottenberg eds.), pp.277-288, 1991.

[5] Givi, P.: "Model-Free Simulations of Turbulent Reactive Flows", Progress in Energy and Combustion Science, Vol.14, 1989.

[6] Launder, B.E. and Spalding, D.B.:" The Numerical Computation of Turbulent Flows", Computer Methods in Applied Mechanics and Engineering, Vol.3, pp.269-289, 1974.

[7] Westbrook, C.K. and Dryer, F.L.:"Simplified Reaction Mechanisms for the Oxidation of Hydrocarbon Fuels in Flames", Combustion Science and Technology, Vol. 27, pp. 31-43, 1981.

[8] Westbrook, C.K. and Dryer, F.L.: "Chemical Kinetics and Modeling of Combustion Process". Eighteenth Symposium (international) on Combustion, The combustion Institute, Pittsburgh, pp. 749, 1980.

[9] Magnussen, B.F. and Hjertager, B.H.: "On Mathematical Modelling of Turbulent Combustion with Special Emphasis on Soot Formation and Combustion", 16th Symposium (Int.) on Combustion, The Combustion Institute, pp. 719-729, 1976.

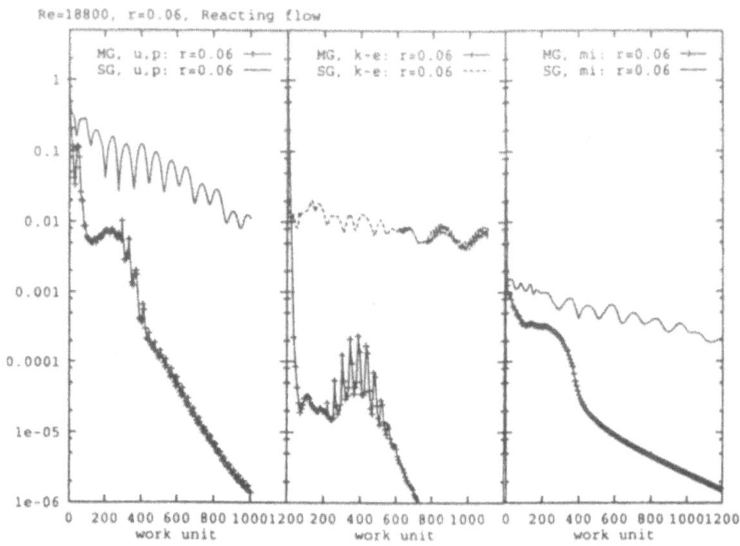

FIGURE 7. Convergence history for *turbulent reacting* flows at $r = 0.06$

[10] Pope, S.B.: "Computations of Turbulent Combustion: Progress and Challenges", 23rd Symposium (Int.) on Combustion, the Combustion Institute,pp. 591-612, 1990.

[11] Wennerberg, D.: "Prediction of Pulverized Coal and Peat Flames", Combustion Science and Technology, Vol. 58, pp. 25-41, 1988.

[12] Hutchinson, P.; Khalil, E.E. and Whitelaw, J. H.: "Measurement and Calculation of Furnace Flow Properties", J. of Energy, Vol.1, No.4, 1977.

[13] Bai, X-S. and Fuchs, L.: "Modeling of Turbulent Reacting Flows Past Bluff Bodies: Assessment of Accuracy and Efficiency". Computer and Fluids, (in print), 1993.

[14] Patankar, S.V.: "Numerical heat transfer and fluid flow", Hemisphere publishing corporation, 1980.

2

On a Multi-Grid Algorithm for the TBA Equations

Alfio Borzì[1] and Anni Koubek[2]

ABSTRACT We analyze a multi-grid algorithm [1] in order to solve numerically the thermodynamic Bethe ansatz equations. This solution method for the system of these non–linear integral equations is particularly important for the investigation of the ultraviolet limit, described by a conformal field theory.

1 Introduction

Massive relativistic field theories can be described on-shell by their scattering matrix. This approach is specially fruitful in two dimensions, where there exists a large class of models which are integrable, and their S-matrix can in principle be computed exactly, being factorizable [2]. Unfortunately there is no general direct method in order to compute the S-matrix of a theory, but usually it is conjectured from general axioms and the underlying symmetries of the corresponding Hamiltonian.

The thermodynamic Bethe ansatz (TBA) was developed in order to provide a means to link a conjectured scattering theory with the underlying field theory [3]. It describes the finite temperature effects of the factorized relativistic field theory, using the S-matrix as an input. If one studies the high temperature limit of the TBA equations, one can identify the conformal field theory (CFT) which governs the ultraviolet (UV) behaviour of the underlying field theory. One should though note, that it is not guaranteed that every consistent S-matrix describes the scattering in some field theoretical model. Therefore the axiomatic bootstrap approach is only of limited value if not linked to field theory by some means, wherefrom the TBA is one of the most powerful ones.

Given the scattering data one can in most cases extract analytically the central charge of the CFT reached in the conformal limit, and in some cases the dimension of the perturbing operator, if the symmetry of the problem is known. Numerical

[1]SISSA, Scuola Internazionale Superiore di Studi Avanzati, Via Beirut 2-4, 34013 Trieste, Italy. Address after Nov.1993: OUCL, Wolfson Building, OX1 3QD Oxford, UK.

[2]Address after Nov. 1993: DAMTP, University of Cambridge, Silver Street, CB3 9EW Cambridge, UK.

calculations on the other hand can solve the TBA equations and therefore extract any measurable quantity.
In [3, 4] the TBA equations were resolved by an iterative method. We propose here a multi-grid algorithm, which is considerable faster particulary if the UV limit is investigated, an important fact if many particles are involved. The heart of the program is the resolution of the coupled integral equations. We specialize our application to the case of diagonal S-matrices, see e.g. [3, 4, 5]. As physical quantities we extract the central charge, the dimension of the perturbing field and the conformal perturbation expansion.

2 The TBA Equations

We briefly review the framework of the TBA, referring to the literature for details ([3] - [6]). Let us investigate an integrable massive scattering theory on a cylinder. Integrability implies factorized scattering, and so one can assume that the wave function of the particles is well described by a free wave function in the intermediate region of two scattering. Consider n particles, and move the kth particle of mass m_k and rapidity β_k, such to scatter all particles and come back to the initial configuration. This implies the following periodic boundary condition,

$$e^{iLm_k \sinh \beta_k} \prod_{j \neq k}^{n} S_{kj}(\beta_k - \beta_j) = -1 \qquad \text{for} \qquad k = 1, 2, \ldots, n \quad . \tag{1}$$

We introduce the phase $\delta_{kj}(\beta_k - \beta_j) \equiv -i \log S_{kj}(\beta_k - \beta_j)$. In terms of these the equation become

$$Lm_k \sinh \beta_k + \sum_{j \neq k} \delta_{kj}(\beta_k - \beta_j) = 2\pi n_k \qquad \text{for} \quad k = 1, 2, \ldots, n \quad , \tag{2}$$

n_k being some integers. These coupled transcendental equations for the rapidities are called the Bethe ansatz equations. One tries to solve these equations in the thermodynamic limit introducing densities of rapidities for each particle species and transferring the equations into integral equations. That is, let $\rho_1^{(a)}(\beta) = \frac{n}{\Delta\beta}$, where we assume that there are n particles in the small interval $\Delta\beta$, be the particle density and $\rho^{(a)}(\beta) = \frac{n_k}{\Delta\beta}$ be the level density corresponding to the particle a, then (2) becomes

$$m_a L \cosh \beta + \sum_{b=1}^{n} \int_{-\infty}^{\infty} \varphi_{ab}(\beta - \beta')\rho_1^{(a)}(\beta')d\beta' = 2\pi\rho^{(a)} \quad . \tag{3}$$

In order to compute the ground state energy one needs to minimize the free energy

$$RLf(\rho, \rho_1) = RH_B(\rho_1) + S(\rho, \rho_1) \quad , \tag{4}$$

where $H_B = \sum_a m_a \int \cosh\beta \rho_1^{(a)} d\beta$ and S denotes the entropy. The extremum condition for a fermionic system[3] takes the form

$$-rM_a \cosh\beta + \epsilon_a(\beta) = \sum_{b=1}^{n} \int_{-\infty}^{\infty} \varphi_{ab}(\beta - \beta') \log(1 + e^{-\epsilon_b(\beta')}) \frac{d\beta'}{2\pi} \quad , \tag{5}$$

where we introduced the so-called pseudo-density $e^{-\epsilon_a} \equiv \frac{\rho_1^{(a)}}{\rho^{(a)} - \rho_1^{(a)}}$, the scaling length $r = Rm_1$ and the rescaled masses $M_a = \frac{m_a}{m_1}$; m_1 is the lightest particle mass. These coupled integral equations are called the TBA equations. The extremal free energy depends only on the ratios $\frac{\rho_1^{(a)}}{\rho^{(a)}}$ and is given by

$$f(r) = -\frac{r}{2\pi} \sum_{a=1}^{n} M_a \int_{-\infty}^{\infty} \cosh\beta \log(1 + e^{-\epsilon_a(\beta)}) d\beta \quad . \tag{6}$$

One can extract several physical quantities from the solution of the TBA-equations ([3, 6, 4]). Since few exact results are known about non-critical systems, it is interesting to examine the equations in the ultraviolet limit, which corresponds to $r \to 0$, where the underlying field theory should become a CFT. The central charge is related to the vacuum bulk energy, and is given by

$$c(r) = \frac{3r}{\pi^2} \sum_{a=1}^{n} M_a \int_{-\infty}^{\infty} \cosh\beta \log(1 + e^{-\epsilon_a(\beta)}) d\beta \quad . \tag{7}$$

Having calculated the central charge one would like to extract the conformal dimension of the perturbing operator. For small r, one expects that $f(r)$ reproduces the behaviour predicted by conformal perturbation theory, which in terms of $c(r)$ reads as

$$c(r) = c - \frac{3f_0}{\pi} r^2 + \sum_{k=1}^{\infty} f_k r^{yk} \quad . \tag{8}$$

The exponent y is related to the conformal dimension of the perturbing field Δ by $y = 2(1 - \Delta)$ if the theory is unitary and by $y = 4(1 - \Delta)$ if it is non-unitary. The coefficients are related to correlation functions of the CFT [3, 4], and even if one cannot read them off directly, this is an ultimate important check of the theory. Note that also non-diagonal S-matrices (see [7]) can be treated, since once one has diagonalized the transfer-matrix also in that case the numerical problem reduces to solving (5). Further quantities to measure can simply be added, and also one can study any range of r.

---

[3] We use the fermionic TBA equations since in diagonal scattering up to now they turned out to be the relevant ones, see e.g.[3] for the general theory

3 The Multi-Grid Method

Although multi-grid (MG) schemes were originally introduced to solve elliptic problems, the same strategy can be applied successfully to many other types of equations, like integral equations [8, 9]. The system of non linear Fredholm integral equations (5) has been solved using iterative methods [3, 4]. However, the number of iterations and corresponding computer process (CPU) time required by these methods to reach a specified precision can become excessively large as the number of grid points N increases. Typically a simple one level relaxation would require $O(N^2 \log N)$ operations. With a multi-grid solution technique the computing time for integral equations is reduced to $O(N^2)$ [9], and in particular cases to $O(N \log N)$ [8], thus justifying the extra effort in programming.

Now we define our numerical problem and we explain how the multi-grid scheme works for solving it. In discretising the TBA equations (5), we use the trapezoidal rule on a grid with mesh size h so that our system yields

$$\epsilon_a(\beta) = rM_a \cosh\beta + \frac{h}{2\pi}\sum_{b=1}^{n}\sum_{\beta'\in\Omega_h} w(\beta')\varphi_{ab}(\beta-\beta')\log(1+e^{-\epsilon_b(\beta')}) \quad , \tag{9}$$

$a = 1, 2, \ldots, n$, $\beta \in \Omega_h$, where Ω_h is the set of grid points with grid spacing h. The weights are $w(\beta) = 1$ unless on the boundary where $w(\beta) = 1/2$. Now let us introduce a sequence of grids with mesh sizes $h_1 > h_2 > \ldots > h_M$, so that $h_{\ell-1} = 2h_\ell$. The system (9) with discretisation parameter h_ℓ will be denoted as

$$\epsilon_a^\ell = K_{ab}^\ell(\epsilon_b^\ell) + f_a^\ell \quad , \quad a = 1, 2, \ldots, n \quad , \tag{10}$$

where a summation over b is intended and where

$$K_{ab}^\ell(\epsilon)(\beta) = \frac{h_\ell}{2\pi}\sum_{\beta'\in\Omega_{h_\ell}} w(\beta')\varphi_{ab}(\beta-\beta')\log(1+e^{-\epsilon(\beta')}) \quad . \tag{11}$$

Following [9] we have applied one Gauss-Seidel iteration to (10), and obtained the approximated solutions $\tilde{\epsilon}_a^\ell$, $a = 1, 2, \ldots, n$. We then transfer them onto the next coarser grid, $\tilde{\epsilon}_a^{\ell-1} = \hat{I}_\ell^{\ell-1}\tilde{\epsilon}_a^\ell$, where $\hat{I}_\ell^{\ell-1}$ is a restriction operator. The coarse grid equations become

$$\hat{\epsilon}_a^{\ell-1} = K_{ab}^{\ell-1}(\hat{\epsilon}_b^{\ell-1}) + \hat{f}_a^{\ell-1} \quad , \quad a = 1, 2, \ldots, n \quad , \tag{12}$$

where

$$\hat{f}_a^{\ell-1} = I_\ell^{\ell-1} f_a^\ell + \tilde{\epsilon}_a^{\ell-1} - K_{ab}^{\ell-1}(\tilde{\epsilon}_b^{\ell-1}) - I_\ell^{\ell-1}(\tilde{\epsilon}_a^\ell - K_{ab}^\ell(\tilde{\epsilon}_b^\ell)) \quad , \tag{13}$$

and with $I_\ell^{\ell-1}$ another fine-to-coarse grid transfer operator not necessarily equal to $\hat{I}_\ell^{\ell-1}$. Having obtained the solution of the coarse grid equation $\hat{\epsilon}_a^{\ell-1}$ the difference $\hat{\epsilon}_a^{\ell-1} - \tilde{\epsilon}_a^{\ell-1}$ is the coarse-grid (CG) correction to the fine-grid solution

$$\tilde{\epsilon}_a^\ell \leftarrow \tilde{\epsilon}_a^\ell + \hat{I}_{\ell-1}^\ell(\hat{\epsilon}_a^{\ell-1} - \tilde{\epsilon}_a^{\ell-1}) \quad , \tag{14}$$

$h_M = 0.02$			$h_M = 0.01$		
Iter. (ν)	*Residual*	*Obs. red.* ($\tilde{\rho}_\nu$)	*Iter.* (ν)	*Residual*	*Obs. red.* ($\tilde{\rho}_\nu$)
1	$0.92 \cdot 10^{-5}$	-	1	$0.27 \cdot 10^{-5}$	-
2	$0.11 \cdot 10^{-10}$	$0.66 \cdot 10^{-7}$	2	$0.95 \cdot 10^{-12}$	$0.58 \cdot 10^{-8}$
3	$0.49 \cdot 10^{-14}$	$0.14 \cdot 10^{-5}$	3	$0.24 \cdot 10^{-14}$	$0.95 \cdot 10^{-6}$
		$\tilde{\rho} = 0.31 \cdot 10^{-6}$			$\tilde{\rho} = 0.74 \cdot 10^{-7}$

TABLE 1. The FAS method.

$a = 1, 2, \ldots, n$, and $\hat{I}^{\ell}_{\ell-1}$ is a coarse-to-fine grid interpolation operator. Finally we perform one relaxation at level ℓ, in order to smoothen errors coming from the interpolation procedure. To solve the system of equations (10) we employ a coarse-grid correction recursively, i.e. equation (12) is itself solved by iteration sweeps combined with a further CG correction.

4 Numerical Investigation

The algorithm described[4] ([1]) above is a non-linear multi-grid (NMGM) method ([9]) with full adaptive scheme (FAS) ([10]). The convergence properties of this scheme can be analyzed using local mode analysis [10]. For, let us consider a simple case where $n = 1$ (and we omit the particle index) and denote with $e^{(\nu)}(\beta) = \epsilon^{(\nu)}(\beta) - \epsilon(\beta)$, $\beta \in G^h = \{jh, j \in \mathrm{Z}\}$, the solution error after ν GS iterations. On G^h we have the decomposition $e^{(\nu)}_j = \sum_\theta E^{(\nu)}_\theta e^{i\theta j}$. If we consider the iterative scheme in terms of the error we find the following reduction factor of the θ component

$$\mu(\theta)_k = \left| \frac{E^{(\nu+1)}_\theta}{E^{(\nu)}_\theta} \right| = \frac{\left| \frac{h}{2\pi} \sum_{j \geq k} \omega_j \varphi_{kj} \frac{e^{-\epsilon_j}}{1+e^{-\epsilon_j}} e^{i\theta(j-k)} \right|}{\left| 1 + \frac{h}{2\pi} \sum_{j<k} \omega_j \varphi_{kj} \frac{e^{-\epsilon_j}}{1+e^{-\epsilon_j}} e^{i\theta(j-k)} \right|} \quad , \tag{15}$$

denoting, let us say $\phi(\beta)$ at $\beta = jh$ simply by ϕ_j. In particular $\varphi_{kj} = \varphi(\beta_k - \beta_j)$. We find that for each k, $\max\{\mu(\theta)_k,\ \ 0 \leq |\theta| \leq \pi\} = \mu(0)_k$, the largest value being approximately $\simeq 0.2$ ($r = 0.1$, $M_1 = 1$, and $h = 0.01$). In the same way, using (15) we obtain a good approximation of the smoothing factor $\mu \simeq 10^{-3}$. Hence after one GS iteration for both pre- and post-smoothing we expect that the convergence factor of the MG cycle is given by $\rho^* = \mu^{\nu_1+\nu_2} \simeq 10^{-6}$. This result is confirmed by numerical experiments as we observe from Table 1, where we report the observed reduction and the mean reduction factor $\tilde{\rho}$ (w.r.t. the maximum norm). Notice the agreement with the predicted [9] behaviour $\tilde{\rho} = O(h^\delta)$, $\delta = 2$ in this case.

[4]The fortran code is available from: CPC Program Library, Queen's University of Belfast, N. Ireland.

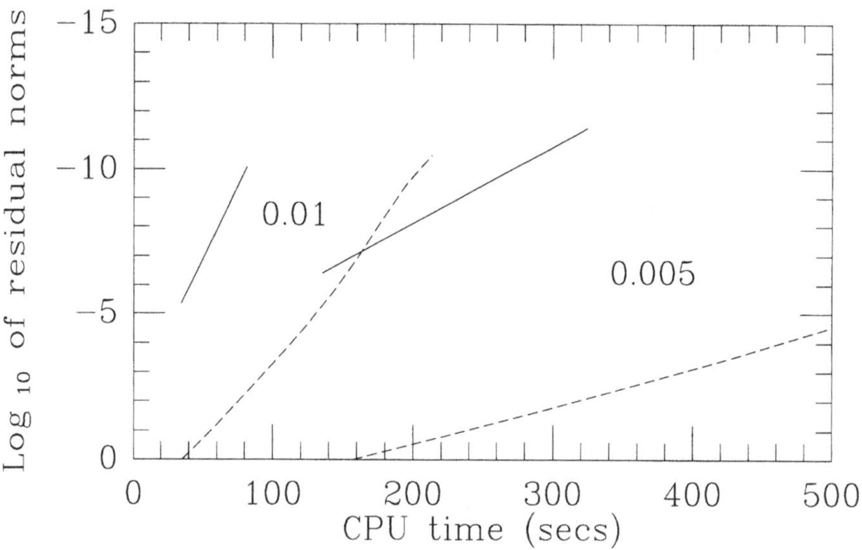

FIGURE 1. Evolution of residual error norm with CPU time for a 1-particle system at $r = 0.1$ for different h_M: solid line for MG, dashed line for iteration only.

In the NMGM scheme with nested iteration, we use an initial approximation which behaves like $rM_a \cosh\beta$, wherefrom the program determines the numerical boundary at which the kernels vanish and verifies that the conditions for the existence of (at least) one solution given by the Schauder's fixed point theorem are satisfied [11].

We compare the performance of the MG and of the Gauss-Seidel iterative schemes in terms of CPU time in Figure 1, there the different initial residual error for MG and iterative scheme is due to the set up of the initial approximated solution in the MG cycle, that is a non-linear nested iteration which uses a MG cycle itself (see [9]). We denote the residuals as $\tau_a(\beta) = (\epsilon_a - K_{ab}(\epsilon_b) - f_a)(\beta)$, $\beta \in \Omega_{h_M}$, and define the norm

$$\| \tau \|_M = \max_{1\leq a\leq n} \sqrt{\sum_{\beta\in\Omega_{h_M}} \tau_a(\beta)^2} \quad . \tag{16}$$

In order to outline how the multi-grid algorithm becomes important as the number of particles increases we give in Table 2 the CPU time required by the two methods to solve the discretized problem to a value of the residual norm $\| \tau \|_M \leq 10^{-14}$.

no. equations	*CPU time (secs)*	
	Relax	Multi-Grid
1	4	3
2	34	22
3	508	331
4	1230	712
5	2530	1320

TABLE 2. A comparison of CPU time required to reach a particular value of the norm, for $r = 0.1, \quad h_M = 0.1$.

5 Numerical Results: An Example

We specifically designed the program for diagonal scattering theories, that is we are concerned with scalar S matrices which in general have the form

$$S_{ab} = \prod_{\alpha_i \in X_{ab}} \frac{\sinh \frac{1}{2}(\beta + i\pi\alpha_i)}{\sinh \frac{1}{2}(\beta - i\pi\alpha_i)} \quad ,$$

X_{ab} is the set of factors f_x appearing in the S-matrix S_{ab} (for a recent review on this subject see [12]). The set of the numbers α, and the masses of the theory are sufficient to resolve the TBA-equations. For example, let us consider the two-particle system $M_1 = 1$ and $M_2 = 2\cos(\pi/5)$, with the S–matrix $S_{11} = f_{\frac{2}{5}} f_{\frac{3}{5}}$, $S_{12} = S_{21} = f_{\frac{1}{5}} f_{\frac{2}{5}} f_{\frac{3}{5}} f_{\frac{4}{5}}$, $S_{22} = f_{\frac{1}{5}} f_{\frac{4}{5}} (f_{\frac{2}{5}} f_{\frac{3}{5}})^2$, which has been conjectured to correspond to the (non-unitary) minimal model $\mathcal{M}_{2,7}$, i.e. $c = \frac{4}{7}$, perturbed by the field with dimension $\Delta = -\frac{3}{7}$.

Using the algorithm [1] we solve the corresponding TBA equations for a set of r close to zero. The above information (masses and indices α) are the only input required by the program. For each r we compute $c(r)$ and using (8) we extract the exponent $y = 2.85714287$ which confirms the conjectured scaling dimension of the perturbing field.

6 Conclusions

We presented a multi-grid scheme for the resolution of the thermodynamic Bethe ansatz equations. The TBA is a means to describe the finite temperature effects of relativistic factorized scattering theories. Our program is specifically designed for theories having a scalar S-matrix. These theories exhibit a unique form, and the only input needed in order to carry out the TBA are the locations of the poles and zeros of the single S-matrix elements.

We calculate the central charge and in the ultraviolet limit the dimension of the perturbing field and the coefficients of the perturbation expansion. These are the most crucial tests in verifying a conjectured S-matrix.

In order to get sensible results for the physical quantities one needs to resolve the integral equations with the highest possible accuracy. Therefore the use of an efficient Multi-Grid algorithm gives the possibility to reach high accuracy in the computation together with a sensible reduction of the CPU time, in confrontation with standard iterative techniques.

References

[1] A. Borzì, A. Koubek, *Computer Phys. Commun.* **75** (1993) 118.

[2] A.B. Zamolodchikov, *JETP Letters* **46** (1987) 160; *Int. Journ. Mod. Phys.* **A3** (1988) 743; *Advanced Studies in Pure Mathematics* **19** (1989) 641.

[3] Al.B. Zamolodchikov, *Nucl. Phys.* **B342** (1990) 695.

[4] T. Klassen, E. Melzer, *Nucl. Phys.* **B350** (1990) 635.

[5] M.J. Martins, *Phys. Lett.* **B240** (1990) 404; P. Christe, M.J. Martins, *Mod. Phys. Lett.* **A5** (1990) 2189; A. Koubek, M.J. Martins, G. Mussardo *Nucl. Phys.* **B368** (1992) 591.

[6] M.J. Martins, *Phys. Rev. Lett.* **67** (1991) 419.

[7] Al.B. Zamolodchikov, *Nucl. Phys.* **B358** (1991) 497.

[8] A. Brandt, A.A. Lubrecht, *J. Comp. Phys.* **90** (1990) 348.

[9] W. Hackbusch, *Multi-Grid Methods and Applications.* Springer-Verlag Berlin, Heidelberg, 1985.

[10] A. Brandt, *Multi-grid techniques: 1984 guide with applications to fluid dynamics.* GMD-Studien. no 85, St. Augustin, Germany, 1984.

[11] W. Pogorzelski, *Integral Equations and their Applications, Vol. I.* PWN-Polish Scientific Publishers, Warsaw, 1966.

[12] G. Mussardo, *Phys. Rep.* **218** (1992) 215.

3

A Multidimensional Upwind Solution Adaptive Multigrid Solver for Inviscid Cascades

L. A. Catalano, P. De Palma, M. Napolitano and G. Pascazio[1]

ABSTRACT A recently developed multidimensional upwind multigrid method is combined with an adaptive grid refinement strategy in order to provide a numerical technique for computing two-dimensional compressible inviscid steady flows accurately and efficiently. A locally nested sequence of mesh refinements is constructed by a quad-tree data-structure, which easily incorporates the multigrid method using compact-stencil space discretization and explicit multi-stage smoother. Computations of flows through channels and cascades are presented which demonstrate the capabilities of the proposed approach.

1 Introduction

Two issues play an ever increasing role in current CFD research: the development of methods suitable for vector and parallel computers, and the application of local refinement strategies as a means for obtaining high quality results for complex flow problems at reasonable costs. Both trends have increased the interest towards employing explicit schemes as smoothers in multigrid methods, as well as towards developing space discretizations based on compact stencils.

Recently, a procedure has been developed for optimizing the coefficients and time step of explicit multi-stage schemes, in order to design an efficient smoother to combine with a multigrid method for multidimensional advection equations [1]. Thanks to its generality, this approach, based on a two-dimensional linear Fourier analysis, has been applied to the optimization of some recently developed genuinely multidimensional upwind schemes [2] characterized by low cross-wind diffusion and thus capable of capturing discontinuities oblique to the mesh very accurately. The efficiency of the resulting smoother combined with the standard FAS multigrid strategy [3], has been proved first for a nonlinear scalar advection equation, using both the finite volume [4] and the recently developed fluctuation

[1]Politecnico di Bari, via Re David 200, 70125 Bari, Italy

splitting [5] schemes. The latter space discretization has been crucial to the development of genuinely multidimensional methods for compressible inviscid flows, based on decomposing the Euler system into an equivalent set of scalar equations with solution-dependent propagation directions, as originally proposed by Roe [6]. Successful applications and improvements of such a methodology [7, 8] have been obtained, thanks to a conservative linearization which can be performed analytically [9]. A major step towards making such a numerical technique robust and efficient has been performed in [5, 10]: the explicit multigrid strategy of [4] has been extended to the Euler equations, using a new wave decomposition model and the fluctuation splitting N-scheme; multigrid acceleration showed to be effective for subsonic, transonic and supersonic flows through channels; and convergence to machine accuracy has been achieved for the first time.

In this paper, the method developed so far is combined with a local grid-refinement strategy in order to solve complex flow configurations at very reasonable computational costs.

2 The multidimensional Euler solver

The governing equations for two-dimensional inviscid non-conducting flows are written in terms of the conservative variables, $q = (\rho, \rho u, \rho v, \rho E)^T$, as:

$$q_t = -\nabla \cdot \mathcal{F} = \mathrm{Res}(q), \qquad \mathcal{F} = (F, G), \tag{1}$$

where $F = (\rho u, \rho u^2 + p, \rho uv, \rho uH)^T$ and $G = (\rho v, \rho uv, \rho v^2 + p, \rho vH)^T$ are the flux-vectors in the x and y directions, respectively, and $\mathrm{Res}(q)$ is the steady-state residual of equation (1).

Classical upwind methods for the solution of the multidimensional Euler equations (1), based on the application of one-dimensional Riemann solvers along grid dependent directions, experience a loss of resolution in presence of discontinuities not aligned with the mesh. For such a reason, a large effort has been recently devoted to the development of numerical methods which contain truly multidimensional features in modelling the propagation phenomena which dominate the behaviour of compressible flows [11, 12, 6]. The approach of [6], of interest here, is based on the application of simple-wave theory and consists in selecting a number N of waves (acoustic, entropy, shear), each having strength α^k and propagation direction $\mathbf{n}^k$. The gradient of the primitive variables $\tilde{q} = (\rho, u, v, p)^T$ can be decomposed as:

$$\nabla \tilde{q} = \sum_{k=1}^{N} \nabla \tilde{q}^k = \sum_{k=1}^{N} \alpha^k \tilde{r}^k \mathbf{n}^{\mathbf{k}}, \tag{2}$$

$\tilde{r}^k$ being the right eigenvector of the Jacobian $(\partial \mathcal{F}/\partial \tilde{q}) \cdot \mathbf{n}^k$ with eigenvalue λ^k; and the steady-state residual, namely, the time derivative of the conservative variable vector, can be obtained by summing up all wave contributions, as follows [6]:

$$q_t = -\sum_{k=1}^{N} \alpha^k r^k \lambda^k = \mathrm{Res}(q). \tag{3}$$

An analytical decomposition can be obtained when employing four orthogonal acoustic, one entropy and one shear waves [6]. Depending on the flow data, namely, $\nabla\tilde{q}$, the intensities of all waves and the propagation directions of the acoustic and entropy ones are provided by equation (2), whereas the direction of the shear remains arbitrary and is the distinctive feature of the various wave models proposed so far in the literature [6, 7, 10, 13]. All results presented in this paper have been obtained by assuming the shear wave front to be parallel to the velocity vector, a choice which has proved robust for a wide range of applications [10, 14].

Concerning the spatial discretization, the use of a cell-vertex grid with linear triangular elements appears to be the most suitable choice for a wave decomposition model based on the flow gradients. Here, in order to combine the basic solver with the quad-tree data-structure used to create locally refined grids [15], the triangular mesh is obtained from a structured quadrilateral one, by subdividing each quadrilateral cell into two triangles. For each triangle T, the global fluctuation, defined as $\Phi_T = \int_S q_t \, dS$, is split into its simple-wave contributions, as follows:

$$\Phi_T = -\int_S \sum_{k=1}^{N} \alpha^k r^k \lambda^k \, dS = -S \sum_{k=1}^{N} \bar{\alpha}^k \bar{r}^k \bar{\lambda}^k = \sum_{k=1}^{N} \Phi_T^k. \tag{4}$$

In equation (4), the cell-averaged values $\bar{\alpha}^k$, $\bar{r}^k$, $\bar{\lambda}^k$ can be evaluated analytically, provided that the parameter vector $z = \sqrt{\rho}(1, u, v, H)^T$ is assumed to vary linearly over each triangle, a feature which is crucial to ensure conservation [9]. A multidimensional upwind residual distribution scheme is then employed to split each wave contribution to the flux balance in each cell, Φ_T^k, among its three vertices, according to the propagation velocity vector $\boldsymbol{\lambda}^k$. In this way, the discrete residual of equation (1) at each cell-vertex v of the grid h is reconstructed as:

$$(q_t^h)_v = \mathrm{Res}_v(q^h) = \frac{1}{S_v} \sum_T \sum_{k=1}^{N} \beta_{T,v}^k \Phi_T^k. \tag{5}$$

In equation (5), $\beta_{T,v}^k$ are the coefficients which define the distribution of the k-th wave component to the vertex v of the triangle T, the area S_v is only a suitable scale factor in the case of steady-state calculations (see, e.g., [2]), and the summation is extended over all triangles having the vertex v in common. Obviously, in order to ensure conservation, for every triangle T and for each wave k, the following condition must be satisfied:

$$\sum_{j=1}^{3} \beta_{T,j}^k = 1. \tag{6}$$

The upwind residual distribution scheme used in this paper is the compact-stencil fluctuation splitting first-order-accurate N-scheme [2].

Boundary conditions, being critical to the accuracy of the solution, have been imposed with particular care. Characteristic boundary conditions are used at subsonic-inlet gridpoints, where the total enthalpy, entropy and flow angle are specified and at subsonic-outlet gridpoints, where the pressure is prescribed. Impermeability at solid boundaries is enforced by extending the symmetry technique with curvature correction, proposed in [16], to the present cell-vertex space discretization: one row of auxiliary cells is used for evaluating the residual at the wall gridpoints, the state at each mirror-image node being computed by imposing no-injection and isentropic simple radial equilibrium [10].

An explicit three-stage Runge-Kutta scheme is used for discretizing the time derivative in equation (5), the predictor coefficients and the time step being chosen so as to optimize the smoothing properties of the scheme. A standard FAS multigrid strategy [3] is used to accelerate convergence to steady-state. See [14] for a more detailed description of the method.

3 Adaptive multigrid strategy

A local refinement technique for the cell-vertex residual distribution method developed so far is proposed in the present section. Starting from a regular structured quadrilateral grid, nested levels of local refinement are created and managed by a quad-tree data-stucture [15], so that the standard multigrid FAS scheme can be applied, with minor changes in the grid transfer operators.

At each level l, the grid Ω^l is composed of an unrefined part Ω^l_c (cells with no *kids*) and a refined part Ω^l_f (cells with *kids* on level l+1), so that Ω^l_f and Ω^{l+1} cover the same region. In such a way, the physical domain is discretized by a grid composed of all unrefined parts Ω^l_c, $l = 0, ..., N$ (*composite* grid with N levels of local refinement).

In order to describe how the conservation property of the basic solver is mantained on the composite grid, a grid with only two nested levels, l and l+1, shown in figure 1, is considered, for simplicity: the boundary points of Ω^{l+1} which do not lie on the physical boundary are called *green* nodes and are denoted by crosses, whereas the internal nodes of Ω^{l+1} are referred to as *interior* nodes and are denoted by dots. The solution at grid level l+1 is firstly obtained by bilinear interpolation of the solution at level l, as done in the standard nested iteration; the fluctuation splitting scheme, namely equations (4) and (5), is then applied on the grid l+1 to reconstruct the residual at all interior nodes; only incomplete contributions are sent to the green nodes, which therefore are not updated in the time integration. In order to apply the coarse grid correction, the solution is then injected from level l+1 to level l, namely:

$$q^l_{i,j} = q^{l+1}_{2i,2j}. \tag{7}$$

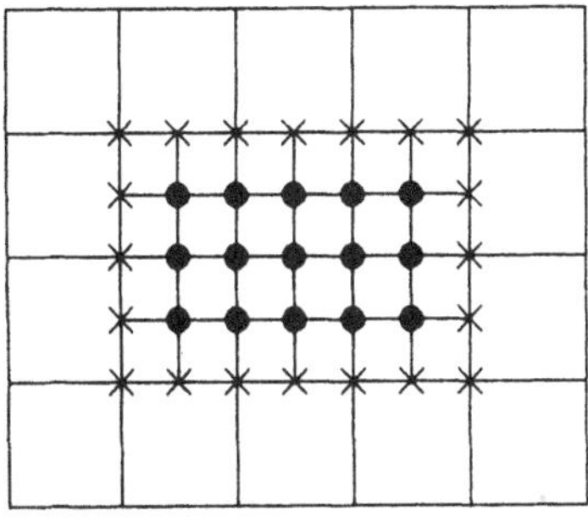

Figure 1: • interior and × green nodes on a composite grid

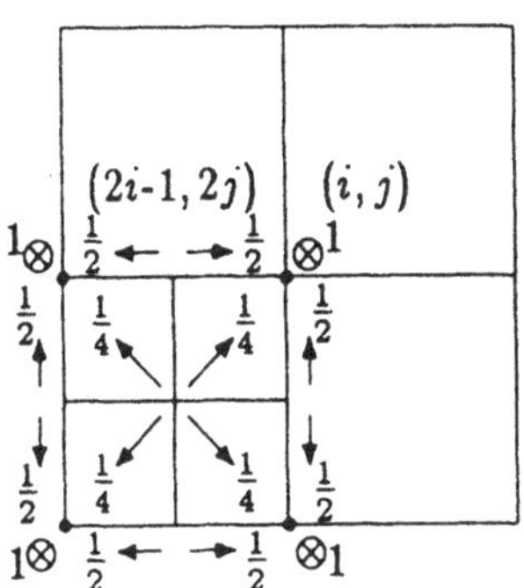

Figure 2: residual collection on the composite grid

A first condition to be satisfied for ensuring conservation on the composite grid is that the flux through each side of $\partial\Omega_f^l$ coincides with the flux through the two corresponding sides on $\partial\Omega^{l+1}$, exactly, or to its local order of accuracy. For the present linear-element cell-vertex discretization, the latter requirement is clearly satisfied when using injection for the restriction of the solution and bilinear interpolation of either the primitive or the conservative variables for the solution at green nodes. Futhermore, thanks to the analytical linearization, the flux conservation can be satisfied exactly by interpolating the parameter vector, which has already been supposed to vary linearly over each cell. Clearly, after integration along $\partial\Omega_f^l$ and $\partial\Omega^{l+1}$, this first condition results in:

$$\oint_{\partial\Omega^{l+1}} \mathcal{F}\cdot\mathbf{n}ds - \oint_{\partial\Omega_f^l} \mathcal{F}\cdot\mathbf{n}ds = 0 \text{ or } \mathcal{O}(h^2), \tag{8}$$

$\mathbf{n}$ being the outward normal; after application of Gauss' theorem, one has:

$$\int_{\Omega^{l+1}} \nabla\cdot\mathcal{F}dS - \int_{\Omega_f^l} \nabla\cdot\mathcal{F}dS = 0 \text{ or } \mathcal{O}(h^2). \tag{9}$$

In the discrete domain, equation (9) corresponds to the following conservation property:

$$\sum_{T\in\Omega^{l+1}}\sum_{v\in T}\sum_{k=1}^{N}\beta_{T,v}^k\Phi_T^k - \sum_{T\in\Omega_f^l}\sum_{v\in T}\sum_{k=1}^{N}\beta_{T,v}^k\Phi_T^k = 0 \text{ or } \mathcal{O}(h^2). \tag{10}$$

Equation (10) can be rewritten in terms of the contributions received by each node, as:

$$\sum_{v\in\Omega^{l+1}} R_v - \sum_{v\in\Omega_f^l} R_v = 0 \text{ or } \mathcal{O}(h^2), \tag{11}$$

R_v being defined as:

$$R_v = \sum_{T \in \hat{\Omega}} \sum_{k=1}^{N} \beta_{T,v}^k \Phi_T^k. \tag{12}$$

In equation (12), the first summation is extended over all triangles of the discrete domain $\hat{\Omega}$ having the vertex v in common.

When using multigrid, the residual computed on the coarser level l has to be corrected by means of the relative local truncation error, τ_{l+1}^l, so that a second condition has to be satisfied for ensuring conservation on the composite grid:

$$\sum_{v \in \Omega^{l+1}} R_v - \sum_{v \in \Omega_f^l} (R_v + (\tau_{l+1}^l)_v S_v) = 0 \text{ or } \mathcal{O}(h^2). \tag{13}$$

Equation (11), combined with equation (13), provides the condition that the relative local truncation errors, τ_{l+1}^l, have to balance each other over the refined domain, Ω_f^l, so that no spurious source terms are introduced at level l:

$$\sum_{v \in \Omega_f^l} (\tau_{l+1}^l)_v S_v = 0 \text{ or } \mathcal{O}(h^2). \tag{14}$$

Equation (14) can be satisfied by choosing the residual collection operator properly: figure 2 provides a sketch of the contributions of each node at level l+1 to the residual collection at coarser-level nodes, denoted by dots. For example, the contribution of the node $(2i-1, 2j)$ to the residual collection in (i, j) is $C_{l+1}^l[R_{2i-1,2j}(q^{l+1})] = 1/2 \ R_{2i-1,2j}(q^{l+1})$. Just like in the standard FAS cycle, the coarse grid correction consists in solving the following equation on level l:

$$q_t^l = Res(q^l) + r^l. \tag{15}$$

In equation (15), the source term r^l corresponds to the local relative truncation error on the refined part of grid l, whereas it is always null on the composite grid:

$$r^l = \begin{cases} 0 & \text{on } \Omega_c^l \\ \tau_{l+1}^l = \dfrac{1}{S_v^l} \left\{ C_{l+1}^l \left[R(q^{l+1}) + r^{l+1} S_v^{l+1} \right] - R_v(q^l) \right\} & \text{on } \Omega_f^l \end{cases} . \tag{16}$$

$R_v(q^l)$ is defined again by equation (12), with $\hat{\Omega} = \Omega_f^l$.

The correction of the solution in Ω_f^l is then prolongated bilinearly and added to the solution in the nodes of Ω^{l+1}; since the same operator is used for the interpolation of both the solution and the correction, the flow variables at the green points need to be interpolated only once, namely in the nested iteration, so that no special treatment is required in the FAS cycle.

Concerning the local refinement strategy, the pressure gradient has been used as the sensor which detects whether a cell must be refined or not. The percentage of cells to be refined on the composite grid is assigned by the user, the threshold value being automatically computed using a subdivision of cells into classes [10]. Thanks to the good capturing properties of the wave decomposition method even on rather coarse grids, the positions of the shocks on the refined grid were always close enough to those on the initial one and, thus, coarsening the grid during the computation has not been found necessary.

4 Results

Subsonic flow through a cosine bump channel (inlet Mach number M_i=0.5 and 20% restriction) [17] has been computed at first, in order to verify if the accuracy of the solution computed on a composite (locally refined) grid (CG) is close enough to that obtained on the standard grid with equal spacing on the finest level (SG). Computations have been performed on four composite grids obtained by adding one to four locally refined grid levels to a 16×4 base grid, 50%, 30%, 25% and 22% of the cells being refined at each level, respectively. Figure 3 provides the value of the Mach number at the top of the bump, M_{max}, and the L_∞-norm of the entropy — computed on the locally refined grids (symbols) and on the corresponding standard ones (lines) — versus the finest mesh size, h. The numerical entropy generation is shown to be almost proportional to the grid spacing for both sets of grids and M_{max} tends to its *exact* value, as $h \to 0$, more regularly for the composite grids than for the standard ones. Figure 4 shows the convergence histories for the four-level composite grid (CG4) — shown in figure 5 — and the corresponding 256×64 standard one (SG4): the logarithm of the L_1-norm of the residual of the mass conservation equation is plotted versus the computational work, one work unit being defined as one residual calculation on the finest level of the standard grid. The computer time is almost proportional to the total number of cells employed.

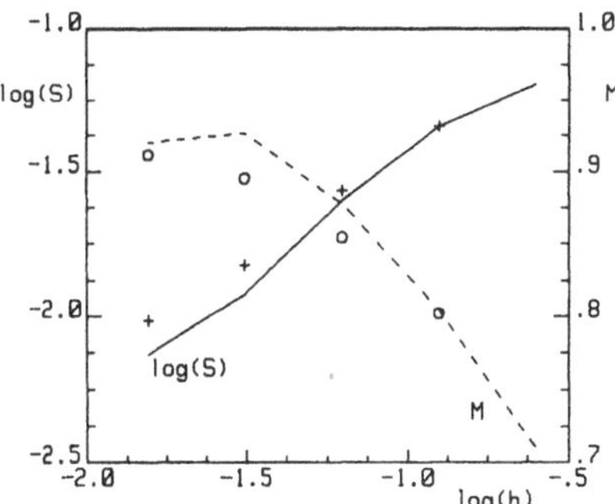

Figure 3: step-size study for the subsonic flow through the cosine bump channel

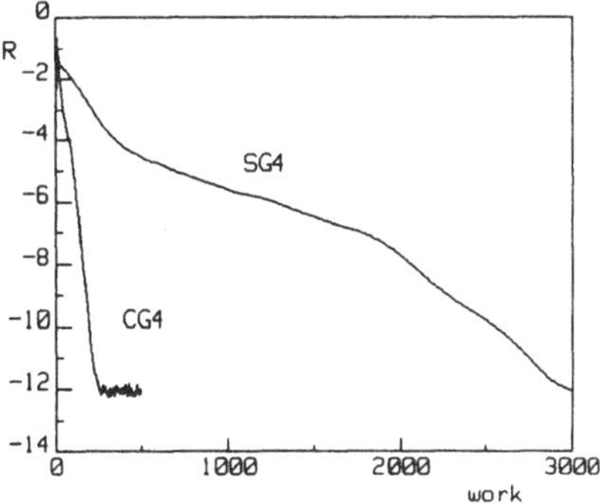

Figure 4: convergence histories for the subsonic flow through the cosine bump channel

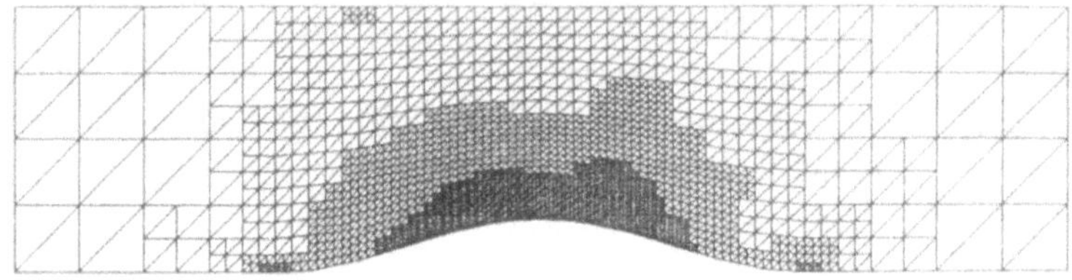

Figure 5: composite grid for the subsonic flow through the cosine bump channel (4 levels — 2080 quadrilateral cells)

Transonic flow through a cascade of NACA-0012 airfoils with pitch to chord ratio equal to 3.6 has been then considered, with M_i=0.8 and incidence angle i=1°. A composite grid with 1689 cells, shown in figure 6 has been obtained at first, starting from a 48 × 8 base grid after two local refinements. The corresponding Mach contours are provided in figure 7, where the shocks are well captured in two-to-three cells. However, a more accurate description of the stagnation point region and a sharper capturing of the shocks are needed. Therefore, a finer composite grid, with 8853 cells, has been obtained by means of two additional local refinements. Such a grid and the corresponding Mach contours are shown in figures 8 and 9, respectively. The marked improvement in the description of the shocks is clearly seen.

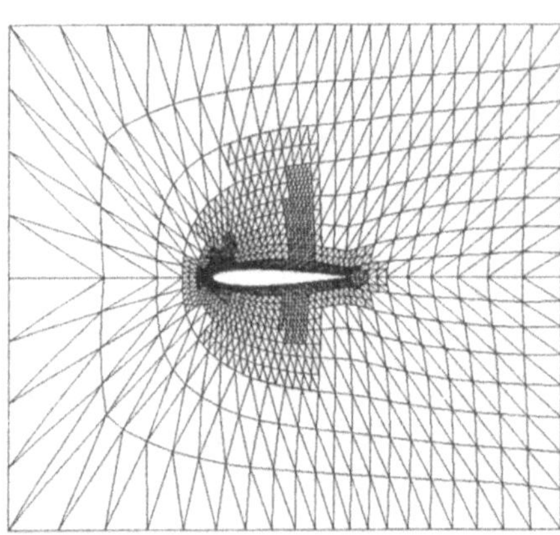

Figure 6: composite grid for the transonic flow through the NACA-0012 cascade (2 levels — 1689 cells)

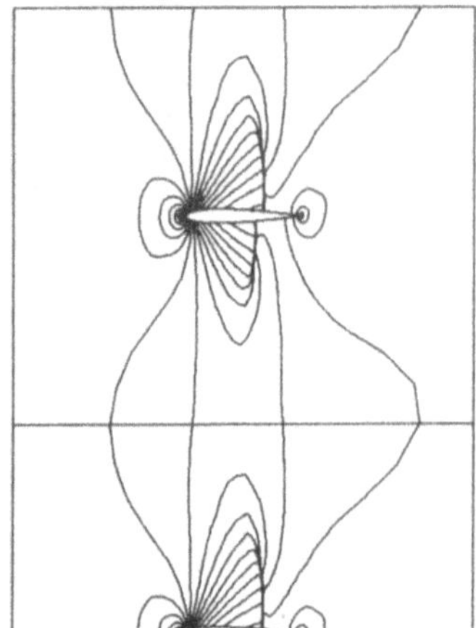

Figure 7: Mach contours for the NACA-0012 cascade (grid of figure 6 — ΔM = .05)

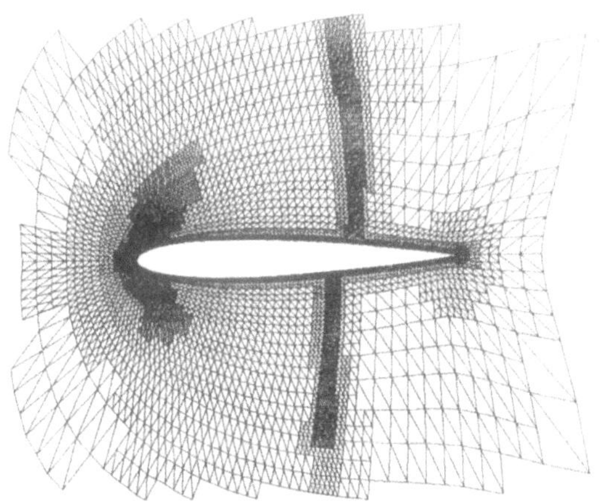

Figure 8: composite grid for the transonic flow through the NACA-0012 cascade (4 levels — 8853 cells)

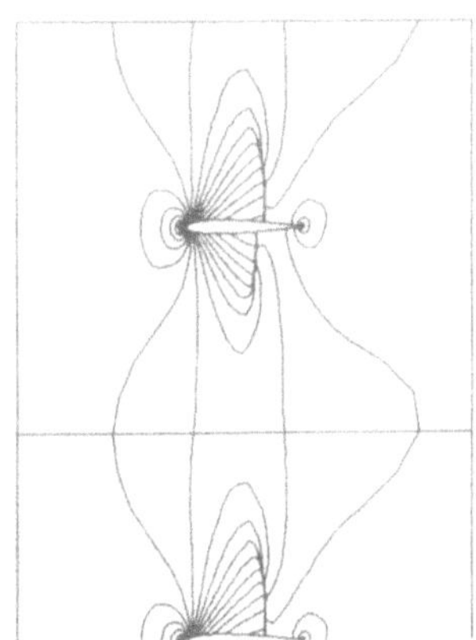

Figure 9: Mach contours for the NACA-0012 cascade (grid of figure 8 — $\Delta M = .05$)

The solution on the 192 × 32 standard grid was also obtained for comparison. Figure 10 provides the surface Mach number distributions computed on such a grid (SG2) as well as those obtained on the composite grids with two (CG2) and four (CG4) refinement levels. The accuracy of the first two solutions, SG2 and CG2, having the same finest-grid size, are almost identical, demonstrating the validity of the proposed adaptive grid approach once more. In this respect, it is noteworthy that the solution on the SG4 grid with 768 × 128 cells, supposedly as accurate as the present CG4 solution, was beyond the available computer resources.

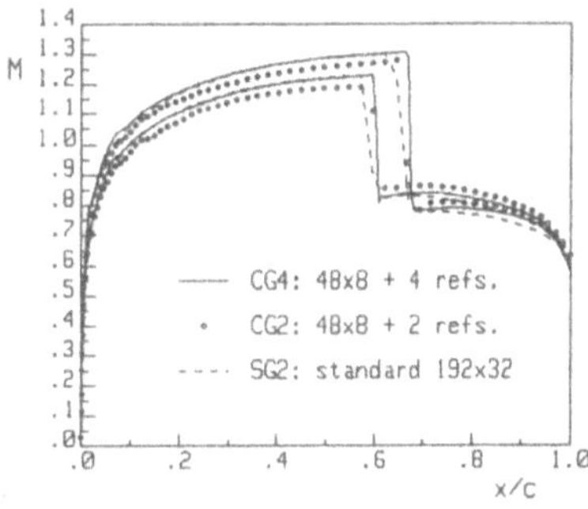

Figure 10: surface Mach distribution for the NACA-0012 cascade

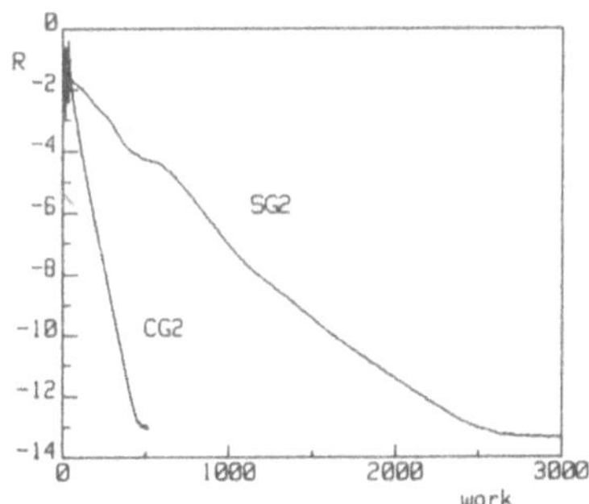

Figure 11: convergence histories for the transonic flow through the NACA-0012 cascade

Furthermore, the convergence histories for the SG2 and CG2 calculations, given in figure 11, demonstrate the reduction in computer time obtained by employing the local refinement strategy.

Finally, the subsonic flow through a high turning cascade (VKI-LS59) has been computed as a rather severe test for the proposed methodology. The outlet isentropic Mach number is equal to 0.81 and the inlet flow angle is 30 degrees. A composite grid with 8383 cells, shown in figure 12, has been obtained from a 128×8 standard grid after three local refinements; the computed Mach contours are presented in figure 13. A comparison between the numerical results (Mach number along the blade profile) and the experimental data provided in [18], is also given in figure 14, and demonstrates the accuracy of the method as well as its capability of handling complex geometries.

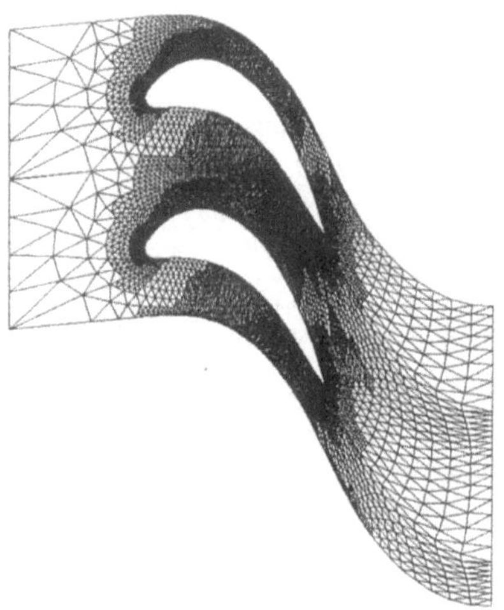

Figure 12: composite grid for the subsonic flow through the VKI-LS59 cascade (3 levels — 8383 cells)

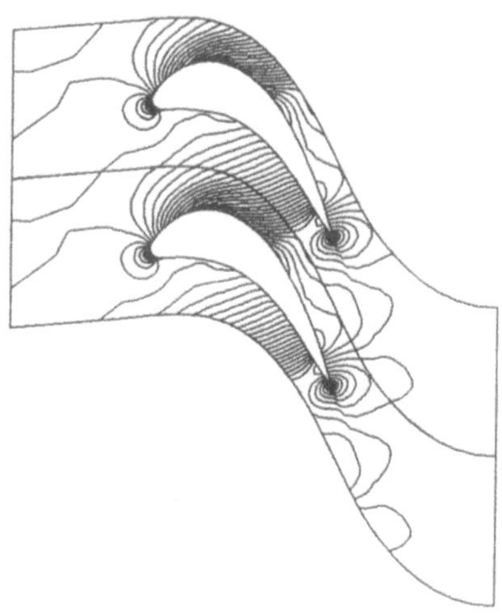

Figure 13: Mach contours for the VKI-LS59 cascade (grid of figure 12 — $\Delta M = .025$)

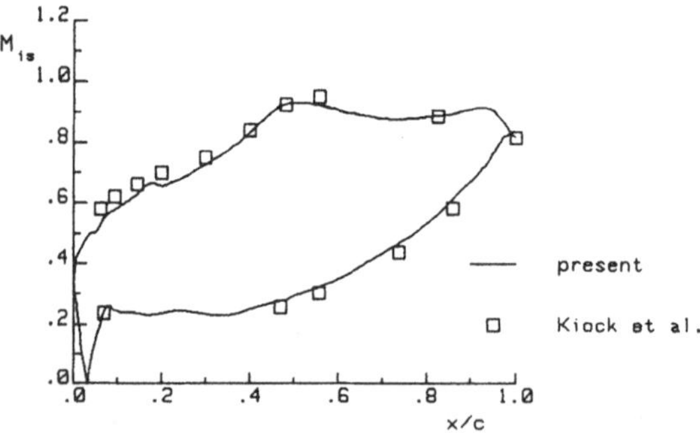

Figure 14: surface Mach distribution for the VKI-LS59 cascade

5 Conclusions

A multidimensional upwind multigrid method recently proposed by the authors has been combined with an adaptive local refinement strategy to provide an efficient tool for computing rather complex two-dimensional compressible inviscid steady flows. The approach has been validated for well-documented channel and cascade flows. The accuracy on the locally refined grids is comparable to that obtained on uniformly refined ones and the gain in efficiency is equal to, or better than, the total reduction of nodes.

ACKNOWLEDGEMENTS

This work has been supported by the EEC under BRITE/EURAM contract No. AER2-CT92-0040, and by MURST, quota 40%..

References

[1] L. A. Catalano, H. Deconinck. Two-dimensional optimization of smoothing properties of multistage schemes applied to hyperbolic equations. *TN-173, von Karman Institute, Belgium,* 1990; also *Multigrid Methods: special topics and applications II, GMD-Studien No. 189, GMD St. Augustin, Germany, pp. 43-55,* 1991.

[2] R. Struijs, H. Deconinck, P. L. Roe. Fluctuation Splitting Schemes for the 2D Euler Equations. *VKI LS 1991-01, Computational Fluid Dynamics, von Karman Institute, Belgium,* 1991.

[3] A. Brandt. *Guide to multigrid development.* Lecture Notes in Mathematics, **960**, pp. 220-312, Springer Verlag, Berlin, 1982.

[4] L. A. Catalano, M. Napolitano, H. Deconinck. Optimal multi-stage schemes for multigrid smoothing of two-dimensional advection operators. *Communications in Applied Numerical Methods,* **8**, *pp. 785-795*, 1992.

[5] L. A. Catalano, P. De Palma, M. Napolitano. Explicit multigrid smoothing for multidimensional upwinding of the Euler equations. *Notes on Numerical Fluid Mechanics,* **35**, *Vieweg, Braunschweig, pp. 69-78*, 1992.

[6] P. L. Roe. Discrete models for the numerical analysis of time-dependent multidimensional gas dynamics. *J. of Comp. Phys.,* **63**, *pp. 458-476*, 1986.

[7] P. De Palma, H. Deconinck, R. Struijs. Investigation of Roe's 2D wave decomposition models for the Euler equations. *TN-172, von Karman Institute, Belgium,* 1990.

[8] R. Struijs, H. Deconinck, P. De Palma, P. L. Roe, K. G. Powell. Progress on multidimensional upwind euler solvers for unstructured grids. *AIAA-91-1550*, June, 1991.

[9] P. L. Roe, R. Struijs, H. Deconinck. A conservative linearization of the multidimensional Euler equations. *J. Comp. Phys.*, to appear.

[10] L. A. Catalano, P. De Palma, G. Pascazio. A multi-dimensional solution adaptive multigrid solver for the Euler equations. *Lecture Notes in Physics,* **414**, *Springer Verlag, pp. 90-94*, 1993.

[11] S. F. Davis. A rotationally biased upwind difference scheme for the Euler equations. *J. of Comp. Phys.*, **56**, *pp. 65-92*, 1984.

[12] H. Deconinck, C. Hirsch, J. Peuteman. Characteristic decomposition methods for the multidimensional Euler equations. *Lecture Notes in Physics,* **264**, *pp. 216-221*, 1986.

[13] P. L. Roe, L. Beard. An improved wave model for multidimensional upwinding of the Euler equations. *Lecture Notes in Physics,* **414**, *Springer Verlag, pp. 135-139*, 1993.

[14] H. Deconinck, R. Struijs, H. Paillere, L. A. Catalano, P. De Palma, M. Napolitano, G. Pascazio. Development of cell-vertex multidimensional upwind solvers for the compressible flow equations. *CWI-Quarterly,* **6**, *No. 1, pp. 1-28, Centrum voor Wiskunde en Informatica, Amsterdam*, 1993.

[15] P. W. Hemker, H. T. M. van der Maarel, C. T. H. Everaars. A data structure for adaptive multigrid computations. *Report NM-R9014, Department of Numerical Mathematics, Centrum voor Wiskunde en Informatica, Amsterdam*, 1990.

[16] A. Dadone, B. Grossman. Surface boundary conditions for the numerical solution of the Euler equations. *ICAM Report 92-10-04, Virginia Tech, Blacksburgh, Virginia*, 1992.

[17] H. T. M. van der Maarel. Adaptive multigrid for the steady Euler equations. *Communications in Applied Numerical Methods,* **8**, *pp. 749-760*, 1992.

[18] R. Kiock, F. Lethaus, N. C. Baines, C. H. Sieverding. The transonic flow through a plane turbine cascade as measured in four European wind tunnels. *Journal of Engineering for gas turbines and power,* **108**, *pp. 277-285*, 1986.

4

Parallel Steady Euler Calculations using Multigrid Methods and Adaptive Irregular Meshes

J. De Keyser and D. Roose[1]

ABSTRACT [2] Solving the Euler equations requires a high spatial accuracy, thus imposing strong demands on the quality of the discretization technique, the numerical solver and the implementation. In this paper we describe the parallel aspects of a steady Euler solver based on solution-adaptive irregular meshes and multigrid. The emphasis is on the run-time load balancing problem that arises in this context, and its solution with a parallel Partitioning by Pairwise Mincut heuristic.

1 Introduction

A nonlinear hyperbolic problem with applications in aerodynamics is defined by the Euler equations. The computation time required to solve such a problem accurately can be reduced by

- *irregular mesh discretization* and *adaptive mesh refinement*, to achieve a prescribed accuracy with a discrete problem that is as small as possible
- *multigrid methods*, giving fast convergence irrespective of problem size
- *distributed memory parallelism*, allowing a high computation rate

Below a steady 2-D Euler solver is described that combines these three acceleration techniques. In order to do so a particular load balancing problem has to be tackled. To this end we construct a cost function describing the multigrid cycle execution time. A parallel Partitioning by Pairwise Mincut heuristic allows to solve the load balancing problem at run-time.

---

[1]Computer Science Department, K. U. Leuven,
Celestijnenlaan 200A, B-3001 Leuven, Belgium

[2]Part of this research took place within the framework of an ESA Project (Development of Parallel Algorithms for Aerothermodynamic Applications, ESA-ESTEC Contract No. AO/1-2420/90/NL/PP). The authors are supported by the Belgian programme on Interuniversitary Poles of Attraction, initiated by the Belgian State – Prime Minister's Service – Science Policy Office. The scientific responsibility is assumed by its authors.

2 Discretization

The state of a compressible fluid in 2-D flow is defined by the *conservative* variables $\mathbf{q} = [\rho \; \rho u \; \rho v \; \rho e]^t$: density ρ, momentum in both coordinate directions ρu and ρv, with u and v the velocity components, and energy $E = \rho e = \frac{p}{\gamma-1} + \rho\frac{u^2+v^2}{2}$, with p the pressure and γ the ratio of specific heats. Other state variables are the enthalpy $h = e + \frac{p}{\rho}$ and the entropy difference with respect to a reference state (p_0, T_0, s_0) : $s - s_0 = C_p \ln T/T_0 - R \ln p/p_0$. The fluid is assumed to be an ideal gas with constant thermal capacities C_p and C_v : $p = \rho RT$, $R = C_p - C_v$, $\gamma = C_p/C_v$ (R is the gas constant and T the absolute temperature). The equations of motion for an inviscid flow field $\mathbf{q}(\mathbf{x}, t)$, $\mathbf{x} \in \Omega, t \in [t_0, \infty)$ are :

$$\frac{\partial \mathbf{q}}{\partial t} + \mathbf{N}\mathbf{q} = \frac{\partial \mathbf{q}}{\partial t} + \frac{\partial \mathbf{f}}{\partial x} + \frac{\partial \mathbf{g}}{\partial y} = \mathbf{0}, \quad \mathbf{f} = \begin{bmatrix} \rho u \\ \rho u^2 + p \\ \rho uv \\ \rho uh \end{bmatrix}, \mathbf{g} = \begin{bmatrix} \rho u \\ \rho uv \\ \rho v^2 + p \\ \rho vh \end{bmatrix}$$

An initial flow field $\mathbf{q}(\mathbf{x}, t_0)$ is given and boundary conditions are defined on $\partial\Omega$.

In our finite volume discretization, $\mathcal{G} = \{\Omega_i\}$ is a collection of polygonal cells covering the domain Ω. Cells Ω_i and Ω_j are said to be *adjacent* if they share a common border. $\mathcal{A}(\Omega_j)$ is the set of neighbors of cell Ω_j. The area of cell Ω_i is denoted by A_i. The interface $\partial\Omega_{ij}$ has length s_{ij}. Two grids $\mathcal{G}^{(k)}$ and $\mathcal{G}^{(l)}$, $k < l$ are *nested* if $\forall\Omega_j^{(l)} : \exists\Omega_i^{(k)} : \Omega_j^{(l)} \subset \Omega_i^{(k)}$. Cell $\Omega_i^{(k)}$ is the *parent cell* of $\Omega_j^{(l)}$; $\mathcal{S}(\Omega_i^{(k)})$ denotes the set of its *subcells*. We will only consider nested grid hierarchies.

$\mathbf{N}$ is a first-order operator that maps flow field $\mathbf{q}(\mathbf{x})$ from the state space E to the residual space $\hat{E}$. The l-th discrete problem in subspaces $E^{(l)} \subset E$ and $\hat{E}^{(l)} \subset \hat{E}$ is — in a first-order approximation with piecewise constant (per cell) grid functions — obtained with the projections :

$$(\mathbf{R}^{(l)}\mathbf{q})_i(t) = \frac{1}{A_i^{(l)}} \int_{\Omega_i^{(l)}} \mathbf{q}(\mathbf{x}, t) \, d\Omega, \qquad (\bar{\mathbf{R}}^{(l)}\mathbf{r})_i = \frac{1}{A_i^{(l)}} \int_{\Omega^{(l)}} \mathbf{r}(\mathbf{x}, t) \, d\Omega.$$

In this finite volume discretization, the flux $\mathbf{f}_{ij}$ through an edge is approximated by the van Leer flux vector splitting [1]. The discrete conservation property $\mathbf{f}_{ij} = -\mathbf{f}_{ji}$ allows to compute this flux once per edge. The conservation laws for Ω_i are :

$$A_i \frac{d\mathbf{q}_i}{dt} + \sum_{\Omega_j \in \mathcal{A}(\Omega_i)} s_{ij}\mathbf{f}_{ij} = A_i \mathbf{r}_i, \quad \text{i.e.} \quad (\mathbf{N}^{(l)}\mathbf{q}^{(l)})_i(t) = \frac{1}{A_i^{(l)}} \sum_{\Omega_j^{(l)} \in \mathcal{A}(\Omega_i^{(l)})} s_{ij}\mathbf{f}_{ij}.$$

With constant prolongation and with the restriction operators

$$(\mathbf{R}^{(l\to k)}\mathbf{q}^{(l)})_i = \frac{1}{A_i^{(k)}} \sum_{\Omega_j^{(l)} \in \mathcal{S}(\Omega_i^{(k)})} A_j^{(l)}\mathbf{q}_j^{(l)} \quad (\bar{\mathbf{R}}^{(l\to k)}\mathbf{r}^{(l)})_i = \frac{1}{A_i^{(k)}} \sum_{\Omega_j^{(l)} \in \mathcal{S}(\Omega_i^{(k)})} A_j^{(l)}\mathbf{r}_j^{(l)}$$

a Galerkin sequence is obtained : $\mathbf{N}^{(l-1)} = \bar{\mathbf{R}}^{(l\to l-1)}\mathbf{N}^{(l)}\mathbf{P}^{(l-1\to l)}$.

A *border cell* Ω_B along each edge of $\partial\Omega$ allows imposing information about state $\mathbf{q}_B$ and flux $\mathbf{f}^*$ at the border.

A parallel polygonal mesh refinement strategy [8] is used. As in the exact Euler solution entropy is constant along streamlines in smooth flow regions and increases over a discontinuity, the streamwise entropy gradient $\mathbf{u}\cdot\nabla s$ is a robust refinement criterion, reflecting the error made in the operator discretization.

3 Multigrid based on explicit time-marching

The spatial semi-discretization described above yields a system of ODEs in time, which can be solved with either explicit or implicit time-integration schemes. We restrict our attention to explicit methods, as they pose the same load balancing challenge as implicit methods, but require less memory.

A first-order explicit time-marching scheme is Forward Euler (FE) :

$$\mathbf{q}^{(l)}(t_{k+1}) = \mathbf{q}^{(l)}(t_k) + \Delta t(\mathbf{f}^{(l)}(t_k) - \mathbf{N}^{(l)}\mathbf{q}^{(l)}(t_k)).$$

Multi-stage Runge-Kutta methods have been developed by choosing the coefficients α_j, $j = 1,\dots,n$ in an n-stage method

$$\mathbf{q}^{[0]} = \mathbf{q}^{(l)}(t_k), \quad \mathbf{q}^{[j]} = \mathbf{q}^{[0]} + \alpha_j\Delta t(\mathbf{f}^{(l)} - \mathbf{N}^{(l)}\mathbf{q}^{[j-1]}), \quad \mathbf{q}^{(l)}(t_{k+1}) = \mathbf{q}^{[n]}$$

so as to improve smoothing properties, e.g. RK4 ($\alpha_1 = 1/4$, $\alpha_2 = 1/2$, $\alpha_3 = 0.55$, $\alpha_4 = 1$) [6]. Local timestepping variants (LT, as opposed to global timestepping, GT) use a different timestep in each cell, giving up time-accuracy in favor of convergence speed. Although explicit time-marching works best in combination with a local preconditioner [9], we do not consider such techniques here as they do not affect the load balancing problem.

Timesteps are expressed by the Courant-Friedrichs-Lewy-number $CFL = \Delta t/\Delta t^{CFL}$ with $\Delta t^{CFL} = \min_{\Omega_i} A_i(\sum_{\Omega_j\in\mathcal{A}(\Omega_i)} s_{ij}\lambda_{ij})^{-1}$ and $\lambda_{ij} = \max\{0, \mathbf{u}_i\cdot\mathbf{n}_{ij} + c_i\}$. The stability limit of FE lies at $CFL \approx 0.7$, that of RK4 at 1.25.

Several methods employ a grid hierarchy to accelerate explicit time-marching. The method proposed by Jameson has been used for both regular and the irregular grid applications [6, 3]. It consists of the FAS-scheme in which the smoother is replaced by a multi-stage RK method. Jespersen showed that it can be regarded as a timestepping method allowing a larger aggregate timestep with less computation. In our approach the refined meshes cover the entire problem domain [3, 5], as opposed to the FAC and MLAT techniques which employ fine grid patches in the refinement regions.

4 Distributed memory parallel implementation

For non-adaptive grid hierarchies a data distribution exists that guarantees load balance and efficient communication, both for the regular and irregular grid case [2, 3]. Adaptive refinement however can result in unpredictable load changes requiring load redistribution at run-time [5]. We have used the two-step approach to load balancing : first the data set is *partitioned*, subsequently the parts are *mapped* onto the parallel machine.

4.1 Distributed data structures and partitioning

The *data parallel* or *Single Program Multiple Data* programming model was adopted. In this approach the term *phase* is used for each data parallel calculation step. A phase is characterized by a *process interaction graph* describing the communication pattern. In the multigrid method there are phases acting on one grid (e.g. smoothing) and phases acting on subsequent grid levels (e.g. prolongation, restriction).

The partitioning depends on the nature of the data dependencies and on the particular data structures used in the multigrid application. We will use a *nested partitioning* of the grid hierarchy. Each grid is partitioned in *parts* $\mathcal{P}^{(l)} = \{p_i^{(l)}\}$. Partitionings $\mathcal{P}^{(l-1)}$ and $\mathcal{P}^{(l)}$ on subsequent levels are *nested* if all subcells of $\Omega_k \in \mathcal{G}^{(l-1)}$ belong to the same fine grid part, and if two fine grid cells with parents in different coarse grid parts do not belong to the same fine grid part. This nestedness property ensures that there is at most one message per fine grid part required during inter-grid operations.

Hierarchical recursive bisection (HRB) is a heuristic that generates nested partitions [5]: the partitioning $\mathcal{P}^{(i)}$ of $\mathcal{G}^{(i)}$ is derived from $\mathcal{P}^{(i-1)}$ by bisecting each part of the latter, and collecting the corresponding subcells. If $\mathcal{G}^{(i)} = \mathcal{G}^{(i-1)} = \ldots = \mathcal{G}^{(0)}$ HRB coincides with recursive bisection for single grid partitioning [10]. *Hierarchical inertial recursive bisection* (HIRB) implicitly tries to minimize the intra-grid communication.

In general, the partitions of a mesh at a given level are not of equal size, as the number of cells depends on the adaptive refinement. There should be sufficiently more parts than processors to allow the mapping step to find a good load distribution. On the other hand, parts should not be too small to avoid a large intra-grid communication volume.

4.2 Formulation of the mapping problem

A balanced workload distribution is obtained by *mapping* : distribute the parts among the processors such as to minimize the execution time for the phase [4]. This problem is known to be NP-hard. Several heuristics have been developed for solving this problem. As the number of processors in the parallel computer and the number of tasks in the task interaction graph increase, such heuristics take progressively more time. Especially in the case of run-time load balancing, one has to resort to parallel heuristics.

This mapping problem is often formulated in terms of graph-theoretic model problems. There are two graphs involved :

- the *process interaction graph* representing the dependencies between parts,
- the *machine graph*, representing the processor interconnection topology.

The graph partitioning problem (GPP) [7, 10] for a digraph $(\mathtt{G}, \rightarrow)$ with weight matrix Λ ($\Lambda_{ii} = 1$, $\Lambda_{ij} = 1$ if $g_i \rightarrow g_j$ and $\Lambda_{ij} = 0$ if $g_i \rightarrow g_j$) consists of finding a P-way partitioning $\{S_k\}$ of $\mathtt{G}$ such that $\lfloor \#\mathtt{G}/P \rfloor \leq \#S_k \leq \lceil \#\mathtt{G}/P \rceil$ and :

$$C = \sum_{g_i, g_j \in \mathtt{G}} \Lambda_{ij} \delta_{ij}$$

is small, with $\delta_{ij} = 0$ or 1 depending on whether g_i and g_j belong to the same partition or not. Such a partition containing all graph nodes that are mapped onto the same processor is a *cluster*. C is a *cost* or *penalty function*, forcing the creation of P equal-sized clusters with minimal inter-cluster connectivity.

The GPP has two limitations. First, all parallel tasks have equal calculation weights Λ_{ii}, and all data dependencies are of equal strength Λ_{ij}. Additionally, the machine architecture is not taken into account, i.e. it is assumed to be fully connected (in that case partitioning and mapping are identical). These limitations may be overcome as follows. Let the speed of the processors be defined by parameter t_{calc} (e.g. floating point multiplication time). Further, assume that the communication time is proportional to message length and communication distance Δ, with a constant of proportionality t_{comm} (e.g. the time to communicate a floating point number). Let $\tau = t_{comm}/t_{calc}$. Additionally, we associate arbitrary calculation and communication weights with the graph nodes. Reasonable estimates of these weights are available in many applications. The resulting generalization of the GPP is equivalent to the mapping problem : for a symmetric Λ, a P-way partitioning must be found such that

$$C_{SG}^2 = \sum_{k=1}^{P} \Big(\sum_{g_i \in S_k} \Lambda_{ii} \Big)^2 + 2\tau^2 \sum_{\substack{k, l = 1 \\ k < l}}^{P} \Big(\sum_{\substack{g_i \in S_k \\ g_j \in S_l}} \Lambda_{ij} \Delta_{kl} \Big)^2$$

is minimal. Δ_{kl} denotes the distance between the processors with clusters S_k and S_l.

The load balancing problem encountered in applications which require remapping at run-time is not always of the nature modeled by C_{SG}. One then looks for a new mapping which is related to the current mapping. A typical example occurs in the parallel implementation of multigrid [5]. Assume that in a full multigrid cycle a solution has been obtained at a given level. After performing adaptive mesh refinement and after partitioning the newly defined fine grid, a mapping of the fine grid must be computed. There is an *inter-grid communication penalty* associated with not mapping a part of a fine grid onto the same processor as its parent part. Let G denote the set of graph nodes which have to be remapped (parts on the fine grid), with weights Λ. Let G′ denote the set of graph nodes whose mapping is given (parts of the coarser grid). With each graph node $g_i \in$ G one associates a graph node $g'_j = t(g_i) \in$ G′ with a penalty T_i for not allocating g_i to the same processor as $t(g_i)$. With $\Delta_{k,t(i)}$ the distance of processor k on which g_i resides to the one holding $t(g_i)$, this is expressed by :

$$C_{MG}^2 = \sum_{k=1}^{P} \Big(\sum_{g_i \in S_k} \Lambda_{ii} \Big)^2 + 2\tau^2 \sum_{\substack{k,l=1 \\ k<l}}^{P} \Big(\sum_{\substack{g_i \in S_k \\ g_j \in S_l}} \Lambda_{ij} \Delta_{kl} \Big)^2 + \tau^2 \sum_{k=1}^{P} \Big(\sum_{g_i \in S_k} T_i \Delta_{k,t(i)} \Big)^2$$

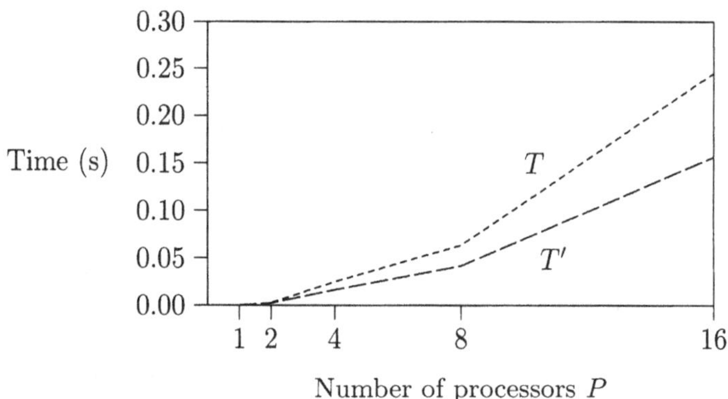

FIGURE 1. Parallel PPM remapping time for the ring problem

4.3 THE PARTITIONING BY PAIRWISE MINCUT HEURISTIC

The Partitioning by Pairwise Mincut heuristic (PPM), originally developed for the GPP [7], is extended here for the generalized problem. PPM yields good solutions as it examines many configurations; nevertheless its parallel complexity is favorable. It is also attractive because it takes advantage of a given initial mapping.

The mincut procedure

The PPM heuristic is based on a *mincut*-procedure, which is applied to every pair of clusters. It examines the effect on the cost function of moving a graph node from one cluster to the other one, and of exchanging graph nodes between both. Move and exchange operations are performed until no further improvement can be obtained. C_{SG} and C_{MG} are such that the effect of moves and exchanges is easy to compute.

The PPM algorithm

The PPM heuristic starts from an initial partitioning and tries to improve it in a number of passes. In each pass the mincut procedure is applied once for every possible pair of partitions. This process continues until no further improvement is achieved. The computational complexity of PPM for P-way partitioning of a graph of degree G with N nodes is determined by :

- the number of passes α_{PPM}, depending on the initial solution and on G
- the number of mincut operations executed in each PPM pass : $P(P-1)/2$
- the average number of passes in the mincut procedure α_m
- the average number of graph nodes involved in each mincut operation : $2N/P$
- the complexity δ of evaluating a move or exchange, depending on G

The sequential complexity is given by :

$$T_{PPM}^{seq} = \alpha_{PPM}\frac{P(P-1)}{2}\alpha_m\frac{2N}{P}\delta = \alpha_{PPM}N(P-1)\alpha_m\delta \, .$$

It has been shown that it is possible to rearrange the pairwise mincut invocations on a P-processor hypercube, such that always $P/2$ mincut operations are performed simultaneously (cf. the algorithm given in [7]). Assuming that communication is negligible, the parallel complexity is :

$$T_{PPM}^{par} = \alpha_{PPM}N\frac{P-1}{P}\,2\alpha_m\delta$$

In the context of run-time remapping the size of the graph is proportional to the number of processors $N = \beta P$. The constant β is the average number of tasks per processor. In typical applications G is independent of the number of processors. The time consumed by run-time parallel PPM remapping is :

$$T_{Remap}(P) = \alpha_{PPM}\beta(P-1)\,2\alpha_m\delta = \alpha_{PPM}\beta(P-1)\mu$$

in which μ depends on the graph degree. The remapping time increases linearly with the number of processors, in spite of the use of parallel remapping techniques.

In order to validate the assumptions on α_{PPM} and α_m, three tests have been performed (with C_{SG}). Let the process interaction graph be a ring of N units, so $G = 2$. We put $\beta = 10$. The initial distribution has precisely N/P units in each processor. As the calculation costs were chosen to increase linearly along the ring from 1.0 to 2.0, the optimal load balance $\lambda = \sum_p \sum_{g_i \in S_p} \Lambda_{ii}/P \cdot \max_p \sum_{g_i \in S_p} \Lambda_{ii}$ is 100 %. The communication costs were (a) zero (b) small (c) of the same order as the calculation costs. We measured the times T and T' to perform Parallel PPM including resp. excluding communication on an Intel iPSC/860 hypercube.

Table 1 demonstrates that PPM finds a perfect load balance in cases (a) and (b) or a good trade-off between calculation and communication in case (c), indicating that the penalty function is well-chosen. Both α_{PPM} and α_m are small. As predicted by the complexity analysis, μ is almost constant. Figure 1 shows for case (c) that the communication overhead takes about 25 % of the parallel remapping time.

P	N	λ	$T'(ms)$	α_{PPM}	α_m	$\mu(ms)$
			a : $\Lambda_{ij} = 0$			
1	10	100	0.0	0	0.0	-
2	20	100	3.1	2	2.0	0.15
4	40	100	8.2	2	1.5	0.14
8	80	100	46.2	4	1.4	0.17
16	160	100	132.1	4	1.5	0.22
			b : $\Lambda_{ij} < \Lambda_{ii}$			
1	10	100	0.0	0	0.0	-
2	20	100	3.0	2	2.0	0.15
4	40	100	8.5	2	1.5	0.14
8	80	100	56.4	5	1.4	0.16
16	160	100	150.9	5	1.4	0.19
			c : $\Lambda_{ij} \approx \Lambda_{ii}$			
1	10	100	0.0	0	0.0	-
2	20	96.4	2.3	2	1.5	0.12
4	40	98.5	16.3	4	1.4	0.14
8	80	94.5	41.5	4	1.3	0.15
16	160	95.5	156.1	6	1.1	0.17

TABLE 1. Analysis of the Parallel PPM algorithm for the ring problem

5 Experimental results

We have implemented the Jameson multigrid algorithm using the data parallel programming library LOCO [4] on the Intel iPSC/860 hypercube. We computed the flow through a channel (horizontal inflow at Mach 2, $\rho = 1.271 kg/m^3$, $p = 101300 Pa$); figure 2 shows the iso-mach lines. The initial grid was taken sufficiently fine to prevent the adaptation process from being misguided. A mesh ratio $\rho = \#\mathcal{G}^{(l)}/\#\mathcal{G}^{(l-1)} \approx 2$ to 3 was used, typical of 2-D irregular mesh hierarchies [3]. The initial mesh contained 444 cells; the fourth multigrid level consisted of 4536

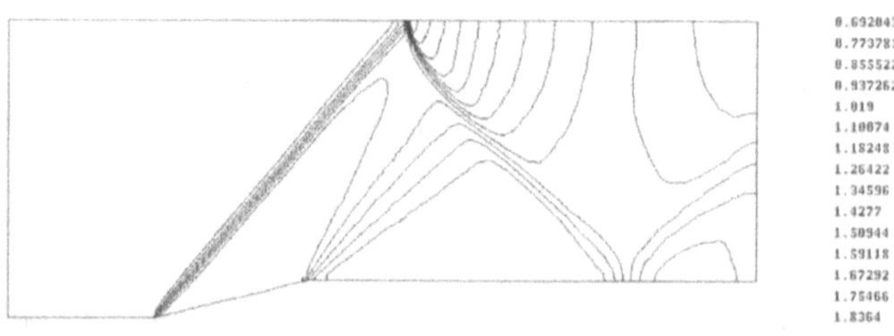

FIGURE 2. Supersonic flow through a channel : iso-mach lines

FIGURE 3. Mesh and partitioning at the 4th multigrid level

Method	seconds/iteration	digits/iteration	relative speed
FE	0.189	0.00052	1.0
MG-FE	1.760	0.036	7.5
MG-RK4	3.387	0.038	4.2

TABLE 2. Relative convergence speed at the fourth multigrid level

cells. HIRB partitioning was used (cf. figure 3 for $P = 4$). The same flow problem solved with a logically rectangular mesh with the same number of cells gave a spatial accuracy which was a factor of 2.5 worse.

Figure 4 compares the convergence of FE (LT, $CFL = 0.7$), a multigrid V-cycle with FE (LT, $CFL = 0.7$, 5 relaxations per cycle), and a multigrid V-cycle based on RK4 (LT, $CFL = 1.25$, 3 relaxations per cycle). Note that the computational cost of an iteration is different for each method. The multigrid methods have a convergence speed which slightly depends on the number of levels as the mesh ratio is not exactly constant, while single-grid convergence slows down as the discrete problem gets larger. A similar behavior was observed for subsonic flow problems, be it that convergence is a lot slower. Table 2 lists the (sequential) asymptotic execution speed relative to FE time-stepping. At the fourth level the V-cycle with FE is ≈ 70 times faster than FE in terms of number of iterations, and

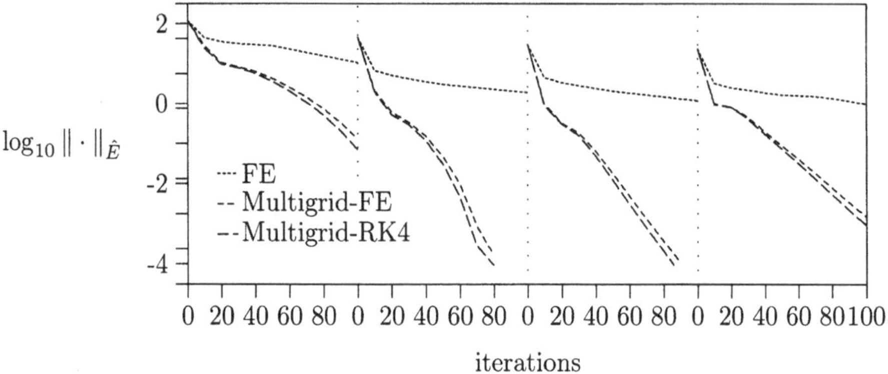

FIGURE 4. Convergence history of explicit multigrid methods

P	N	G	λ	α_{PPM}	α_m
1	1	1	100	-	-
2	9	6	97	2	1.6
4	17	8	95	3	1.2
8	26	11	94	5	1.4
16	33	11	91	4	1.3

TABLE 3. PPM parameters at the 4th multigrid level

7.5 times faster in terms of sequential computation time. The relative performance of multigrid will even be better for larger problem sizes.

The remapping procedure took substantially less time than the grid refinement process. Table 3 lists the PPM parameters for solving the load balancing problem at the fourth multigrid level. The remapping procedure is based on the weight matrix Λ, giving information about the time involved in calculation and communication operations :

$$\Lambda_{ii} = a_i t_{calc}, \qquad \forall p_i \in \mathcal{A}(p_j) : \Lambda_{ij} = b_{ij} t_{comm}, \qquad \forall p_i \in \mathcal{S}(p_j) : T_i = c_j t_{comm}$$

a_i is the number of floating point operations required for part p_i during one multigrid cycle (proportional to the number of cells in p_i and to the number of smoothing steps). Intra-grid communication requires exchanging b_{ij} numbers between p_i and p_j (proportional to the number of edges along the interface between both parts, and to the number of smoothing steps). The intergrid transfer message length to or from parent part p_j is c_j (proportional to the number of cells in parent p_j, and to the number of coarse grid corrections). Based on these cost estimates, a load balance between 90-100 % is always obtained when $N/P > 4$. A difficulty is the inaccuracy of the estimates : e.g. imposing boundary conditions may take an unpredictable amount of time. Note that C_{MG} does not try to optimize λ, but is prepared to accept a worse load balance if this can save intergrid communication overhead. This is the case on the coarsest multigrid levels, where there may be only enough work to keep a few processors busy (an effect known as *agglomeration*).

The parallel efficiency of an algorithm is defined as : $\epsilon_p = T(1,S)/P \cdot T(P,S)$, in which $T(P,S)$ denotes the time to apply the algorithm to a problem of size S on a machine with P processors. For the V-cycle with FE, our current implementation achieved $\epsilon_p \approx 75$ % for $P = 16$. Parallel efficiency losses are due to load imbalance (0-10 %), data exchange communication (5-10 %), and the double calculation of fluxes for edges along part interfaces (5-15 %) : these are calculated twice, once for each cell on either side of part interfaces. Communication and double calculation losses decrease as the problem size per processor is larger.

6 Conclusion

Three acceleration techniques have been combined in one Euler solver. Incorporating adaptivity and multigrid in a distributed memory parallel code poses a particular load balancing problem. We have extended the original Partitioning by Parallel Pairwise Mincut algorithm to allow the solution of this problem. For this application, mappings of good quality are obtained in a reasonable time. The time taken by parallel PPM increases approximately linearly with the number of processors. In spite of this asymptotic behavior, the remapping overhead in our application remains small compared to the time invested in grid refinement.

The proposed code proves to be effective on medium-sized parallel computers. For a model problem on a 16 processor machine, a global acceleration factor of the order of $(2.5)^2 \times (7.5) \times (16 \times 0.75) \approx 600$ has been observed for Jameson multigrid with Forward Euler time-stepping. When larger problems are solved, each of the three acceleration techniques will be more effective, leading to a larger global acceleration factor.

References

[1] W. K. Anderson, J. L. Thomas, and B. van Leer. A comparison of finite volume flux vector splittings for the Euler equations. *AIAA Paper No. 85-0122, presented at the AIAA 23rd Aerospace Sciences Meeting, Reno, Nevada,* 1985.

[2] G. Chesshire and A. Jameson. FLO87 on the iPSC/2 : A parallel multigrid solver for the Euler equations. In J. Gustafson, editor, *Proceedings of the Fourth Conference on Hypercubes, Concurrent Computers and Applications,* pages 957–966. Golden Gate Enterprises, May 1990.

[3] R. Das, D. Mavriplis, J. Saltz, S. Gupta, and R. Ponnusamy. The design and implementation of a parallel unstructured Euler solver using software

primitives. In *Proceedings of the 30th Aerospace Sciences Meeting and Exhibit.* AIAA-92-0562, 1992.

[4] J. De Keyser and D. Roose. A software tool for load balanced adaptive multiple grids on distributed memory computers. In *Proceedings of the 6th Distributed Memory Computing Conference*, pages 122–128. IEEE Computer Society Press, 1991.

[5] J. De Keyser and D. Roose. Incremental mapping for solution-adaptive multigrid hierarchies. In *Proceedings of the Scalable High Performance Computing Conference '92*, pages 401–408. IEEE Computer Society Press, 1992.

[6] A. Jameson. Solution of the Euler equations for two dimensional transonic flow by a multigrid method. *Applied Math. Comp.*, 13:327–355, 1983.

[7] P. Sadayappan, F. Ercal, and J. Ramanujam. Parallel graph partitioning on a hypercube. In J. Gustafson, editor, *Proceedings of the 4th Distributed Memory Computing Conference*, pages 67–70. Golden Gate Enterprise, CA, 1990.

[8] R. Struijs, P. Van Keirsbilck, and H. Deconinck. An adaptive grid polygonal finite volume method for the compressible flow equations. In *Proceedings of the AIAA 9th CFD Conference*, 1989.

[9] B. van Leer, W.-T. Lee, and P.L. Roe. Characteristic time-stepping or local preconditioning of the Euler equations. In *AIAA 10th Computational Fluid Dynamics Conference*, 1991.

[10] R. D. Williams. Performance of dynamic load balancing algorithms for unstructured mesh calculations. *Concurrency: Practice and Experience*, 3:457–481, 1991.

5

Multigrid Methods for Steady Euler Equations Based on Multi-stage Jacobi Relaxation

Erik Dick and Kris Riemslagh[1]

1 Introduction

First order accurate upwind methods of flux-difference type applied to steady Euler equations generate a set of discrete equations of positive type. This set can be solved by any classic relaxation method in multigrid form. The set of discrete equations generated by a higher order accurate form does not have this property and cannot be solved in the same way. The common approach is then to use defect correction [1, 2, 3]. In this procedure the multigrid method is applied to the first order accurate form and constitutes an inner iteration for a higher order correction only made on the finest grid. The defect correction proves to work well in many applications. The speed of convergence is however largely determined by the outer iteration and sometimes is found to be rather dissapointing, especially when the first order and the higher order solutions differ significantly. It can be expected that if the higher order approximation also could be used in the multigrid itself a better performance could be obtained. A second difficulty is that often convergence cannot be obtained unless a suitable initial flow field is specified, i.e. there is a risk of choosing an initial approximation which is out of the attraction region of the iterative method.

In principle, both difficulties can be avoided by using time stepping methods on the unsteady equations instead of relaxation methods on the steady equations. The higher order accurate discretization can then be used on any grid so that the defect correction becomes unnecessary and convergence is guaranteed starting from any initial field due to the hyperbolicity of the equations with respect to time. Many multi-stage time stepping methods with optimization strategies for the smoothing have been proposed for this purpose in recent years. We cite the methods of Van Leer et al. [5], Catalano and Deconinck [6], among others.

The drawback of time stepping is that, even if local time stepping is used,

[1]Universiteit Gent
Sint-Pietersnieuwstraat 41, 9000 Gent, Belgium

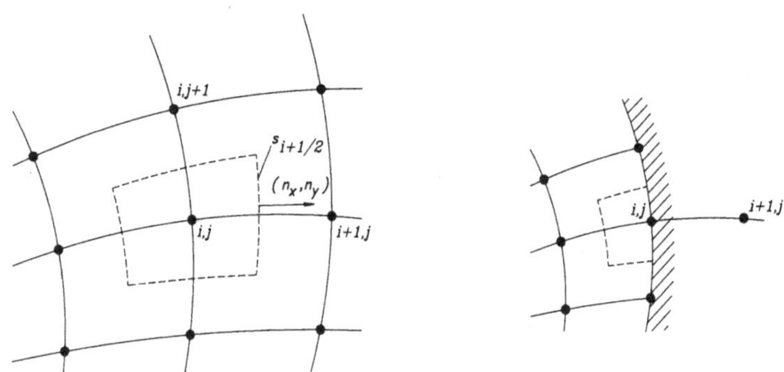

FIGURE 1. Control volumes in the interior and on a solid boundary.

the smoothing only can be tuned well for the fastest wave components in the flowfield. This results in a rather poor multigrid performance. As a remedy to this, we propose to use Jacobi relaxation as a basic algorithm, equivalent to single stage time stepping, and to bring in multi-staging in the same way as single stage time stepping is transformed into multi-stage time stepping. This procedure has the advantage that all wave components are first scaled so that, so to speak, they move all with the same CFL-number. This guarantees optimal tuning for all wave components. Nevertheless, the hyperbolicity with respect to the relaxation direction, i.e. the ficticious time, is not lost. The principle of combining Jacobi relaxation and time stepping was first suggested by Morano et al. [7], but not worked out. The present authors made a preliminary analysis of the possible multigrid performance in [8] and an analysis of the performance for the linear $\kappa = 1/3$ scheme in [9]. Here, we analyse the performance for the non-linear TVD-scheme.

2 The flux-difference splitting method

The discretization is based on the vertex-centred finite volume method. Figure 1a shows a control volume centred around a vertex (i, j) in the grid. The control volumes are formed by connecting the centres of gravity of the surrounding cells.

The flux-difference over the surface $S_{i+1/2}$ is written as

$$F_{i+1} - F_i = s_{i+1/2} A_{i+1/2} (U_{i+1} - U_i), \tag{1}$$

where U stands for the vector of conserved variables, $s_{i+1/2}$ is the length of the surface and $A_{i+1/2}$ is the discrete flux-Jacobian. The first order upwind flux is defined by

$$F_{i+1/2} = \frac{1}{2}(F_i + F_{i+1}) - \frac{1}{2} s_{i+1/2} (A^+_{i+1/2} - A^-_{i+1/2})(U_{i+1} - U_i), \tag{2}$$

where $A^+_{i+1/2}$ and $A^-_{i+1/2}$ are the positive and the negative parts of $A_{i+1/2}$. The upwind flux can also be written as

$$F_{i+1/2} = F_i + s_{i+1/2} A^-_{i+1/2} (U_{i+1} - U_i). \tag{3}$$

This flux expression shows the incoming wave components. To determine the flux-Jacobians, we use here the polynomial flux-difference splitting. Details on this method are given in [3, 4]. The technical form of the splitting is however not relevant for the method we describe here. The second order accurate flux is defined using the flux-extrapolation technique. Details on this technique, using the minmod-limiter, are again given in [3, 4].

The resulting flux expression is

$$F_{i+1/2} = F_i + s_{i+1/2} A^-_{i+1/2} (U_{i+1} - U_i) + F.C., \tag{4}$$

where $F.C.$ denotes the flux-correction for higher order accuracy.

3 Boundary conditions

The examples to follow are channel flows. These internal type flows have solid boundaries, inlet and outlet boundaries. For inflow and outflow boundaries, the classic extrapolation procedures are used. At solid boundaries, impermeability is imposed by setting the convective part of the flux equal to zero. This requires a modification of the flux expression (4). For (i, j) a point on the boundary, the point $(i+1, j)$ does not exist (see figure 1b). This can be introduced in (4) by setting the term $F.C.$ to zero and by taking the values of the variables in the ficticious node $(i+1, j)$ equal to the values of the variables in the node (i, j). The matrix $A^-_{i+1/2}$ in (4) is then calculated with the values of the variables in the node (i, j). Of course, since the difference of the variables is zero, the first order difference part in (4) is also zero. The impermeability is introduced trough replacing F_i by $F_i - F_{i'}$, where $F_{i'}$ is the convective part of the flux. The term $-F_{i'}$ can be seen as a new flux correction term $F.C.$ As will be discussed in the next section, the matrix $A^-_{i+1/2}$ at a solid boundary plays an important role in the relaxation method, although it is multiplied with a zero term.

4 The multi-stage Jacobi relaxation

In earlier multigrid formulations for steady Euler equations the Gauss-Seidel relaxation method was always used [3, 4]. Gauss-Seidel relaxation was preferred to Jacobi relaxation because of its much better smoothing properties (effectiveness associated to the coarse grid correction) and much better speed of convergence (effectiveness associated to the relaxation method itself). A simultaneous relaxation method, like the Jacobi relaxation has the advantage of being easily vectorizable and parallelizable. The only drawback is that a simultaneous relaxation method, at least in its basic form, is much less effective than a sequential method.

To repair this, we bring multi-staging into the Jacobi method in the same way as multi-staging is used for time stepping methods and we use the optimization results known for time stepping schemes.

For the time-dependent Euler equations, the discrete set of equations associated to the node (i,j) reads

$$Vol_{i,j}\frac{dU_{i,j}}{dt}+\sum_{k}A_k^-(U_k-U_{i,j})s_k+\sum_{k}F.C.=0, \tag{5}$$

where the index k loops over the faces of the control volume and the surrounding nodes. A single stage time stepping method on (5) gives

$$\left(\frac{Vol_{i,j}}{\Delta t}\right)\delta U_{i,j}+\sum_{k}A_k^-(U_k^n-U_{i,j}^n)s_k+\sum_{k}F.C.^n=0. \tag{6}$$

The Jacobi-relaxation applied to the steady part of (5) reads

$$\sum_{k}A_k^-(U_k^n-U_{i,j}^{n+1})s_k+\sum_{k}F.C.^n=0. \tag{7}$$

Using increments $\delta U_{i,j}=U_{i,j}^{n+1}-U_{i,j}^n$, this gives

$$\left(-\sum_{k}A_k^-s_k\right)\delta U_{i,j}+\sum_{k}A_k^-(U_k^n-U_{i,j}^n)s_k+\sum_{k}F.C.^n=0. \tag{8}$$

The 4x4 matrix coefficient of $\delta U_{i,j}$ in (8) is non-singular. In the expressions (6) to (8), the matrices A_k^- are on the time or relaxation level n. The difference between (single stage) Jacobi relaxation (8) and single stage time stepping (6) is seen in the matrix coefficient of the vector of increments $\delta U_{i,j}$.

In the time stepping method, the coefficient is a diagonal matrix. In the Jacobi method, the matrix is composed of parts of the flux-Jacobians associated to the different faces of the control volume. The collected parts correspond to waves incoming to the control volume. In the time stepping, the incoming waves contribute to the increment of the flow variables all with the same weight factor. In the Jacobi relaxation the corresponding weight factors are proportional to the wave speeds. As a consequence, Jacobi relaxation can be seen as a time stepping in which all incoming wave components are scaled to have the same effective speed, i.e. all have a CFL-number equal to unity. Using terminology already in use nowadays, time stepping can be referred to as scalar time stepping while Jacobi relaxation can be referred to as matrix time stepping.

For a node on a solid boundary, an expression similar to (8) is obtained provided that for a face on the boundary the flux expression (4) is used and that the difference in the first order flux-difference part is introduced as $U^n_{i,j} - U^{n+1}_{i,j}$, similar to the term $U^n_k - U^{n+1}_{i,j}$ which is used for a flux on an interior face. So, in order to avoid a singular matrix coefficient of the vector of increments in (8), this treatment at boundaries is necessary. A boundary node can then be updated in the same way as a node in the interior. The solid boundary treatment is different from the treatment used earlier [3, 4].

To bring in multi-staging is now very simple. For instance, a three-stage modified Runge-Kutta stepping is given by

$$\begin{aligned} U^0 &= U^n_{i,j} \\ U^1 &= U^0 + \alpha_1\,\delta U^0 \\ U^2 &= U^0 + \alpha_2\,\delta U^1 \\ U^3 &= U^0 + \alpha_3\,\delta U^2 \\ U^{n+1}_{i,j} &= U^3, \end{aligned}$$

where δU is the increment obtained from single stage time stepping or single stage Jacobi relaxation. The last coefficient in the stepping series (here α_3) has the significance of a CFL-number for time stepping. We refer to this coefficient as the step size or simply as the CFL-number.

5 Optimization of the multi-stage parameters

We follow here the Fourier-representation method for operators and solution methods used e.g. by Van Leer et al. [5].

Figure 2a shows the Fourier-symbols of the first order upwind scheme (U1), the second order upwind scheme (U2), the second order central scheme (C2) and the $\kappa = 1/3$ scheme (K3). Figure 2b shows the contours of the amplification factor for three-stage stepping with optimum smoothing for the first order upwind scheme, according to Van Leer et al. [5]. Figure 2c shows the contours for the K3-scheme, optimized in the same way. Figure 2d shows the contours corresponding to three consecutive single stage steppings with relaxation factor 0.5. Interpreted as a three stage stepping, the corresponding coefficients are 1/6, 1/2, 3/2. The three stage steppings illustrated in figure 2b, c and d are designed to optimize the smoothing of a particular linear discretization scheme. We consider now also steppings suitable for use with the non-linear TVD-scheme. For these steppings, stability for both the central discretization C2 and the upwind discretization U2 is necessary. Figure 2e shows contour levels of a three stage stepping stable for both the C2- and U2-schemes and with maximum step size. Figure 2f shows the contour levels of a similar five stage stepping. Figure 2g shows the contour levels for a five stage stepping stable for the U2- and C2-schemes and with optimum smoothing for the U2-scheme. Figure 2h shows a three stage stepping scheme stable for the K3-scheme, but not for the U2- and C2-schemes, with maximum step size.

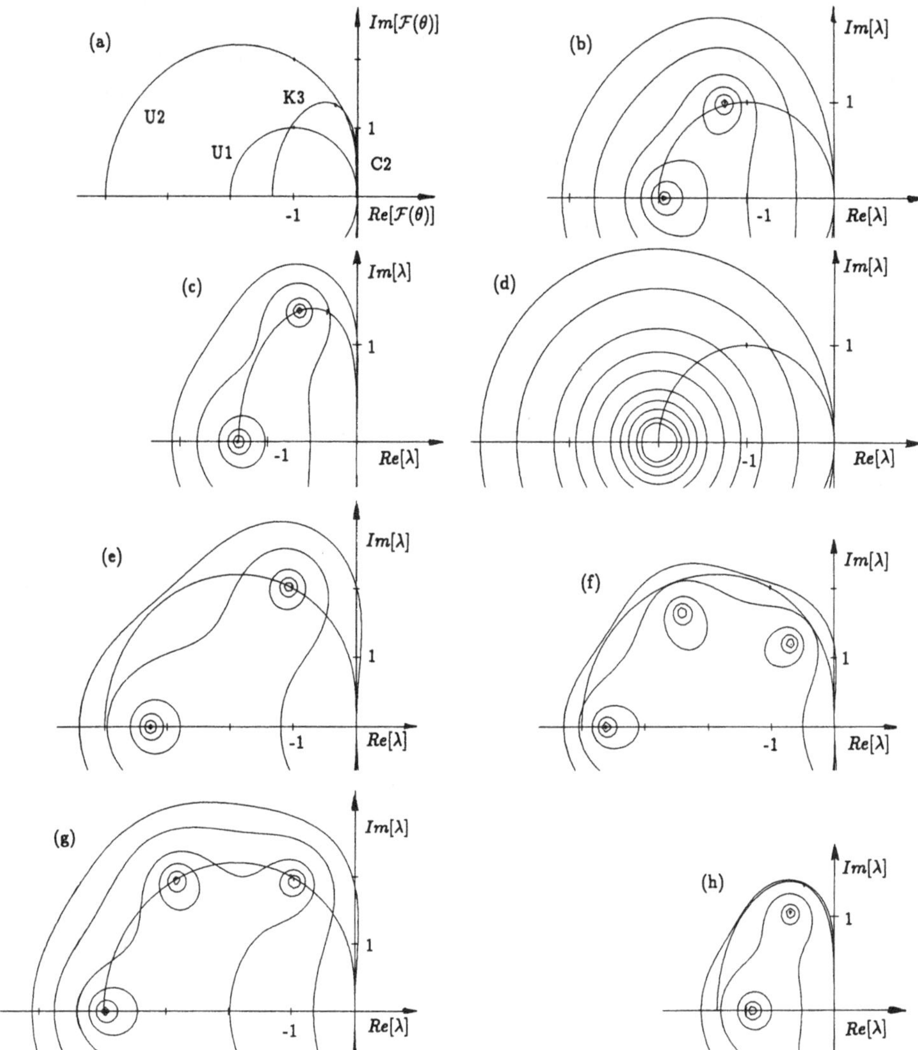

FIGURE 2. a: Fourier symbols of the basic schemes. Symbols from $\theta = 0$ to $\theta = -\pi$. The dot corresponds to $|\theta| = \pi/2$; b: Amplification factor for three stage stepping with optimized smoothing for U1 (coefficients 0.223, 0.60, 1.50) [5]; c: Idem for the K3-scheme (coefficients 0.382, 0.664, 1.325) [5]; d: Idem for a three stage scheme lineary equivalent to three consecutive Jacobi relaxations (coefficients 0.167, 0.50, 1.50); e: Amplification factor for three stage stepping stable for U2 and C2 and with maximum step size (coefficients 0.185, 0.443, 0.712); f: Idem for five stage stepping (coefficients 0.102, 0.248, 0.476, 0.893, 1.55); g: Amplification factor for five stage stepping stable for U2 and C2 and with maximum smoothing for U2 (coefficients 0.0867, 0.202, 0.366, 0.633, 1.14); h: Amplification factor for three stage stepping with maximum step size for the K3-scheme (coefficients 0.521, 0.857, 1.82); Amplification levels shown : 1, 0.5, 0.2, 0.1, 0.05, 0.02, ...

The results shown in figure 2 apply to the one-dimensional convection equation. As is known, the set of Fourier symbols for a two-dimensional convection equation lies inside the Fourier symbol of a one-dimensional equation. So, two-dimensional stability is guaranteed for the equation when one-dimensional stability for the equation is obtained. It is difficult to make precise statements about smoothing.

6 Performance analysis

6.1 Test problems

A channel with a circular perturbation in the lower wall is used. The grid has 32 by 96 cells. Four consecutive grids are used. The coarser grids have 16 x 48, 8 x 24 and 4 x 12 cells. The height of the channel is equal to the length of the perturbation. The height of the circular perturbation is 4.2% of its length. The grids used have an almost uniform distribution of the mesh-size. The test geometry is the same as in [8, 9]

The same multigrid structure as in [3, 4, 8, 9] is used. The W-cycle is employed. On each level there is one pre- and postrelaxation consisting of either three Gauss-Seidel relaxations or a multi-stage Jacobi relaxation. The defect restriction operator is full weighting. The computation starts on the coarsest grid. To evaluate the work, the number of relaxation steps or stages are counted and the number of defect corrections and defect calculations. One basic operation on the finest grid is considered as 1 work unit. So the work unit corresponds to 3201 point-relaxation operations. A defect correction or a defect calculation is somewhat less expensive than a relaxation operation. Nevertheless these operations are given the same weight to compensate for the neglect of the work involved in the grid transfer. Since precisely the defect operations are connected with grid transfer this is believed to be fair. A relaxation operation for the first order (U1), for the third order (K3) and for the TVD operator are counted to be equivalent. This is not completely correct. The third order and the TVD operator are slightly more expensive.

Two flow fields are considered : a transonic flow field corresponding to an outlet Mach number of 0.79 and a supersonic flow field corresponding to an inlet Mach number of 1.39. The quality of the solutions is not particularly good due to the rather low resolution in the shock-regions (results not shown).

6.2 Defect correction

Figure 3 shows the convergence results for the transonic and the supersonic test cases using defect correction. Gauss-Seidel relaxation is compared to multi-stage Jacobi relaxation. Convergence results are expressed by the log_{10} of the L_∞-norm of the defect as function of the number of work units. For the Gauss-Seidel, three

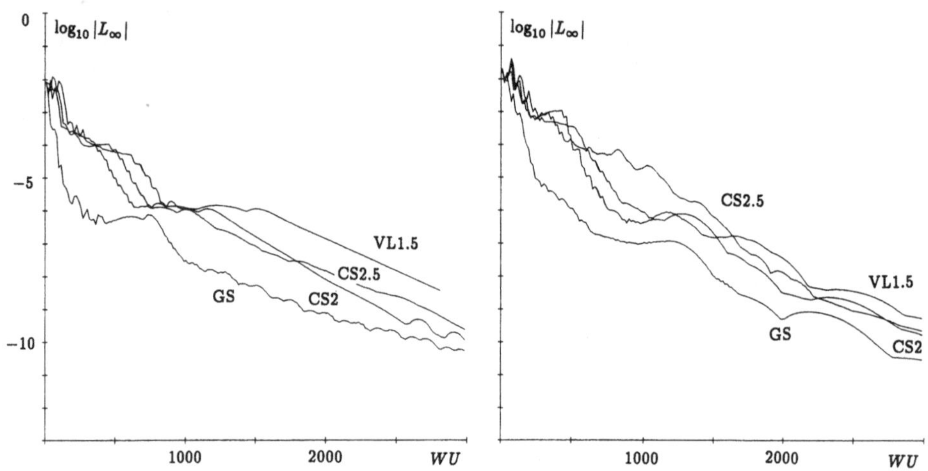

FIGURE 3. Convergence behaviour for defect correction. Transonic (left) and supersonic (right) test cases. Gauss-Seidel relaxation (GS). Three stage Jacobi relaxation with Van Leer coefficients (VL1.5) and consecutive coefficients (CS2 and CS2.5).

relaxations are done per level. The relaxation factor is 0.95 for the transonic case and 1.0 for the supersonic case. The ordering is lexicographic. In the first sweep the relaxation starts in the left bottom corner, goes up in the first column, then in the second column and so forth up to the right top corner. In the second sweep the ordering is reversed. The third sweep has again the ordering of the first sweep. The convergence history is also shown for the three stage stepping with the Van Leer coefficients (set of figure 2b) and with the *consecutive* coefficients with different step sizes. With a consecutive scheme we mean a three stage Jacobi relaxation lineary equivalent to three consecutive single stage Jacobi relaxations. We denote such a scheme by CS followed by the step size. The amplification factors for the CS1.5 scheme are shown in figure 2d. The coefficients are (1/9, 1/3, 1) multiplied with step size.

The three stage Jacobi schemes with optimum smoothing (the convergence behaviour of CS1.5 is not shown but is almost identical to the behaviour of VL1.5) do not perform as good as a three stage scheme with a somewhat larger step size. A step size around 2 seems to be the best. This proofs that smoothing is to be sacrificed a bit in favour of convection speed. Sufficient smoothing is necessary but not optimum smoothing. A step size of 3 is the stability limit. Even a step size of 2.5 performs better than the step size of 1.5 for optimum smoothing.

6.3 Mixed discretization

By mixed discretization we mean that the second order TVD-operator is used in all relaxations on the finest level but that an other (linear) operator is used on the

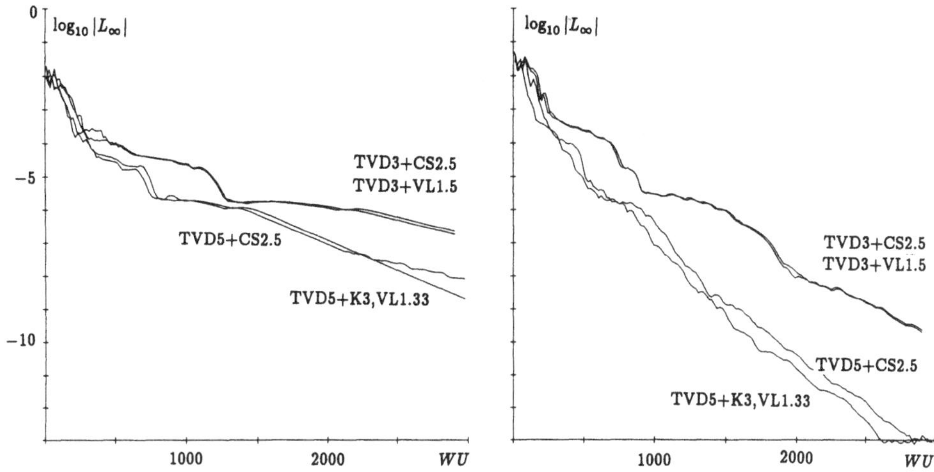

FIGURE 4. Convergence behaviour for mixed discretization. Transonic (left) and supersonic (right) test cases. TVD on finest level, U1 or K3 on coarser levels.

coarser levels. Figure 4 shows the convergence behaviour for the TVD-operator on the finest level with three stage stepping with maximum CFL (coefficient set of figure 2e) and with five stage stepping with maximum CFL (coefficient set of figure 2f) combined with several three stage steppings for the first order operator.

The performance is not very sensitive to the choice of the coefficient set on the coarser levels. So, again optimum smoothing is not necessary. The five stage stepping performs best.

In the transonic case the best obtained performance, i.e. TVD5+CS2.5, does not compete with the performance of the Gauss-Seidel defect correction (figure 3), but in the supersonic case the mixed discretization TVD5+CS2.5 performs better than the Gauss-Seidel defect correction. This easily can be understood. In the transonic test case the shock is largely aligned with the grid. As a consequence, the second order TVD-solution does not differ very much from the first order solution. For the supersonic test case the difference between the second order and the first order solutions is rather large. This makes defect correction a much more effective procedure in the transonic test case than in the supersonic test case. The better performance of the mixed discretization in the supersonic test case also shows that it pays off to bring the second order operator in the multigrid formulation in those cases where second and first order solutions differ significantly. We further illustrate this on figure 4 where also the convergence behaviour is shown for mixed discretization but with the K3 operator on the coarser levels. The convergence improves somewhat, since the K3 operator is closer to the TVD operator than the first order upwind operator U1 is.

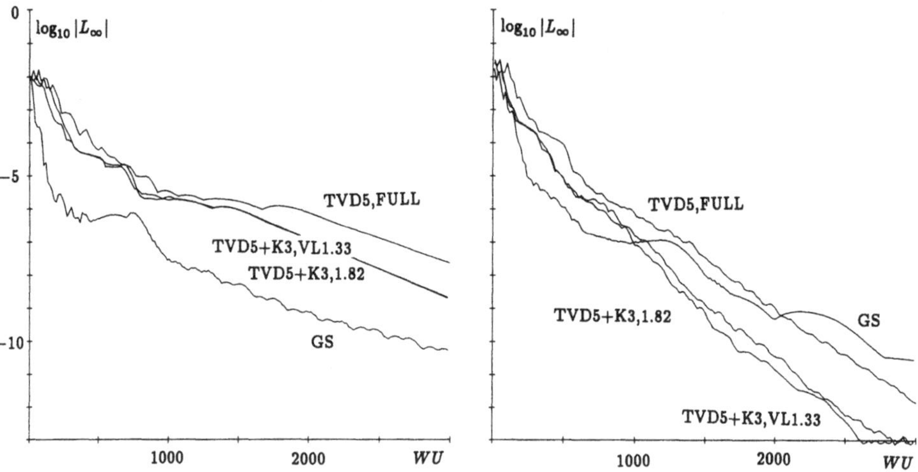

FIGURE 5. Convergence behaviour for full second order formulation (FULL) with the TVD operator on all levels. Transonic (left) and supersonic (right) test cases. Convergence behaviour for mixed discretization with the K3 operator on coarser levels, explicit and implicit residual weighting. Comparison with Gauss-Seidel (GS).

6.4 Full second order with implicit residual weighting

We illustrate now the performance for the TVD-operator used on all levels. Since the TVD-operator changes from the central to the upwind scheme and vice versa, depending on the solution, no smoothing can be obtained for this operator. In order to make the multigrid method possible, the restriction of a smooth residual must be obtained by supplementary means. The technique of explicit and implicit residual smoothing is well known for use with time stepping schemes. In analogy with the residual smoothing we bring here implicitness in the weighting. The usual full weighting as restriction is already a residual smoother of explicit type. An implicit version of it can be much more efficient. In one dimension an explicit residual weighting (ERW) on the same grid gives

$$(1+2\epsilon)\tilde{R}_i = R_i + \epsilon(R_{i-1} + R_{i+1}), \tag{9}$$

where $\epsilon = 0$ corresponds to injection and $\epsilon = 1/2$ corresponds to full weighting.

Figure 6 compares the amplification factor for both types of residual weighting.

A corresponding implicit residual weighting (IRW) is given by

$$(1+2\epsilon)\tilde{R}_i - \epsilon(\tilde{R}_{i-1} + \tilde{R}_{i+1}) = R_i. \tag{10}$$

The obtained weighted residuals $\tilde{R}_i$ still have to be injected to the coarser grid. By enlarging the value of the weight ϵ, the smoothing of the implicit residual weighting (IRW) increases. Maximum smoothing does not correspond to optimum

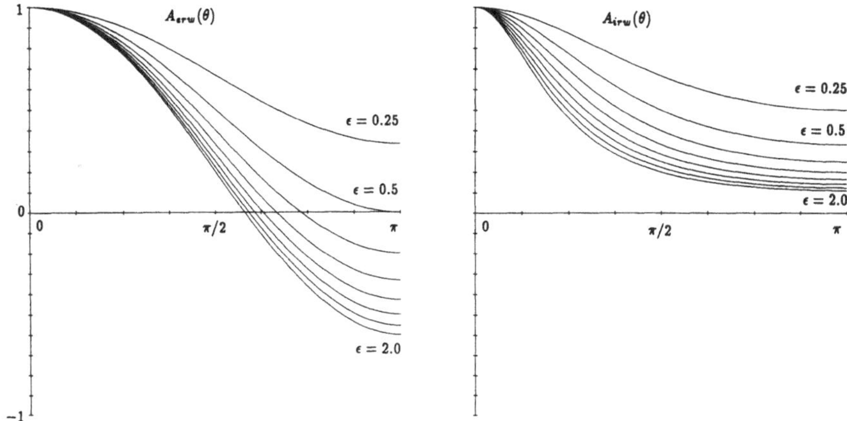

FIGURE 6. Amplification factor for explicit (left) and implicit (right) versions of residual weighting.

multigrid performance. The optimum is a compromise between the reduction of high frequency components, i.e. diminishing of the alaising in the fine to coarse grid transfer, and leaving as much as possible intact the low frequency components, i.e. the components that have to be treated by the coarse grid. In two dimensions it was found that $\epsilon = 2$ was optimum for a nine point stencil of form

$$\begin{aligned}(1+6\epsilon)\tilde{R}_{i,j} &- \epsilon(\tilde{R}_{i-1,j} + \tilde{R}_{i+1,j} + \tilde{R}_{i,j-1} + \tilde{R}_{i,j+1}) \\ &- \frac{\epsilon}{2}(\tilde{R}_{i-1,j-1} + \tilde{R}_{i+1,j-1} + \tilde{R}_{i-1,j+1} + \tilde{R}_{i+1,j+1}) = R_{i,j}.\end{aligned}$$

Figure 5 shows the convergence behaviour for a full second order formulation, i.e. using the TVD operator on all levels, with implicit residual weighting, compared to mixed discretization using the K3 operator on the coarser levels. Five stage stepping with maximum CFL is used for the TVD operator (coefficients of figure 2f). Three stage steppings with optimum smoothing coefficients with ERW (coefficients of figure 2c) and with maximum step size with IRW (coefficients of figure 2h) are used for the K3 operator.

In the full second order formulation, it does not pay off to change the coefficient set from a set not corresponding to maximum step size. One could try to introduce smoothing for the U2-operator (coefficient set of figure 2g). This does not help since smoothing never can be obtained for the C2 operator. As can be seen in figure 5, the mixed discretization works better than the full second order formulation. The best mixed discretization is the one with the K3 operator on the coarser levels. It does not pay off to enlarge the step size for the K3 operator, since smoothing with the operator itself is then lost and has to be introduced by implicit residual weighting. The resulting performance is not better than the performance with a coefficient set corresponding to smoothing and explicit residual weighting.

7 Conclusion

By the combination of Jacobi relaxation and multi-stage stepping, multigrid methods for Euler equations can be constructed that are more general than defect correction procedures. With implicit residual smoothing it is even possible to use the TVD operator on all grid levels. The best performance is however obtained with a mixed discretization formulation with the K3 operator on the coarser levels. In the case where the second order solution and the first order solution differ considerably, this formulation is more efficient than a defect correction formulation based on Gauss-Seidel relaxation.

ACKNOWLEDGEMENTS

The research reported here was granted under contract IT/SC/13, as part of the Belgian National Programme for Large Scale Scientific Computing and under contract IUAP/17 as part of the Belgian National Programme on Interuniversity Poles of Attraction, both initiated by the Belgian State, Prime Minister's Office, Science Policy Programming.

References

[1] P. W. Hemker, Defect correction and higher order schemes for the multigrid solution of the steady Euler equations, *Proc. 2nd European Conf. on Multigrid Methods*, Lecture Notes in Mathematics *1228* (Springer, 1986) 149-165.

[2] B. Koren, Defect correction and multigrid for an efficient and accurate computation of airfoil flows, *J. Comp. Phys. 77* (1988) 183-206.

[3] E. Dick, Second-order formulation of a multigrid method for steady Euler equations trough defect-correction, *J. Comput. Appl. Math. 35* (1991) 159-168.

[4] E. Dick, Multigrid methods for steady Euler and Navier-Stokes equations based on polynomial flux-difference splitting, *Proc. 3rd European Conf. on Multigrid Methods, Multigrid Methods III*, Int. Series on Numerical Mathematics, *98* (Birkhauser Verlag, Basel, 1991) 1-20.

[5] B. Van Leer, C. H. Tai and K. G. Powell, Design of optimally smoothing multi-stage schemes for the Euler equations, *AIAA-paper 89-1933*, 1989.

[6] L. A. Catalano and H. Deconinck, Two-dimensional optimization of smoothing properties of multi-stage schemes applied to hyperbolic equations, *Proc. 3rd European Conf. on Multigrid Methods* (GMD-Studien 189, Bonn, 1991).

[7] E. Morano, M.-H. Lallemand, M.-P. Leclerq, H. Steve, B. Stoufflet and A. Dervieux, Local iterative upwind methods for steady compressible flows, *Proc. 3rd European Conf. on Multigrid Methods* (GMD-Studien 189, Bonn, 1991).

[8] E. Dick and K. Riemslagh, Multi-stage Jacobi relaxation as smoother in a multigrid method for steady Euler equations, *Proc. ICFD Conf. on Numerical Methods for Fluid Dynamics*, (Reading, 1993) to appear.

[9] E. Dick and K. Riemslagh, Multi-staging of Jacobi relaxation to improve smoothing properties of multigrid methods for steady Euler equations. *J. Comput. Appl. Math.*, (1993) to appear.

6

Multigrid and Renormalization for Reservoir Simulation

Michael G Edwards [1] and Clive F Rogers[2]

ABSTRACT We present a new approach to multigrid for the case of strongly varying equation coefficients which arise in the reservoir simulation pressure equation. Renormalization (hierarchical rescaling) is incorporated into the cell centred multigrid method of Wesseling et.al. and the new method is applied to the pressure equation. Significant improvement in multigrid performance is obtained with the new scheme for typical cases of randomly varying permeability distributions of finite correlation length. A new 9-point scheme is described which is flux continuous both for diagonal and full permeability tensors. Results from the new scheme are presented.

1 Introduction

The flow equations of reservoir simulation are a coupled system of hyperbolic conservation laws for fluid transport and parabolic/elliptic equation for pressure (elliptic for incompressible flow). The coupling between the equations is provided by Darcy's law which defines the fluid velocity to be proportional to the medium permeability and pressure gradient. The pressure equation is generally of the form [2],

$$\nabla\cdot(K \nabla \phi) = f$$

where K is a full matrix of tensor permeabilities and f is the source/sink distribution dependent on the wells. The solution of the pressure equation typically consumes between 50% and 90% of the net cpu time used in reservoir simulations. Thus the development of fast elliptic solvers remains an active area of simulation research. To date most commercial simulator pressure solvers employ some variant of a pre-conditioned conjugate gradient technique, where the pre-conditioning is

[1] Current address: Dept.of Petroleum Engineering, CPE, University of Texas at Austin, Austin TX 78712, USA.

[2] BP Exploration Operating Co Ltd, BPX Technology Provision, Chertsey Road, Sunbury-on-Thames, Middlesex TW16 7LN, United Kingdom

based on an approximate LU factorisation.

Multigrid for the reservoir simulation pressure equation with a diagonal permeability tensor has been investigated by several authors [3,4,5,6,7,8]. The main attraction of multigrid is the theoretical and observed convergence rate which is proportional to O(N), where N is the number of grid blocks. In contrast, conjugate gradient methods have a convergence rate $O(N^{\alpha})$, where α is greater than 1 and depends on problem dimension, pre-conditioning and ordering of the unknown pressures. Superior performance of multigrid over conjugate gradient methods for two dimensional problems greater than 33*33 have been reported [3].

One of the difficulties of applying multigrid to reservoir simulation is in determing how to treat the strongly varying coefficients which arise due to the permeability distribution. The strategies described in the literature fall into two camps:

(a) Cell vertex discretisations where multigrid interpolation must be operator dependent to ensure MG convergence, following Alcouffe et.al. [1]

(b) Cell centred discretisations where polynomial interpolation is found to be sufficient, following Wesseling et.al. [9,10]

In this paper we investigate the Wesseling method.

The purpose of this paper is two fold; first to report a modification to Wesseling's multigrid method which exploits spatial renormalization and demonstrates enhanced multigrid performance for some realistic permeability distributions. Secondly to introduce the notion of flux continuity for diagonal and full permeability tensors within a nine point scheme framework. While our focus here is in two dimensions, we do not anticipate any fundamental difficulty in extending our operators to three dimensions.

2 Multigrid methods for reservoir simulation

A key issue in the construction of reservoir simulation multigrid schemes is the development of operators which can cope with rapidly varying coefficients with large jumps of orders of magnitude and large numbers of interfaces which can occur in a randomly varying reservoir rock permeability maps.

Reservoir simulation multigrid literature has in the main addressed the diagonal tensor pressure equation using either cell vertex or cell centred formulations [3-8].

Cell vertex multigrid schemes have been based on the work of Alcouffe et. al. The permeability is defined at cell centres and flow variables including pressure are defined at the cell vertices or corner nodes. A control volume is constructed around a node and a face value permeability (transmissibility) is defined by an appropriate average of the cell centred permeabilities. Definition of coarse grid transmissibilities involves weighted averages of fine grid face values using homogenisation theory [1]. Authors of other multigrid schemes based on this approach do not explicitly discuss the definition of coarse grid transmissibilities, but leave it defined implicitly through the use of the Galerkin operator.

A common key ingredient in these schemes involves the definition of the prolongation operator which is constructed such that flux continuity is maintained. This construction is crucial to obtaining multigrid convergence for cell vertex schemes according to Alcouffe et al. and is employed by [3-8]. While all authors use a Galerkin formulation, and construct restriction operators from the adjoint of the resulting flux continuous prolongation operator, significantly the adjoint of the standard polynomial interpolation is found to be sufficient for difficult cases by Alcouffe et al. [1].

Cell centred multigrid schemes for elliptic equations with diagonal tensors and large variations in coefficients have been proposed by Kahil & Wesseling. Permeability and pressure are both defined at the cell centre (control volumes are the actual grid cells), and the face coefficients are defined by harmonic means of adjacent permeabilities for the pressure equation ensuring flux continuity on the finest (top level) grid, which is precisely the standard reservoir simulator scheme. In this multigrid scheme, coarse grid pressure locations are not embedded within fine grid locations, which is an important distinction between this approach and the cell vertex formulation. Coarse grid transmissibilities are obtained by a simple mean of the fine grid face values figure 1(a) as part of the Galerkin formulation. Namely,

$$\mathrm{K}^x_{i,j} = \frac{1}{2}(\mathrm{K}^x_{2i,2j} + \mathrm{K}^x_{2i,2j-1})$$

The prolongation operator is defined by polynomial approximation (a fundamental difference to the cell vertex approach), and restriction is defined by the adjoint of bilinear interpolation. However, since a coarse grid value lies at the centre of the corresponding four fine grid cells, it is possible to use piecewise constant prolongation and not violate flux contiuity constraints, as no neighbour information is required. We therefore conclude that this scheme is consistent with that of Alcouffe et al. since the prolongation operator does not violate flux continuity.

3 Renormalization

Fluid and rock properties such as permeability and porosity are measured on a very fine scale, generally ten orders of magnitude smaller than a typical reservoir grid block in two dimensions. The aim of renormalization is to replace fine scale properties by effective properties on a coarse grid cell. This is achieved by a hierarchical rescaling of the rock properties. The initial fine grid domain Ω is considered to comprise a set of local sub-domains Ω_m with boundary $\partial\Omega_m$, where Ω_m is of dimension $m \times m$. Each subdomain Ω_m is replaced by a single cell, by solving for the pressure field over Ω_m, subject to local boundary conditions on

$\partial\Omega_m$. The effective permeability for Ω_m is defined such that the flux through the outlet boundary of $\partial\Omega_m$ is equal to the product of the effective permeability and the global pressure gradient across Ω_m. The process is repeated in a hierarchical fashion after identifying the next level of subdomains of dimension $m \times m$.

For a diagonal tensor the choice $m = 2$ corresponds to the closed form analytic solution derived by a resistor network analogy [11]. This method has been well tried and proves to be extremely economic in deriving effective permeabilities at the reservior grid block scale. This 2*2 cell renormalization technique provides a natural permeability restriction operator both for grid adaptivity [12] and for use in multigrid.

4 Renormalisation coupled with multigrid

The multigrid scheme of Wesseling is the natural choice for Reservoir simulation with a cell centred discretisation. Use of a harmonic mean of neighbouring permeabilities ensures flux continuity on the finest grid. Piecewise constant prolongation does not violate flux continuity and use of the adjoint of bilinear interpolation as restriction for the Galerkin operator ensures that the necessary regularity condition ($m_p + m_r > 2m$) for convergence is obeyed. A certain parallel can be drawn with the scheme of Alcouffe et al., where it is observed that only prolongation need be flux continuous. However, the performance of this scheme can sometimes deteriorate with complexity in the media permeability, although convergence is still obtained.

Renormalization, with a Ω_m (2×2 subdomain), provides a natural mobility restriction operator, which fits neatly within the Wesseling scheme. By using renormalization prior to the multigrid solution algorithm effective permeability fields are determined on all grid levels. This facilitates the use of a flux continuous discrete operator on all grid levels via a harmonic mean of the respective neighbouring cell permeabilities for each level figure1(b). Namely

$$\mathrm{K}^x_{i,j} = \frac{2K^x_{i,j}K^x_{i+1,j}}{K^x_{i,j} + K^x_{i+1,j}}$$

5 Diagonal tensor results

Geostatistics and reservoir description are playing an increasingly important role in reservoir simulation. Well measurements, seismic and outcrop studies often provide the only hard facts, which are used as a basis for generating the most likely realization for a reservoir description via geostatistics technique. The resulting permeability fields are generated on a fine scale, far smaller than can be modelled by a reservoir simulator. Consequently, coarse scale effective permeabilities need to be derived by renormalization techniques and the test cases chosen are taken from realizations on a range of length scales.

The standard cell centred multi-grid scheme of Wesseling et al. is compared with renormalized multigrid employing local cell renormalization for a range of diagonal tensor test cases. Problems ranging from model reservoir simulation cases described in the literature to general cases generated by means of geostatitical techniques are presented.

To aid comparison and avoid complications from different applied boundary conditions all the test cases are solved with two isolated sources or wells in a quarter five spot configuration. In some case the appropriate physical problem may more realistically correspond to a vertical cross-section with a different source distribution. A fixed rate production well at the top right hand corner grid block and a pressure constrained injection well at the bottom left hand grid block are used in each of the examples. Consequently, f takes the form:

$$f = \delta_i \lambda(\phi - \phi_{bh}) - \delta_p q_p$$

where δ_i and δ_p are 1 in well blocks and zero elsewhere. λ and ϕ_{bh} are parameters describing the well connection factor and pressure respectively. This converts the pressure equation into a helmholtz equation with λ added onto the diagonal of the well block equation. The singular nature of the Neumann problem is lifted, but importantly for the success of the multi-grid method the implied global constraint must be implimented in discrete form. Since the discrete form of the diffusion operator sums to zero over the whole grid, the constant to be added to the fine grid solution after each interation [1] is given by,

$$-(f_{1,1} + f_{N,N} + \lambda\phi_{1,1})/\lambda$$

In this way the solution level is fixed without frustrating the smoothing process. The multgrid method used performs ten V-cycles with 1 pre-restriction and 2 post-prolongation smoothing sweeps. In all cases alternating line Gauss-Seidel is used as smoothing process.

The first two test cases are similar to standard test problems used for pressure solution comparisons [13,1]. Namely a modification of stones problem and the so-called staircase problem. The permeability fields are represented by a number of discontinuous regions. Both cases are given below and are represented in figure2.
Problem 1 : Stone's problem

$$K_A = \begin{pmatrix} 2 & 0 \\ 0 & 2 \end{pmatrix} K_B = \begin{pmatrix} 1 & 0 \\ 0 & 100 \end{pmatrix} K_C = \begin{pmatrix} 100 & 0 \\ 0 & 1 \end{pmatrix} K_D = \begin{pmatrix} 0 & 0 \\ 0 & 0 \end{pmatrix}$$

Problem 2 : Staircase problem

$$K_A = \begin{pmatrix} 100 & 0 \\ 0 & 100 \end{pmatrix} \quad K_B = \begin{pmatrix} 1 & 0 \\ 0 & 5 \end{pmatrix}$$

A summary of the convergence behaviour for the standard cell-centred multigrid scheme compared with renormalised multigrid are given in figure 4. The performance of both schemes is similar for Stone's problem. The simple variation in permeability offers little opportunity for renormalization to have any effect. For the staircase problem the renormalized multigrid offers an advantage.

The next three problems are taken from geostatistical applications with a randomly varying permeability field. A summary of the statistical properties is given below,

Problem 3 : Monet problem

$$K_{av} = \begin{pmatrix} 1.752 & 0 \\ 0 & 1.626 \end{pmatrix} \quad \mathrm{var}(K) = \begin{pmatrix} 3.002 & 0 \\ 0 & 2.596 \end{pmatrix}$$

Problem 4 : Cross-bed problem

$$K_{av} = 161.3 \qquad \mathrm{var}(K) = 0.6107 \times 10^5$$

Problem 5 : Channel sand problem

$$K_{background} = \begin{pmatrix} 1 & 0 \\ 0 & 1 \end{pmatrix} \quad K_{channel} = \begin{pmatrix} 10 & 0 \\ 0 & 10 \end{pmatrix} \quad \text{net to gross} = 0.6$$

The permeability distributions are also represented in figure 3. Problems 3 and 4 provide a useful comparison pair in that the permeability distributions correspond to a modest variance example and a more extreme choice. Problem 4 is taken from a cross-bed application.

The final example has been developed as a channel sand description and exhibits the features of meandering high permeability channels through a relatively low permeability background.

Figure 4 reports the corresponding convergence histories for each of the final problem examples. A significant improvement in convergence performance is achieved by the renormalization multigrid scheme compared to the standard cell centred scheme for problems 3 to 5. The permeability maps for these cases differ distinctly from each other while all involve a large number of discontinuities.

6 Full tensor equation and a new 9-point scheme

The need for a full tensor pressure equation arises since although at fine scales the permeability tensor is expected to be diagonal the process of re-scaling to practical reserviour simulation scales is expected to introduce off-diagonal terms in the effective permeability tensor representing fine scale cross flow in a coarse grid cell. Discretisation of the resulting cross-derivative terms in the general tensor equation requires an increase in the support of the scheme from five cells to nine cells. Extensions to a full permeability tensor has been the subject of number of recent contributions [14,15,16]. Except for [16] these schemes reduce to a five-point scheme in the diagonal tensor limit. The latter case is an unconventional scheme with interface pressures as unknowns and reducing to a seven-point scheme in the diagonal case.

Further motivation for a nine point scheme arises in the case of

(1) a diagonal tensor pressure equation, when modelling unfavourable mobility ratio floods. Numerical predictions of the resulting unstable flow, computed with a five point scheme can exhibit grid dependence (grid-orientation), which has been shown to be reduced by use of more accurate nine point schemes [17,18,19,20].

(2) solving the diagonal/full tensor equation on an arbitrary nonorthogonal quadrilateral grid, where use of a standard five point scheme(for the diagonal tensor) introduces an O(1) error in the flux for severely non-orthogonal grids.

To date, a general flux continuous nine point scheme which embodies these cases has not been derived. The new scheme described below achieves flux continuity for both diagonal and full tensor pressure equation by imposing pointwise pressure and flux continuity in the framework of a locally conservative 9-point finite volume formulation. The scheme is applicable for an arbitrary nonorthogonal quadrilateral grid. Here we illustrate the construction only for a uniform grid. Refering to figure 5(a) the scheme is derived by introducing four mean interface pressures at s, e, n and w at distances q_x and q_y (measured in local co-ordinate units $h_x/2$ and $h_y/2$) from the common vertex of the four neighbouring cells. The interface values are expressed in terms of the cell centred pressures ϕ_1, ϕ_2, ϕ_3 and ϕ_4, by introducing a piecewise linear approximation for pressure over each of the resulting triangle (shaded) in the figure and demanding pointwise continuity of flux and pressure at the four interface positions s, e, n and w. Since pressure gradients are piecewise constant with values in the notation of figure5(b) given by

$$\partial_x \phi = \frac{2}{h_x}(q_y\phi_j + (1-q_y)\phi_i - \phi_k)/d$$

$$\partial_y \phi = \frac{2}{h_y}(q_x\phi_j + (1-q_x)\phi_k - \phi_i)/d$$

where $d = 1-(1-q_x)(1-q_y)$. Each pointwise flux can be expressed as a linear combination of the four cell pressures. In practice, the elimination of the interface pressures is readily obtained by Gaussian elimination. Repetition of this procedure for each group of four cells in the grid leads to a nine point scheme which is flux continuous for both the diagonal and full tensor pressure equations.

In general a family of 9-point stencils is obtained dependent on q_x and q_y, from which the standard five point scheme is recovered with the choice $q_x = q_y = 1$.

The choice $q_x = q_y = 1/2$ produces the optimal sixth order accurate approximation for Laplace's equation.

$$\begin{bmatrix} 1/6 & 2/3 & 1/6 \\ 2/3 & -10/3 & 2/3 \\ 1/6 & 2/3 & 1/6 \end{bmatrix}$$

The final numerical result illustates the multigrid solution of the pressure equation for a full tensor problem in figure 6. The permeability fields are obtained by rotating the Monet problem by 45°. For this symmetric permeability tensor case the standard cell centred scheme is used with the full Galerkin construction of the coarse grid equations.

Renormalization must also be formulated to incorporate rescaling of the off diagonal components. A study of appropriate renormalization schemes for incorporation within a nine point multigrid scheme applied to the full tensor equation will be presented in a future paper together with further details of the flux continuous nine point scheme.

7 Conclusions

- Local cell renormalization is incorporated into the cell centred multigrid scheme of Wesseling et al, enabling flux continuity to be maintained on all grid levels.
- Significant improvement in convergence is obtained by the new multigrid scheme combined with renormalization, for a wide range of randomly varying discontinuous permeability fields.
- Piecewise constant prolongation preserves flux continuity and therefore provides some justification for use of polynomial interpolation in the cell centred multigrid method.
- A new flux continuous locally conservative nine point scheme is presented which applies to both diagonal and full tensor pressure equations on an arbitrary quadrilateral grid.

References

[1] R.E. Alcouffe, A. Brandt, J.E. Dendy Jr., and J.W. Painter, The Multi-grid method for the diffusion equation with strongly discontinuous coefficients. SIAM J. Sci. Stat. Comput., 2(430-454) (1981)

[2] K. Aziz and A. Settari, Petroleum reservoir simulation. Applied Science Publishers, London(1979)

[3] A. BEHIE and P.A. FORSYTH Jr., Multi-grid solution of the pressure equation in reservoir simulation. Soc.Pet.Eng.J., 23,623-632(1983)

[4] A. BEHIE and P.A. FORSYTH Jr., Multi-grid solution of three-dimensional problems with discontinuous coefficients. Appl.Math. Comp., 13,229-240(1983)

[5] T. SCOTT, Multi-grid methods for oil reservoir simulation in two and three dimensions. J.Comp.Phy.,59,290-307(1985)

[6] J.E. DENDY, S.F. MCCORMICK, J.W. RUGE, T.F. RUSSELL and S. SCHAFFER, Multigrid methods for three-dimensional petroleum reservoir simulation. SPE paper 18409(1989)

[7] F. BRAKHAGEN and T.W. FOGWELL, Multigrid methods in modelling porous media flow

[8] R. TEIGLAND and G.E. FLADMARK, Multilevel methods in porous media flow

[9] M. KHALIL, P. WESSELING, Vertex-centered and cell-centered multigrid methods for inteface problems. Report 88-42, Delft University of Technology.

[10] P. WESSELING, Cell-centered multigrid for interface problems. J. Comp. Phy.,79,85-91(1988)

[11] P.R. KING, The use of renormalization for the calculating effective permeability. Transport in Porous Media 4, 37-58(1989).

[12] M.G. EDWARDS and M.A. CHRISTIE, Dynamically adaptive godunov schemes with renormalisation in reservoir simulation. SPE Paper 25268, 413(1993).

[13] H.L. STONE, Iterative solution of implicit approximation of multidimensional partial differential equations. SIAM J. Numer. Anal., 5(3),530-558(1968)

[14] I. FAILLE, Control Volume Method to Model Fluid Flow on 2D Irregular Meshing, 2nd European Conference on the Mathematics of Oil Recovery,Arles,September 11-14,1990.

[15] G.I. SHAW, Discretisation and multigrid solution of pressure equations in reservoir simulation. Oxford University Computing Laboratory Report,(Sept.1992)

[16] M.J. KING, Application and Analysis of Tensor Permeability to Crossbedded Reservoirs. SPE Paper 26118(May 1993)

[17] J.L. YANOSIK and T.A. MCCRACKEN, A Nine-point, Finite-Difference reservoir simulator for realistic prediction of adverse mobility ratio displacements. Soc.Pet.Engr.J.,253-262(Aug.1979)

[18] K.H. COATS and A.B. RAMESH, Effects of grid type and difference scheme on pattern steamflood simulation results. SPE Paper 11079,(Sept.1982)

[19] K.H. COATS and A.D. MODINE, A consistent method for calculating transmissibilities in nine-point difference equations. SPE Paper 12248(1983)

[20] J.B. BELL, C.N. DAWSON and G.R. SHUBIN, An unsplit, higher order Godunov method for scalar conservation laws in multiple dimensions. J.Comp. Phys., 74,1-24(1988)

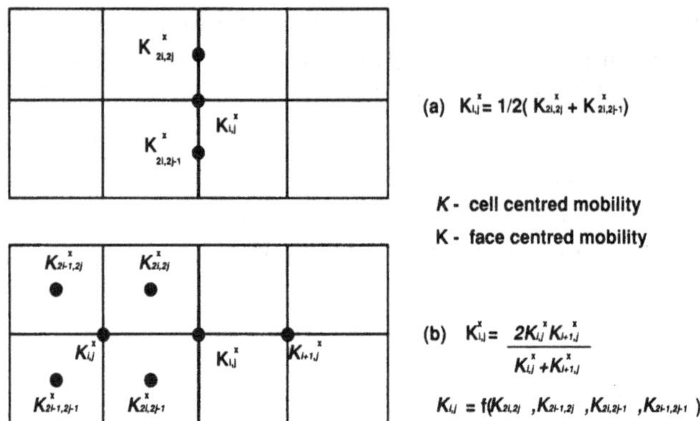

Figure 1 : Mobility Restriction

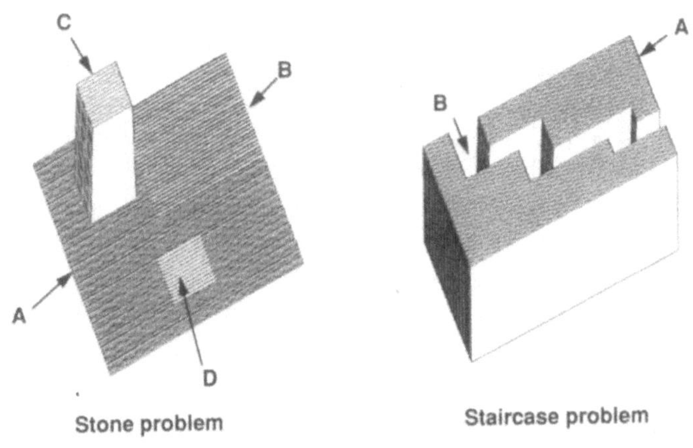

Figure 2:Permeability distribution

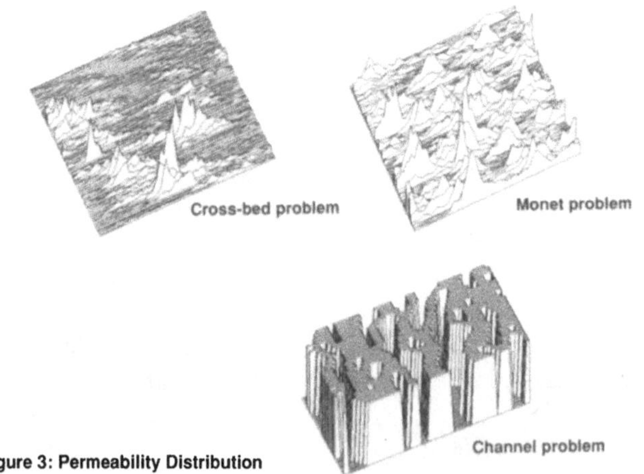

Figure 3: Permeability Distribution

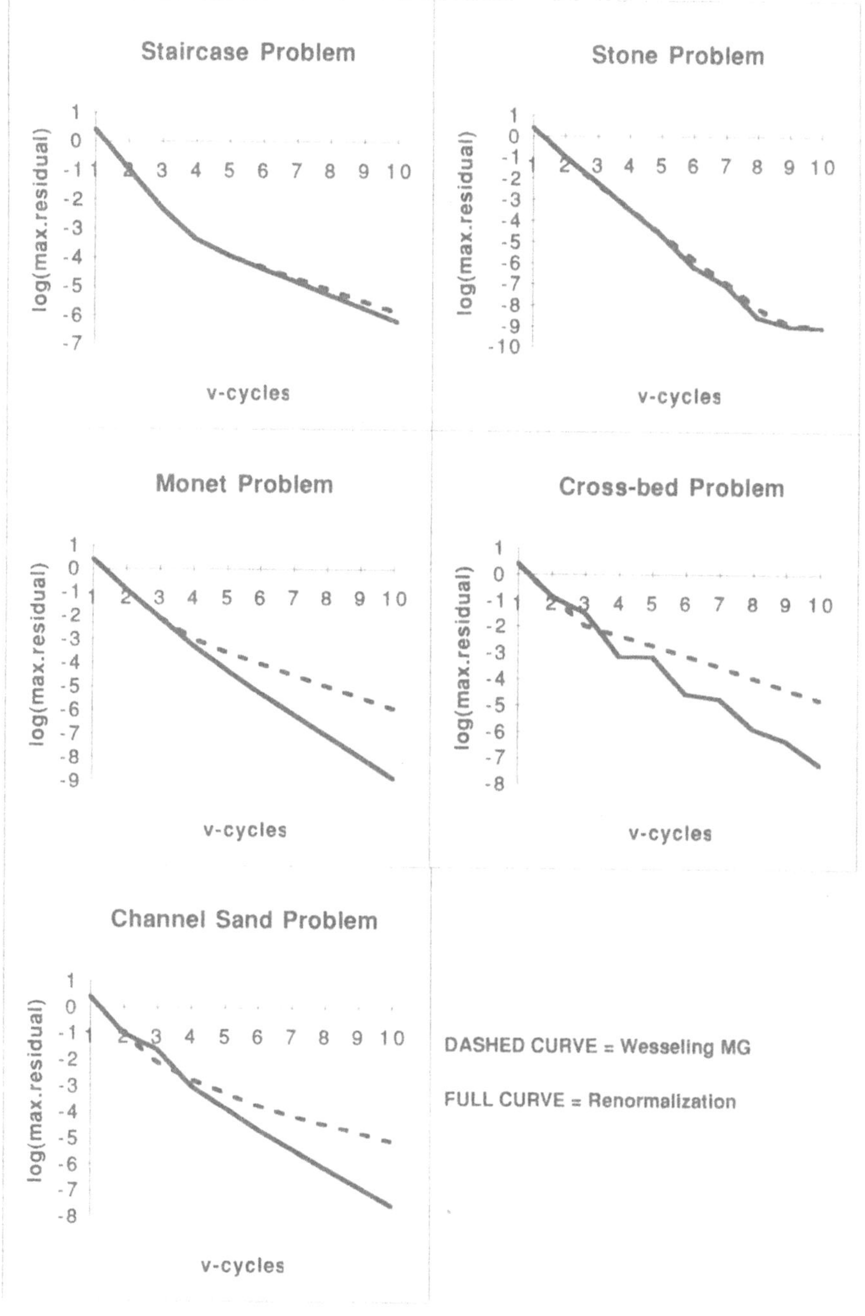
Multigrid Convergence
Staircase Problem
log(max.residual)
1
0
-1
-2
-3
-4
-5
-6
-7
1 2 3 4 5 6 7 8 9 10
v-cycles
Stone Problem
log(max.residual)
1
0
-1
-2
-3
-4
-5
-6
-7
-8
-9
-10
1 2 3 4 5 6 7 8 9 10
v-cycles
Monet Problem
log(max.residual)
1
0
-1
-2
-3
-4
-5
-6
-7
-8
-9
1 2 3 4 5 6 7 8 9 10
v-cycles
Cross-bed Problem
log(max.residual)
1
0
-1
-2
-3
-4
-5
-6
-7
-8
1 2 3 4 5 6 7 8 9 10
v-cycles
Channel Sand Problem
log(max.residual)
1
0
-1
-2
-3
-4
-5
-6
-7
-8
1 2 3 4 5 6 7 8 9 10
v-cycles
DASHED CURVE = Wesseling MG
FULL CURVE = Renormalization

Figure 4

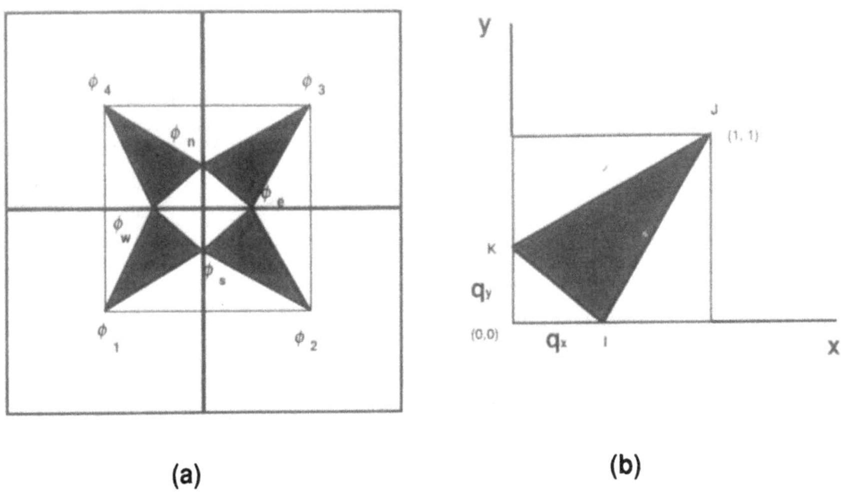

Figure 5 : Interpolation triangles for pointwise continuous 9-point scheme

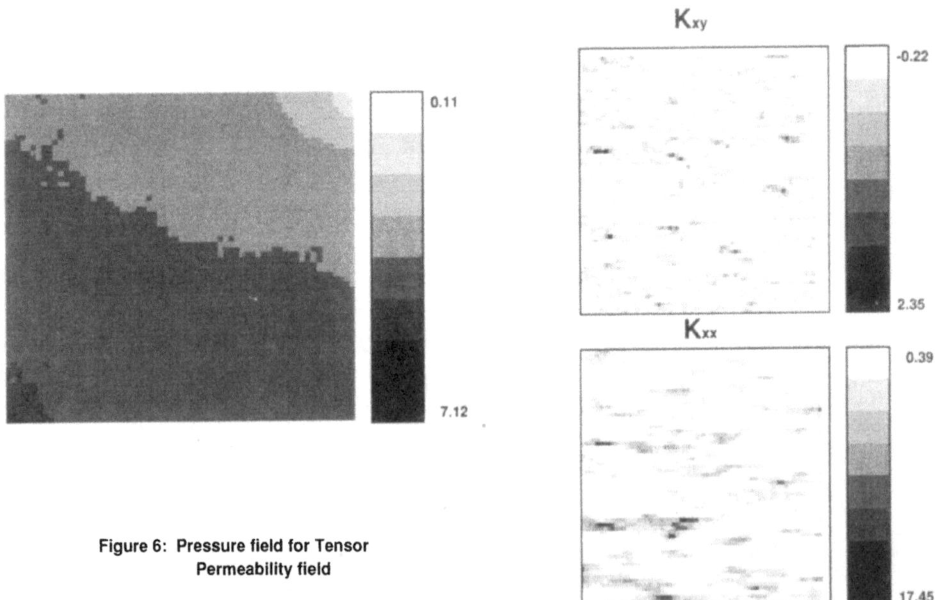

Figure 6: Pressure field for Tensor Permeability field

7

Interpolation and Related Coarsening Techniques for the Algebraic Multigrid Method

G. Golubovici and C. Popa[1]

1 Introduction

Let A be a symmetric and positive definite matrix and $b \in R^r$. We consider the system

$$Au = b, \tag{1}$$

with the (unique) exact solution $u \in R^r$. For $q \geq 2$ let $C_1, C_2, \ldots, C_q$ be a sequence of nonvoid subsets of $\{1, \ldots, r\}$ such that

$$\{1, \ldots, r\} = C_1 \supset C_2 \supset \cdots \supset C_q, \tag{2}$$

$$|C_m| = n_m, \quad m = 1, \ldots, q, \tag{3}$$

$$r = n_1 > n_2 > \ldots > n_q \geq 1, \tag{4}$$

where by $|C_m|$ we denoted the number of elements in the set C_m. Furthermore, for $m = 1, 2, \ldots, q-1$ we consider the matrices $A^1 = A$ and A^{m+1} and the linear operators

$$I_{m+1}^m : R^{n_{m+1}} \rightarrow R^{n_m}, \quad I_m^{m+1} : R^{n_m} \rightarrow R^{n_{m+1}}, \tag{5}$$

with the properties: I_{m+1}^m has full rank,

$$I_m^{m+1} = (I_{m+1}^m)^t, \tag{6}$$

$$A^{m+1} = I_m^{m+1} A^m I_{m+1}^m. \tag{7}$$

We also define the coarse grid correction operators T^m by

[1]University of Constantza, Department of mathematics, Bd. Mamaia, nr. 124, Constantza-8700, Romania

$$T^m = I_m - I^m_{m+1}(A^{m+1})^{-1}I^{m+1}_m A^m, \tag{8}$$

and the smoothing process

$$u^m_{new} = G^m u^m_{old} + (I_m - G^m)(A^m)^{-1}b^m, \tag{9}$$

where I_m is the identity and

$$A^m u^m = b^m \tag{10}$$

are the systems corresponding to the coarse levels.

With all the above defined elements we consider a classical V-cycle type algorithm (with one smoothing step performed after each coarse grid correcting step, see e.g. [5], [11]).

In what follows, in order to simplify the notations, we shall write n, p, I^n_p, I^p_n, C_p, A, A_p instead of n_m, n_{m+1}, I^m_{m+1}, I^{m+1}_m, C_m, A^m, A^{m+1}, respectively (where $m \in \{1,\dots,q-1\}$ is arbitrary fixed). We shall also suppose (without loss of generality) that the 'coarse grid' C_p is given by

$$C_p = \{n-p+1,\ n-p+2,\dots,n\} \tag{11}$$

Accordingly to (11) we consider the following block decomposition of A,

$$A = \begin{bmatrix} A_1 & B \\ B^t & A_2 \end{bmatrix}, \tag{12}$$

where we suppose that A_1 is symmetric and positive definite and A_2 is symmetric and invertible.

In the papers [1], [2], [12] methods are presented for preconditioning the systems (10), starting from a decomposition like (12). These methods use (at least at the end, when some conditions have to be fulfilled) the fact that the system (1) originates from a finite element or finite difference discretization of a partial differential equation (i.e. some 'geometric' information like: the discretization of a domain, the finite element basis functions etc.). For this reason it is very hard or even impossible to apply them for general ('algebraic') systems (1). On the other hand it is very hard to construct an efficient preconditioning method (which modifies the condition number of the matrix A) for arbitrary 'purely algebraic' problems (1). In the present paper we will present a preconditioning method (for general systems (1)), which does not modify the condition number of the matrix A, but ensures the fulfillment of the approximation assumption (see [3], [11] and relation (15) in the next section) in the algebraic multigrid V-cycle type algorithm presented above. In Section 2 we describe the method and we present some theoretical results (under the additional condition (30)). In Section 3 we present three particular cases for the preconditioning matrix. In Section 4 we describe a coarsening algorithm which ensures the fulfillment of the condition (30). Section

5 presents numerical examples with a 5-grid, V-cycle type algebraic algorithm for plane Dirichlet, Helmholtz and anisotropic Poisson problems. Beside the classical, variational (Galerkin) variant (relations (6)-(7)) we also used a truncated (non-Galerkin) version of the 5-grid algorithm, suggested by the fill-in process observed in the coarse grids matrices (beginning with the third level).

2 The preconditioning method

We shall consider on the level n (with the notation as in Section 1) the inner products

$$< u, v >_0 = < Du, v >; < u, v >_1 = < Au, v >; < u, v >_2 = < D^{-1}Au, Av >; \quad (13)$$

together with their corresponding norms $|| \cdot ||_i$, $i = 0, 1, 2$, (where D =diag (A), $<, >$ is the Euclidian inner product and $|| \cdot ||$ the Euclidian norm). Let v be an approximation of the exact solution u and $e = v - u$ the corresponding error. From [3], [11] we know that the convergence of the V-cycle type algebraic multigrid described in Section 1 (with a convergence factor, in the energy norm $|| \cdot ||_1$, independent of the dimension of the initial system (1)), is governed by the following two assumptions

$$||Ge||_1^2 \leq ||e||_1^2 - \alpha ||e||_2^2, \quad (14)$$

$$||Te||_1^2 \leq \beta ||e||_2^2, \quad (15)$$

where constants α, β exists independently of the level n and e.

REMARKS 1. Condition (14) is called *the smoothing assumption* and is fulfilled by the classical relaxation schemes (see [3], [8], [11]).

2. Condition (15), *the approximation assumption*, cannot be easily obtained (some results are known for M-matrices of positive type, see [3], [11]).

Let us now start with the decomposition (12) of A and with another arbitrary symmetric and positive definite matrix $\bar{A}^1$. Let

$$A_1 = L_1 L_1^t, \quad \bar{A}^1 = \bar{L}_1 \bar{L}_1^t, \quad (16)$$

be the Cholesky decompositions of A_1 and $\bar{A}_1$ (with L_1, $\bar{L}_1$ lower triangular). We define the matrix $\bar{\Delta}_1$ by

$$\bar{\Delta}_1 = \begin{bmatrix} \bar{L}_1 L_1^{-1} & 0 \\ 0 & I_2 \end{bmatrix}, \quad (17)$$

and we precondition the system (1) in the following way

$$(\bar{\Delta}_1 A \bar{\Delta}_1^t) \quad (\bar{\Delta}_1^{-t} u) = \bar{\Delta}_1 b \tag{18}$$

Thus, the system (1) becomes

$$\bar{A}\bar{u} = \bar{b} \tag{19}$$

where

$$\bar{A} = \bar{\Delta}_1 A \bar{\Delta}_1^t = \begin{bmatrix} \bar{A}_1 & \bar{B} \\ \bar{B}^t & A_2 \end{bmatrix}, \quad \bar{u} = \bar{\Delta}_1^{-t} u, \quad \bar{b} = \bar{\Delta}_1 b, \tag{20}$$

and the $(n-p) \times p$ matrix $\bar{B}$ is given by

$$\bar{B} = \bar{L}_1 \bar{L}_1^{-1} B. \tag{21}$$

REMARKS 1. The preconditioned matrix $\bar{A}$ is also symmetric and positive definite; thus we can consider inner products like in (13) and the associated norms which we shall denote by $|||\cdot|||_i$, $i = 0, 1, 2$.

2. We shall use the notations I_1, I_2, $u = [u_1, u_2]$ for the identities on R^{n-p}, R^p and for $u \in R^n = R^{n-p} \oplus R^p$, respectively. We shall also denote by $\rho(S)$, $\lambda_{\min}(S)$ the spectral radius and the smallest eigenvalue, respectively, for a positive definite matrix S. Defining the interpolation I_p^n by

$$I_p^n = \begin{bmatrix} -\bar{A}_1^{-1}\bar{B} \\ I_2 \end{bmatrix} = \begin{bmatrix} -\bar{L}_1^{-t} L_1^{-1} B \\ I_2 \end{bmatrix}, \tag{22}$$

we observe that it has full rank and we obtain the following result.

PROPOSITION 1. *(i) The coarse grid matrix A_p is independent of the preconditioning matrix $\bar{A}_1$ and is given by*

$$A_p = A_2 - B^t A_1^{-t} B \tag{23}$$

(i.e. the Schur's complement of A).

(ii) If $\bar{e} = [\bar{e}_1, \bar{e}_2]$ is the error for the preconditioned system (19), after the correction step, then

$$|||\bar{e}|||_1^2 \geq \lambda_{\min}(\bar{A}_1)||\bar{e}||^2. \tag{24}$$

PROOF. (i) We first observe from (22), (20) and (6)-(7) that

$$I_n^p \bar{A} = [0; \tilde{A}_2], \tag{25}$$

with

$$\tilde{A}_2 = A_2 - \bar{B}^t \bar{A}_1^{-1} \bar{B} \tag{26}$$

and

$$A_p = I_n^p \bar{A} I_p^n = \tilde{A}_2. \tag{27}$$

Then, using (16) and (21), the equality (23) follows from (26)-(27).

(ii) For the error after the correction step we have (see e.g. [5])

$$I_n^p \bar{A} \bar{e} = 0. \tag{28}$$

From (28), using (25) and (27), we obtain

$$A_p \bar{e}_2 = 0 \Rightarrow \bar{e}_2 = 0, \tag{29}$$

because A_p is invertible. Then (24) is obvious and the proof is complete. □

At this moment we shall formulate the following assumptation: there exists a constant $\gamma > 0$, independently of the dimension n of A such that

$$\lambda_{\min}(A_1) \geq \gamma, \quad \lambda_{\min}(\bar{A}_1) \geq \gamma. \tag{30}$$

THEOREM 1. *If the conditions (30) are satisfied, for every vector $e \in R^n$ we have*

$$|||Te|||_1^2 \leq \beta \cdot |||e|||_2^2, \tag{31}$$

with $\beta > 0$ given by

$$\beta = \left[\gamma \cdot \max_{1 \leq i \leq n} \sum_{j=1}^{n} |a_{ij}|\right]^{-2} \cdot \frac{\max\limits_{1 \leq i \leq n-p} \bar{a}_{ii} \cdot \min\limits_{1 \leq i \leq n-p} \bar{a}_{ii}}{\gamma \cdot \min\limits_{n-p+1 \leq i \leq n} a_{ii}}, \tag{31'}$$

where we denote by $a_{ij}, \bar{a}_{ij}$ the elements of the matrices $A_1, \bar{A}_1$, respectively.

PROOF. It is not hard to observe that, if (30) holds, then we have

$$||\bar{A}_1^{-1} \bar{B}|| \leq C, \tag{32}$$

with $C > 0$ given by

$$C = \left[\gamma \cdot \max_{i \leq i \leq n} \sum_{j=1}^{n} |a_{ij}|\right]^{-1}. \tag{33}$$

Then, if

$$\bar{e} = [\bar{e}_1, \bar{e}_2] = Te \tag{34}$$

is the error after the correction step, using (24) and (29) we obtain

$$|||\bar{e}|||_1^2 \geq \lambda_{\min}(\bar{A}_1) \quad ||e_1||^2 = \lambda_{\min}(\bar{A}_1) \quad ||\bar{e}||^2 \geq$$

$$\geq \frac{\lambda_{\min}(\bar{A}_1)}{\min\limits_{1\leq i\leq n-p} \bar{a}_{ii}} \cdot |||\bar{e}|||_0^2. \tag{35}$$

But, a simple calculation using the Cauchy-Schwarz inequality, yields (using again (29))

$$|||\bar{e}|||_1^2 \leq |||\bar{e}|||_2 \cdot |||\bar{e}|||_0. \tag{36}$$

From (34)-(36) we obtain

$$|||Te|||_1^2 \leq \frac{\min\limits_{1\leq i\leq n-p} \bar{a}_{ii}}{\lambda_{\min}(\bar{A}_1)} \cdot |||Te|||_2^2. \tag{37}$$

From (8) and the symmetry of $\bar{A}$ we observe that

$$\bar{A}T = T^t\bar{A}. \tag{38}$$

Then, if we denote $\bar{D} = diag(\bar{A})$ and E is the matrix

$$E = \bar{D}^{1/2}T\bar{D}^{-1/2}, \tag{39}$$

we obtain

$$|||\bar{e}|||_2^2 = < \bar{D}^{-1}\bar{A}Te, \bar{A}Te > \leq \rho(EE^t)|||e|||_2^2. \tag{40}$$

But, if K is the matrix

$$K = \bar{D}_1^{1/2}\bar{A}_1^{-1}\bar{B}\bar{D}_2^{-1/2}, \tag{41}$$

we observe that

$$EE^t = \begin{bmatrix} I_1 + KK^t & 0 \\ 0 & 0 \end{bmatrix}. \tag{42}$$

Then, using (32)-(33), we have

$$\rho(EE^t) \leq 1 + ||K||^2 \leq \frac{\max\{\bar{a}_{ii},\ 1 \leq i \leq n-p\}}{\min\{a_{ii},\ n-p+1 \leq i \leq n\}} \cdot C^2. \tag{43}$$

From (40) and (43) we obtain

$$|||Te|||_2^2 \leq \frac{\max\{\bar{a}_{ii},\ 1 \leq i \leq n-p\}}{\min\{a_{ii},\ n-p+1 \leq i \leq n\}} \cdot C^2 \cdot |||e|||_2^2. \tag{44}$$

From (37) and (44) results (31) with β from (31′). □

REMARKS 1. If (30) holds, we can see that β from (31′) depends only on the elements of the matrices A and $\bar{A}_1$. Thus, the approximation assumption (31) for the preconditioned system (19) holds.

2. If A is weakly diagonally dominant, accordingly to the decomposition (12), condition (30) holds if we ask for in the case of the arbitrary matrix $\bar{A}_1$ and if A_1 is relatively sparse (see an algorithm for obtaining this in Section 4).

3. According to (31) the approximation assumption (15) holds for the preconditioned system (19). It can be proved (see [9]) that if the smoothing assumption (14) holds for the initial system (1) with a constant $\alpha > 0$, then it also holds (with the same constant α) for the preconditioned one (19).

3 Three particular cases

CASE I. $\bar{A}_1 = A_1$. In this case $\bar{A} = A$, thus no preconditioning occurs. Condition (30) will hold if, e.g. A_1 is strictly diagonally dominant, i.e.

$$\nu_i = a_{ii} - \sum_{j \neq i} |a_{ij}| > 0, \quad i = 1, \ldots, n-p. \tag{45}$$

Then, γ from (30) can be taken as

$$\gamma = \min\{\nu_i, \ i = 1, \ldots, n-p\}. \tag{46}$$

The interpolation operator I_p^n will be given by

$$I_p^n = \begin{bmatrix} -A_1^{-1} B \\ I_2 \end{bmatrix}, \tag{47}$$

and it is possible to obtain the product $A_1^{-1}B$ without inverting the matrix A_1 (see [7]).

CASE II. $\quad \bar{A}_1 = diag(d_1, d_2, \ldots, d_{n-p})$ with $d_i > 0$, $i = 1, \ldots, n-p$. (48)

Then

$$\bar{L}_1 = \bar{L}_1^t = diag(\sqrt{d_1}, \sqrt{d_2}, \ldots, \sqrt{d_{n-p}}), \tag{49}$$

and

$$I_p^n = \begin{bmatrix} -\bar{L}_1^{-1} L_1^{-1} B \\ I_2 \end{bmatrix}. \tag{50}$$

In order to obtain the product $L_1^{-1}B$ (with L_1 the Cholesky factor of A_1, from (16)) we make a Gaussian elimination (without pivoting and making 1 on the diagonal) on the first $n-p$ rows of A. In this way we obtain the matrix

$$\tilde{A} = \begin{bmatrix} \tilde{A}_1 & \tilde{B} \\ B^t & A_2 \end{bmatrix}, \tag{51}$$

where

$$A_1 = \tilde{L}_1 \tilde{A}_1 \tag{52}$$

is an LU-decomposition of A_1 ($\tilde{A}_1$ upper triangular and with 1 on its diagonal) and

$$\tilde{B} = \tilde{L}_1^{-1} B. \tag{53}$$

Then, if $\tilde{D}_1 = diag(\tilde{L}_1) = diag(\tilde{l}_{11}, \tilde{l}_{22}, \dots, \tilde{l}_{n-p,n-p})$ it is obvious that

$$L_1^t = \tilde{D}_1^{1/2} \tilde{A}_1. \tag{54}$$

The elements of the matrix $\tilde{D}_1$ can be recursively obtained by the formulas

$$a_{11} = \tilde{l}_{11}, \quad a_{ii} = \tilde{l}_{ii} + \sum_{k=1}^{i-1} \tilde{l}_{kk} \tilde{a}_{ki}^2, \quad i = 2, \dots, n-p, \tag{55}$$

(where $\tilde{a}_{ij}$ are the elements of $\tilde{A}_1$). Then we have

$$L_1^{-1} B = \tilde{D}_1^{1/2} \tilde{B}. \tag{56}$$

The constant γ from (30) can be taken as

$$\gamma = \min\{\nu_i, d_i, \ i = 1, \dots, n-p\}. \tag{57}$$

Case III. $\bar{A}_1 = A_1 + R_1$, where

$$A_1 = \bar{A}_1 - R_1 \tag{58}$$

is an incomplete Cholesky decomposition of A_1 (if A_1 is supposed to be an M-matrix, cf. [6]). The factors $\bar{L}_1, \bar{L}_1^t$ are obtained during this decomposition. We know from [6] that

$$\rho = \rho(\bar{A}_1^{-1} R_1) < 1. \tag{59}$$

The constant γ from (30) will depend on the number ρ (the case $\rho = 0$ means $\bar{A}_1 = A_1$ i.e. our particular case I).

4 The coarsening algorithm

In this section we will come back at the decomposition (11)-(12) and formulate the following question: if the matrix A is weakly diagonally dominant, how to obtain the splitting (11)-(12) of A such that A_1 be strictly diagonally dominant and A_2 be the matrix corresponding to the connections between the coarse grid points (satisfying the general rules formulated in [3], [11]). In what follows we shall briefly describe an algorithm which tries to answer at this question. Details can be found in [10]. We shall use the following notations: for $i \in \{1, \ldots, n\}$

$$N(i, A) = \{j \in \{1, \ldots, n\} \mid j \neq i, a_{ij} \neq 0\},$$

$$N^t(i, A) = \{j \in \{1, \ldots, n\} \mid i \in N(j, A)\},$$

and for $\Delta \subset \{1, \ldots, n\}$ a nonempty subset

$$N(i, \Delta, A) = N(i, A) \cap \Delta, \quad N^t(i, \Delta, A) = N^t(i, A) \cap \Delta.$$

Then we have

ALGORITHM $\mathcal{C}$

Step 1. Set $k = 1$ and $\Delta^k = \{1, \ldots, n\}$.

Step 2. Set $\alpha = 1$, $C^k = \emptyset$, $F^k = \emptyset$, $\Delta_\alpha = \Delta^k$.

Step 3. For all $i \in \Delta_\alpha$ set

$$\lambda_i = |N^t(i, \Delta_\alpha, A)|, \ \lambda = \max\{\lambda_i, \ i \in \Delta_\alpha\}.$$

Step 4. a. If $\lambda = 0$ set $F^k = \Delta_\alpha$ and go to step 6.
b. If $\lambda \neq 0$ define the set $I_\alpha \subset \Delta_\alpha$ by $I_\alpha = \{i \in \Delta_\alpha, \lambda_i = \lambda\}$.
c. If $I_\alpha \subset F^k$ set $\bar{\Delta} = \Delta^k$, $k = k + 1$, $\Delta^k = \Delta \backslash C^k$ and go to step 2.
d. If $I_\alpha \not\subset F^k$ set $I_\alpha = I_\alpha \backslash F^k$ and for each $i \in I_\alpha$ do:
–if $i \in F^k$ go to another index from I_α.
–if $i \notin F_k$ set

$$C^k = C^k \cup \{i\},$$

$$F^k = F^k \cup N^t(i, \Delta_\alpha \backslash C^k, A).$$

Step 5. Set $\bar{\Delta} = \Delta_\alpha$, $\alpha = \alpha + 1$, $\Delta_\alpha = \bar{\Delta} \backslash C^k$ and go to step 3.

Step 6. Set $C = C^1 \cup C^2 \cup \ldots \cup C^k$

$$F = F^k$$

and Stop.

After the application of the algorithm $\mathcal{C}$ we obtain a partition $\{C, F\}$ of the set $\{1, \ldots, n\}$. If we reorder the indices in $\{1, \ldots, n\}$ beginning with those from the set F, then the set C will become C_p from (11) and the submatrix of A defined by the connections between the F-points will be A_1 from (12).

REMARKS 1. We have to observe (see also [10]) that no information concerning the symmetry of A was needed. Thus we can apply the algorithm $\mathcal{C}$ to arbitrary matrices and we can use the idea of strong - connections (from [3], [11]).

2. We used the above algorithm (with some improvements) for obtaining the coarse grids in the examples presented in Section 5.

5 Numerical examples. The non-Galerkin approach

We consider the following plane problems

$$\text{Dirichlet: } \begin{cases} -\Delta u = f & in \quad \Omega, \\ u = 0 & on \quad \partial\Omega, \end{cases}$$

$$\text{Helmholtz: } \begin{cases} \Delta u + k^2 u = f & in \quad \Omega, \\ u = 0 & on \quad \partial\Omega, \end{cases}$$

$$\text{Anisotropic Poisson: } \begin{cases} -\epsilon \dfrac{\partial^2 u}{\partial x^2} - \dfrac{\partial^2 u}{\partial y^2} = f & in \quad \Omega, \\ u = 0 & on \quad \partial\Omega, \end{cases}$$

with $\Omega = (0,1)^2 \subset R^2$, discretized by classical 5-points stencils (see e.g. [5]). We used two different initial (finest grid) discretizations (corresponding to mesh sizes h=1/14 and h=1/32) and a 5-grid V-cycle type algebraic multigrid. We applied the preconditioning from cases I and II (Section 3) with coarsening made by algorithm $\mathcal{C}$ (Section 4) with some improvements (see [10]). Initially we used the Galerkin approach (6)-(7). The results are presented in tables for TOL=0.0. We observed a fill-in process starting at the 3rd-level. Thus, we decided to change the Galerking approach in the following way: for coarsening on level 3 we only took into account 'connections' (in absolute value) ≥ 0.1. After that we truncated the matrices A_1 and B from (12) (corresponding to the 3rd level) such that we kept only the elements (in absolute value) larger than a parameter TOL. We constructed the 'truncated' interpolation operator (in cases I and II see Section 3) using this 'truncated' matrices. After that we defined the fourth level matrix by (7) but with the above 'truncated' version of interpolation and restriction. The fifth level matrix was constructed in the same way starting with the 'truncated' version of the fourth level matrix. The results are presented in Tables 1-6 for different values of the truncation parameter TOL. We also indicated the worst step reduction factor ρ for obtaining a desired precision (Euclidean norm of the error $\leq 10^{-6}$).

TOL	0.0	10^{-4}	10^{-3}	10^{-2}	10^{-1}	0.5	1.
ρ for case I	0.051	0.051	0.051	0.055	0.1	0.34	0.35
ρ for case II	0.19	0.19	0.19	0.195	0.22	0.4	0.4

TABLE 1. The Dirichlet problem, h=1/14.

TOL	0.0	10^{-4}	10^{-3}	10^{-2}	10^{-1}	0.5	1.
ρ for case I	0.078	0.078	0.078	0.16	0.53	0.82	0.82
ρ for case II	0.40	0.40	0.41	0.46	0.65	0.84	0.84

TABLE 2. The Dirichlet problem, h=1/32.

TOL		0.0	10^{-3}	10^{-2}	10^{-1}
ρ for case I	$k^2 = 4$	0.054	0.054	0.058	0.13
	$k^2 = 19$	0.058	0.059	0.061	0.8
	$k^2 = 25$	0.09	0.09	0.06	0.81
	$k^2 = 30$	0.37	0.37	0.38	0.39
ρ for case II	$k^2 = 4$	0.21	0.21	0.22	0.26
	$k^2 = 10$	0.27	0.27	0.27	0.35
	$k^2 = 25$	0.48	0.49	0.5	0.51
	$k^2 = 30$	0.74	0.74	0.77	0.68

TABLE 3. The Helmholtz problem, h=1/14.

References

[1] AXELSSON, O., VASSILEVSKI, P.S. Algebraic multilevel preconditioning methods. Part I: Numer. Math., 56, 157-177 (1989); Part II: SIAM J. Numer. Anal., 27, no. 6, 1569-1590 (1990); Part III: Report 9045, Dept. of Math., Catholic Univ. Nijmegen (1990).

TOL		0.0	10^{-4}	10^{-3}	10^{-2}	10^{-1}
ρ for case I	$k^2 = 19$	0.077	0.076	0.22	0.81	>1
	$k^2 = 55$	0.34	0.34	0.34	0.44	>1
	$k^2 = 100$	0.83	0.83	0.81	0.82	>1
ρ for case II	$k^2 = 19$	0.56	0.56	0.63	0.9	>1
	$k^2 = 55$	0.80	0.80	0.80	0.69	>1
	$k^2 = 100$	0.97	0.97	0.96	0.95	>1

TABLE 4. The Helmholtz problem, h=1/32.

TOL		0.0	10^{-4}	10^{-3}	10^{-2}	10^{-1}
ρ for case I	$\epsilon = 10^{-1}$	0.052	0.052	0.052	0.056	0.11
	$\epsilon = 10^{-2}$	0.052	0.052	0.052	0.057	0.11
	$\epsilon = 10^{-6}$	0.054	0.054	0.054	0.06	0.15
ρ for case II	$\epsilon = 10^{-1}$	0.19	0.19	0.19	0.20	0.23
	$\epsilon = 10^{-2}$	0.20	0.20	0.20	0.20	0.24
	$\epsilon = 10^{-6}$	0.23	0.23	0.23	0.23	0.29

TABLE 5. The Poisson problem, h=1/14.

TOL		0.0	10^{-4}	10^{-3}	10^{-2}	10^{-1}
ρ for case I	$\epsilon = 10^{-1}$	0.078	0.078	0.079	0.16	0.55
	$\epsilon = 10^{-2}$	0.078	0.078	0.079	0.17	0.56
	$\epsilon = 10^{-6}$	0.079	0.078	0.079	0.2	0.63
ρ for case II	$\epsilon = 10^{-1}$	0.41	0.41	0.41	0.47	0.66
	$\epsilon = 10^{-2}$	0.41	0.41	0.42	0.48	0.67
	$\epsilon = 10^{-6}$	0.42	0.43	0.43	0.51	0.73

TABLE 6. The Poisson problem, h=1/32.

[2] AXELSSON, O., NEYTCHEVA, M. Algebraic multilevel iteration method for Stieltjes matrices. Report 9102, Dept. of Math., Catholic Univ. Nijmegen (1991).

[3] BRANDT, A. Algebraic multigrid theory: the symmetric case. Preprint, Weizmann Inst. of Science, Rehovot (1983).

[4] BRANDT, A., TA'ASSAN, S. Multigrid method for nearly singular and slightly indefinite problems. In "Multigrid methods II", ed. W. HACKBUSCH and U. TROTTENBERG, Lecture Notes in Math., vol. 1228, Springer Verlag, Berlin (1986).

[5] HACKBUSCH, W. Multigrid methods and applications. Springer Verlag, Berlin (1985).

[6] MEIJERINK, J.A., VAN DER VORST, H.A. An iterative solution method for linear systems of which the coefficient matrix is a symmetric M-matrix, Math. of Comput., 31 (1977) 148-162.

[7] POPA, C. ILU decomposition for coarse grid correction step on algebraic multigrid. Paper presented at the 3rd European Conference on Multigrid Methods, Bonn, 1990. In GMD-Studien, nr. 189, 1991.

[8] POPA, C. On smoothing properties of SOR relaxation for algebraic multigrid method. (Studii si Cerc. Mat., Ed. Academiei Române, 5 (1989) 399-406.

[9] POPA, C. Preconditioning for the fulfillment of the approximation assumption in the algebraic multigrid method (unpublished paper, presented as an internal communication at Institut für Informatik und Praktische Mathematik, Univ. of Kiel, January 1993).

[10] POPA, C. Coarsening algorithms for the algebraic multigrid method (as [9]).

[11] RUGE, J., STÜBEN, K. Algebraic multigrid. In "Multigrid methods", S. MCCORMICK ed., SIAM, Philadelphia, 1987.

[12] YSERENTANT, H. Hierarchical bases of finite-element spaces in the discretization of nonsymmetric elliptic boundary value problems. Computing, 35 (1985) 39-49.

8

Parallel Point-oriented Multilevel Methods

Michael Griebel[1]

ABSTRACT Instead of the usual nodal basis, we use a generating system for the discretization of PDEs that contains not only the basis functions of the finest level of discretization but additionally the basis functions of all coarser levels of discretization. The Galerkin-approach now results in a semidefinite system of linear equations to be solved. Standard iterative GS-methods for this system turn out to be equivalent to elaborated multigrid methods for the fine grid system.
Beside Gauss-Seidel methods for the level-wise ordered semidefinite system, we study block Gauss-Seidel methods for the point-wise ordered semidefinite system. These new algorithms show basically the same properties as conventional multigrid methods with respect to their convergence behavior and efficiency. Additionally, they possess interesting properties with respect to parallelization. Regarding communication, the number of setup steps is only dependent on the number of processors and not on the number of levels like for parallelized multigrid methods. The amount of data to be communicated, however, increases slightly. This makes our new method perfectly suited to clusters of workstations as well as to LANs and WANs with relatively dominant communication setup.

1 Introduction

Recently, see [4], [6], a new concept for the development of multigrid and BPX-like multilevel algorithms has been presented. There, instead of a basis approach on the finest grid and the acceleration of the basic iteration by MG-coarse grid correction or a BPX-type preconditioner, a generating system is used to allow a non-unique level-wise decomposed representation of the solution. The degrees of freedom are associated to the nodal basis functions of all levels under consideration. With this non-unique multilevel decomposed representation of a function, the Galerkin-approach leads to a semidefinite linear system with unknowns on all levels. Its solution is non-unique but in some sense equivalent to the unique solution of the

[1]Institut für Informatik, Technische Universität München, Arcisstraße 21, D-80290 München, email: griebel@informatik.tu-muenchen.de
This work is supported by the Bayerische Forschungsstiftung via FORTWIHR - Bayerischer Forschungsverbund für technisch- wissenschaftliches Hochleistungsrechnen.

standard problem on the finest grid.

Furthermore, it has been shown that traditional iterative methods for the semidefinite system are equivalent to modern elaborated multilevel methods applied to the standard system which exhibit optimal convergence properties. The conjugate gradient method (with appropriate diagonal scaling) for the semidefinite system is equivalent to the BPX-conjugate gradient method for the fine grid system. Gauss-Seidel-type iterations for the semidefinite system are equivalent to certain multigrid methods. For details, see [4], [6]. These methods are grid- or level-oriented and can be considered as level block techniques. An outer iteration switches from level to level, and an inner iteration operates on the specific grid.

Now, we consider the semidefinite system from a different point of view. We group together all unknowns which are associated to the same grid point. This results in a point-oriented method and can be considered as a point block technique. Now, an outer iteration switches from grid point to grid point. The local system that belongs to all basis functions of different levels centered in the same grid point can be solved either directly or by an inner iteration that runs over all levels that are associated to the grid point under consideration. Furthermore, grid points can be grouped together to form subdomains. In this sense, we get some sort of simple domain decomposition method which exhibits MG-like convergence properties. Compare also [3], [5], [6].

In contrast to the parallelization of a multilevel method where communication has to take place on all levels, our point block approach needs substantially less setup steps for the communication due to its domain decomposition qualities. In this sense, our new method is superior to other parallel multigrid and multilevel methods and very well suited to clusters of workstations or LANs and WANs with relatively dominant communication setup.

2 The semidefinite system

In this section, we introduce a generating system that replaces the usual finite element basis in the discretization of a boundary value problem of a partial differential equation. We then derive the associated semidefinite linear system and discuss its properties.

Consider a partial differential equation in d dimensions with a linear, second-order operator L on the domain $\Omega = (0,1)^d$, d=1,2,...,

$$Lu = f \text{ on } \Omega, \tag{1}$$

with appropriate boundary conditions and corresponding solution u. For reasons of simplicity, we restrict ourselves to homogeneous Dirichlet boundary conditions. Given an appropriate function space V, the problem can be expressed equivalently in its variational form: Find a function $u \in V$ with

$$a(u,v) = (f,v) \quad \forall\, v \in V. \tag{2}$$

(In the case of homogeneous Dirichlet boundary conditions, V would be the Sobolev space $H_0^1(\Omega)$.) Here, a: $V \times V \to \mathbb{R}$ is a bounded, positive-definite, symmetric bilinear form and (.,.) is the linear form for the right-hand side. Let $\|.\|_a := \sqrt{a(.,.)}$ denote the induced energy norm. We assume that V is complete with respect to $\|.\|_a$, which is true if $a(.,.)$ is H_0^1-elliptic. The Lax-Milgram lemma then guarantees the existence and uniqueness of the solution of (2). If we consider directly the functional $J(u) := 1/2 \cdot a(u,u) - (f,u)$, the problem can be stated alternatively as minimization of $J(u)$ in V.

2.1 Spaces, bases, and the generating system

Assume we are given a sequence of uniform, equidistant, and nested grids

$$\Omega_1 \subset \Omega_2 \subset ... \subset \Omega_{k-1} \subset \Omega_k \tag{3}$$

on $\bar{\Omega}$ with respective mesh sizes $h_l = 2^{-l}$, $l=1,..,k$, and an associated sequence of spaces V_l of piecewise d-linear functions,

$$V_1 \subset V_2 \subset .... \subset V_{k-1} \subset V_k, \tag{4}$$

with dimensions

$$n_l := dim(V_l) = (2^l - 1)^d, l = 1, .., k. \tag{5}$$

Here, 1 denotes the coarsest and k the finest level of discretization. Consider also the sequence

$$N_1 \subset N_2 \subset ... \subset N_{k-1} \subset N_k \tag{6}$$

of sets of inner grid points $N_l = \{x_1, ..., x_{n_l}\}$ of the grid Ω_l, $l=1,..,k$.

The standard finite element basis, that spans V_l on the equidistant grid Ω_l, is denoted by B_l. It contains the nodal basis functions $\phi_i^{(l)}$, $i=1,..,n_l$, which are defined by

$$\phi_i^{(l)}(x_j) = \delta_{i,j}, \quad x_j \in N_l. \tag{7}$$

Now, any function $u \in V_k$ can be expressed uniquely by

$$u = \sum_{\phi \in B_k} u_\phi \cdot \phi \tag{8}$$

with the vector $u_k := (u_\phi)_{\phi \in B_k}$ of nodal values associated to some given ordering of the functions of B_k.

In contrast to this conventional basis approach we now consider the set of functions E_k defined as the union of all the different nodal bases B_l for the levels $l=1,..,k$,

$$E_k = B_1 \cup B_2 \cup .... \cup B_{k-1} \cup B_k. \tag{9}$$

Obviously, being a linearly dependent set of functions, E_k is no longer a basis for V_k, but merely a generating system. See Figure 1 for a simple 1D example.

FIGURE 1. Functions of the generating system E_3 in the 1D case.

In any case, an arbitrary function $u \in V_k$ can be expressed in terms of the generating system by

$$u = \sum_{\phi \in E_k} w_\phi \cdot \phi \tag{10}$$

with the vector $w_k^E := (w_\phi)_{\phi \in E_k}$ associated to some given ordering of the functions of E_k.

Here and in the following, we denote representations in terms of the generating system E_k by the superscript E. The length of w_k^E is

$$n_k^E = \sum_{l=1}^{k} n_l, \tag{11}$$

which is in the 1D case about twice, in the 2D case about 4/3 times, and in the 3D case about 8/7 times as large as the length of the vector u for the basis representation (8). This is due to the geometric rate of decrease of the number of grid points from fine to coarse levels with factors of 1/2, 1/4, and 1/8 for 1D, 2D, and 3D, respectively. The generalization of this concept to higher dimensions is straightforward.

Note that the representation of u in terms of E_k is not unique. In general, there exists a variety of level-wise decompositions of $u \in V_k$. However, for a given representation w_k^E of u in E_k, we can easily compute its unique representation u_k with respect to B_k. This involves in 2D the bilinear interpolation which can be expressed and implemented by MG-prolongation operators. For details, see [6], [7].

2.2 Galerkin-approach and linear systems

Using the nodal basis B_k, the Galerkin-approach results in the discrete variational problem for $u \in V_k$

$$\forall \phi_i \in B_k : a(u, \phi_i) = f(\phi_i) \tag{12}$$

and the equivalent linear system of equations for the vector of nodal values

$$L_k u_k = f_k, \tag{13}$$

where

$$\begin{array}{rcll} (L_k)_{i,j} & := & a(\phi_j^{(k)}, \phi_i^{(k)}), & 1 \le i,j \le n_k \\ (f_k)_i & := & (f, \phi_i^{(k)}), & 1 \le i \le n_k \end{array} \tag{14}$$

for some appropriate ordering of the functions of B_k.

For the generating system E_k, the Galerkin-approach leads to the discrete variational problem for $u \in V_k$

$$\forall \phi_i \in E_k : a(u, \phi_i) = f(\phi_i) \tag{15}$$

and, with representation (10), to the linear system

$$L_k^E w_k^E = f_k^E, \tag{16}$$

where

$$\begin{array}{rcll} (L_k^E)_{i,j} & := & a(\phi_j, \phi_i), & 1 \le i,j \le n_k^E \\ (f_k^E)_i & := & (f, \phi_i), & 1 \le i \le n_k^E \end{array} \tag{17}$$

for some appropriate ordering of the functions of E_k.

The system $L_k^E w_k^E = f_k^E$ has the following properties. The matrix L_k^E is semidefinite and has the same rank as L_k. Thus $n_{k-1}^E = n_k^E - rank(L_k)$ eigenvalues of L_k^E are zero. The system is solvable because the right-hand side is constructed in a consistent manner, i.e. $rank(L_k^E) = rank(L_k^E, f_k^E)$. It has not just one unique solution, but a variety of different solutions. However, the evaluation of two different solutions $w_k^{E,1}$ and $w_k^{E,2}$ with respect to their representation in B_k by means of MG-prolongation operators results in the unique solution u_k of the system $L_k u_k = f_k$. Therefore, it is sufficient to compute just one solution of the enlarged semidefinite system to obtain, via interpolation, the unique solution of the system $L_k u_k = f_k$. Note that the enlarged matrix L_k^E contains the submatrices L_l that arise from the use of the standard basis B_l, $l = 1, .., k$. A similar property holds for the right-hand sides.

3 Gauss-Seidel-type iterative methods

Now, multilevel-type algorithms are easy to construct. In [4], [6], it is shown that (after appropriate scaling) the conjugate gradient method for the semidefinite system $L_k^E w_k^E = f_k^E$ is equivalent to CG with the preconditioner of Bramble, Pasciak, and Xu [1] for the system $L_k u_k = f_k$. Furthermore, the Gauss-Seidel iteration for the semidefinite system is equivalent to the multigrid-method with Gauss-Seidel smoother applied to $L_k u_k = f_k$. Here, depending on the ordering of the unknowns of the semidefinite system, different MG-cycle strategies can be modeled easily. See the discussion in [4] and [6] for further guidance.

In the following, we focus on Gauss-Seidel-type iterations for the semidefinite system only. First, we consider level-oriented Gauss-Seidel iterations and state that they are equivalent to certain multigrid methods. An outer iteration switches from level to level, and an inner iteration operates on the respective grids. Then, we study point-oriented Gauss-Seidel iterations. There, the unknowns of the semidefinite system that belong to basis functions centered at the same grid point are grouped together. Now, an outer iteration switches from grid point to grid point, and an inner iteration or a direct solver works on the local subsystem of the semidefinite problem belonging to the respective grid point. In the same way, domain-oriented Gauss-Seidel methods can be constructed for the semidefinite system.

Note that, in both cases, it is not necessary to assemble the matrix L_k^E and the right-hand side f_k^E explicitly. It is possible to use MG-prolongation and MG-restriction operators and the fine grid discretizations L_k, f_k to express L_k^E and f_k^E in a certain product form. For an example for the level-oriented approach, see (35) in [4] or the more detailed explanations in [6]. Furthermore, by storing and updating certain parts of the current residual in (at least two) additional vectors, it is possible to realize, for appropriate traversal orderings, the implementation of level-, point- and domain-oriented Gauss-Seidel methods to need $O(n_k^E) = O(n_k)$ operations per iteration step, only. Especially for the point- and domain-oriented block Gauss-Seidel methods, this is quite technical. A description of implementation details will be given in [7]. Altogether, the required storage and the number of operations to perform a Gauss-Seidel- or block Gauss-Seidel-step is proportional to the number of grid points employed.

3.1 Grid-oriented Gauss-Seidel algorithms

We consider Gauss-Seidel iterations for the semidefinite system $L_k^E w_k^E = f_k^E$. As usual, we decompose the semidefinite matrix L_k^E by $L_k^E = D_k^E + F_k^E + (F_k^E)^T$, where D_k^E and F_k^E denote the diagonal and strictly lower triangular parts of L_k^E, respectively. Then, the Gauss-Seidel iteration (GS) is expressed by using $C_k^{E,GS} := (D_k^E + F_k^E)^{-1}$, and its symmetric counterpart SGS is expressed by using $C_k^{E,SGS} := (D_k^E + F_k^E)^{-T} D_k^E (D_k^E + F_k^E)^{-1}$ in the iteration

$$w_k^{E,(\kappa+1)} := w_k^{E,(\kappa)} + C_k^E (f_k^E - L_k^E w_k^{E,(\kappa)}). \tag{18}$$

Note that $D_k^E + F_k^E$ is generally not symmetric, but it is positive definite and invertible.

Of course, the lower triangular part of L_k^E depends on the ordering of the unknowns. For practical reasons, i.e. to maintain $O(n_k)$ operations per iteration step, not all orderings are advisable. Here, we order the unknowns grid-wise (starting for example with the finest grid). This leads to a grid-oriented block partition of L_k^E. The unknowns of each level can be ordered lexicographically or, in the 2D case, in a four color manner.

Note that this level-wise ordering corresponds to a level-wise decomposition

$$V_k = \sum_{l=1}^{k} V_l = \sum_{l=1}^{k} \sum_{i=1}^{n_l} V_{l,x_i} \tag{19}$$

of V_k, where $V_{l,x_i} = span\{\phi_i^{(l)}\}$.

In [4], [6], it was shown that, for a level-wise ordering of the unknowns, the Gauss-Seidel iteration for the semidefinite system corresponds to the multigrid method for the standard system. The switching from grid to grid in the multigrid method corresponds to an outer (block) Gauss-Seidel iteration for the semidefinite system, whereas the MG-smoothing steps resemble an inexact solver for each block by inner Gauss-Seidel iterations. Especially the multigrid V-cycle with one pre- and post-smoothing step by Gauss-Seidel iterations becomes the SGS-method, which is also known as Aitken's double sweep. But other MG-cycle types can be modelled by different orderings of the block GS-traversal. The case of multiple smoothing steps corresponds to multiple inner iterations. Furthermore, other smoothers can be incorporated in the block GS-method as inner iterations.

These algorithms on the semidefinite system (16) can be interpreted alternatively as subspace correction methods [12]. The relaxation of an iterate $u^{(\kappa)} \in V_k$ then takes place with respect to a $\phi \in E_k$ by

$$u^{(\kappa+1)} := u^{(\kappa)} + \lambda \cdot \phi \quad \text{with} \quad a(u^{(\kappa+1)}, \phi) = f(\phi). \tag{20}$$

The different MG-type GS-algorithms for the generating system involve the cyclic application of (20) for a corresponding sequence of functions of E_k.

3.2 Point-oriented block Gauss-Seidel algorithms

Now, we partition the unknowns of the semidefinite system into groups and perform a block Gauss-Seidel iteration on the associated block-partitioned system. Thus, we decompose the semidefinite matrix L_k^E by $L_k^E = \mathcal{D}_k^E + \mathcal{F}_k^E + (\mathcal{F}_k^E)^T$, where $\mathcal{D}_k^E$ and $\mathcal{F}_k^E$ now denote the block diagonal and strictly lower block triangular parts of L_k^E, respectively. Then, the block GS-iteration is expressed by using $\mathcal{C}_k^{E,GS} := (\mathcal{D}_k^E + \mathcal{F}_k^E)^{-1}$, and its symmetric counterpart SGS is expressed by using $\mathcal{C}_k^{E,SGS} := (\mathcal{D}_k^E + \mathcal{F}_k^E)^{-T} \mathcal{D}_k^E (\mathcal{D}_k^E + \mathcal{F}_k^E)^{-1}$ in the iteration

$$w_k^{E,(\kappa+1)} := w_k^{E,(\kappa)} + \mathcal{C}_k^E (f_k^E - L_k^E w_k^{E,(\kappa)}). \tag{21}$$

Note that $\mathcal{D}_k^E + \mathcal{F}_k^E$ is invertible only for certain choices of partitions, where each group contains unknowns associated to linear independent functions. Nevertheless, if this is not fulfilled, an iterative method still produces a non-unique solution for the corresponding semidefinite subsystem.

Alternatively, one block Gauss-Seidel step can be interpreted as a subspace correction method. The relaxation of an iterate $u^{(\kappa)} \in V_k$ now takes place simultaneously with respect to a set $\Phi \subset E_k$ by

$$u^{(\kappa+1)} := u^{(\kappa)} + \sum_{\phi \in \Phi} \lambda_\phi \cdot \phi \quad \text{with} \quad \forall \phi \in \Phi : a(u^{(\kappa+1)}, \phi) = f(\phi). \tag{22}$$

For the computation of the values of $\lambda_\Phi = \{\lambda_\phi : \phi \in \Phi\}$, the system

$$L_\Phi \lambda_\Phi = r_\Phi \tag{23}$$

with

$$\begin{array}{rcll} (L_\Phi)_{i,j} & := & a(\phi_j, \phi_i), & \phi_i, \phi_j \in \Phi \\ (r_\Phi)_i & := & (f_\Phi)_i - a(u^{(\kappa)}, \phi_i) & \\ (f_\Phi)_i & := & (f, \phi_i), & \phi_i \in \Phi, \end{array} \tag{24}$$

has to be solved. The block Gauss-Seidel method for the semidefinite system is the cyclic application of (22) for a given sequence of (disjoint) subsets of E_k. This type of iteration is only convergent if the union of all involved sets of functions span V_k. Note that the case of non-disjoint subsets can still be denoted in terms of (22), but not more by (21).

Of course, this approach is heavily dependent on the chosen sequence of subsets of E_k. In the following, we suggest a point-oriented approach. We group all unknowns together that are associated to the same grid point. This results in point-oriented methods and can be considered as a point-block technique. Now, an outer iteration switches from grid point to grid point. The local system that belongs to all basis functions of different levels centered in the same grid point can be solved either directly or by an inner iteration running over all unknowns associated to the grid point under consideration.

To be more specific, we consider as blocks the unknowns associated to all functions that are centered in the same grid point $x \in N_k$:

$$P_x := \{\phi \in E_k : \phi(x) = 1\} \tag{25}$$

This corresponds to a point-oriented decomposition

$$V_k = \sum_{x \in N_k} \sum_{l : x \in N_l} V_{l,x} \tag{26}$$

of the space V_k. Note that in comparison with the level-wise decomposition (19) just the summations are exchanged.

Now, we can step through the set N_k of grid points and relax simultaneously the unknowns that belong to the same grid point. This results in systems of linear equations (23) with $\Phi = P_x$, $x \in N_k$, that form the block diagonal matrix $\mathcal{D}_k^E$ involved in (21). Note that the size of the systems belonging to $x \in N_l \setminus N_{l-1}, l = k, .., 2$, is $k - l + 1$. The size of the system belonging to the center point $x = (0.5, 0.5)$ is k. For the case of an operator with constant coefficients, the point-block matrices L_{P_x} are full, definite and symmetric Toeplitz matrices. It turns out that their condition number is $O(1)$. See also the results in [6]. Thus, the solution of (23) for $x \in N_l \setminus N_{l-1}$ can be obtained by some appropriate iterative method in $O((k-l+1)^2)$ operations. The number of operations necessary for solving all arising point-block subsystems is in 2D

$$C\frac{3}{4}n_k\left(1^\gamma + \frac{1}{4}2^\gamma + \frac{1}{16}3^\gamma + \frac{1}{64}4^\gamma + \dots\frac{1}{4^{k-l}}(k+1-l)^\gamma + \dots + \frac{1}{4^{k-1}}k^\gamma,\right) \tag{27}$$

where $\gamma = 2$ for SGS and $\gamma = 3$ for direct solution by Gaussian elimination. Altogether, this results in $O(n_k)$ operations. The coupling between two point-blocks is described by the respective submatrix entries of $\mathcal{F}_k^E$, or, in (24), by $a(u^{(\kappa)}, \phi_i), i = 1,..,|P_x|$, with a non-unique representation (10) of $u^{(\kappa)}$. It is easy to see that by this coupling information is exchanged on all respective levels of discretization simultaneously.

In practice, however, not all traversal orderings through the set of grid points N_k are advisable. We restrict ourselves to the sequence of point-blocks where the grid points of $N_k \setminus N_{k-1}$ are ordered first, for example in a three-color fashion. Second, the remaining grid points of N_{k-1} are ordered recursively in the same way. Then, neither the point-block matrix nor the off-diagonal blocks have to be assembled explicitly, but can be expressed by means of prolongation and restriction operators. However, the efficient solution of the diagonal-block problems and the block Gauss-Seidel iteration is still tricky to implement. It can be shown that one point-block Gauss-Seidel step can be implemented to need $O(n_k)$ operations. Details will be reported in [7].

Note that it is in general not necessary to compute the exact solution for each point-block problem. Usually, a few GS- or SGS-iterations are sufficient. In the extreme case of one GS step only, we obtain the GS iteration for the semidefinite system (16) with just a special point-oriented ordering.

3.3 Domain-oriented block Gauss-Seidel algorithms

The point-oriented approach can be easily generalized. We may allow an arbitrary domain decomposition of Ω into K non-overlapping subdomains with associated decomposition

$$N_k = \bigcup_{i=1}^{K} N_k^i, \quad N_k^i \cap N_k^j = \{\} \text{ for } i \neq j \tag{28}$$

of the grid points N_k. Now, we group together the unknowns of the semidefinite system that are associated to functions of E_k whose center points are situated in the same N_k^i by

$$P_{N_k^i} := \bigcup_{x \in N_k^i} P_x. \tag{29}$$

The resulting block Gauss-Seidel algorithm now switches from subdomain to subdomain in some prescribed order. Thus, we obtain some sort of simple domain decomposition method which exhibits MG-type convergence properties.

Note that, in contrast to the point block approach, the arising subproblem matrices are now in general not longer invertible, since they can be semidefinite. However, an iterative method still is able to produce a non-unique solution for the corresponding subsystem. Alternatively, a direct solver for any definite subsystem with full rank can be used.

For practical purposes, a nested dissection-like decomposition [2] of the grid points into subdomains is advisable. Compare also Figure 7. Then, by using multigrid prolongation and restriction operators, the submatrices for the subdomains have not to be assembled explicitly, and one overall block Gauss-Seidel iteration can be performed with $O(n_k)$ operations.

3.4 Convergence properties

Based on Xu [12], Zhang [14] and [6], it can be shown that the above mentioned plain Gauss-Seidel methods, the point-oriented Gauss-Seidel methods and to some extend also the domain-oriented Gauss-Seidel methods working on the semidefinite system converge *independent* of k or n_k^E without any regularity assumption.

In contrast to Xu [12], the use of the generation system gives us the possibility to express the estimates for the convergence rate directly in terms of the semidefinite matrix itself, which is similar to conventional convergence estimates for classical Gauss-Seidel and SOR methods, c.f. [6].

Interestingly, the arising estimate for the plain Gauss-Seidel method (as well as for the point-oriented algorithm) is *independent* of the ordering of the generating system and gives an upper bound for *all* possible orderings. Thus, it also holds for plain Gauss-Seidel method with level-oriented traversal ordering, that corresponds to a multigrid V-cycle with either one pre- or one post-smoothing step with any Gauss-Seidel smoother (lexicographic, red-black, four-color) on each level. Furthermore, since the symmetric Gauss-Seidel method consists just of two Gauss-Seidel iterations with one in reversed order, we immediately obtain the estimate $\rho^{SGS} \leq (\rho^{GS})^2 \leq c < 1$. for the convergence rate of the symmetric Gauss-Seidel method (i.e. V-cycle with one pre- and one post-smoothing step), which is also independent of k and n_k^E.

In addition, not only level-wise orderings have to be considered. We can give up the level-fixed approach that is inherent in usual multigrid methods and gain the freedom to perform Gauss-Seidel relaxations in a point- or domain-oriented ordering as well without loosing the property that the convergence rate is independent of k and n_k^E. For example, the point- and domain-oriented Gauss-Seidel methods where the point- and domain-subsystems are not solved exactly but only iterated by one Gauss-Seidel step reduce to a simple Gauss-Seidel method for the overall semidefinite system with just a special point- or domain-oriented traversal ordering. Therefore, their convergence rate is also independent of k and n_k^E.

Similar observations can also be made for the point- and for the domain-oriented methods with exact inner solver. For further details, we refer to [6].

4 Parallelization properties

So far, we have mentioned that point-oriented Gauss-Seidel methods or domain-oriented Gauss-Seidel methods (with one Gauss-Seidel step in each subdomain) possess a convergence rate that is independent of k just like level-oriented Gauss-Seidel methods (multigrid). Note that the convergence rates measured in numerical experiments for point- and domain-oriented methods, c.f. [5], are slightly worse than that for sophisticated level-oriented methods. We obtain convergence factors of 0.1-0.2. Furthermore, the amount of operations to perform one iteration step is, at least for the present implementation, somewhat larger.

Thus, we can ask the question whether we need the point- and domain-oriented method at all. The answer gets clear if we consider the parallelization properties of the different methods on most presently available MIMD computers and especially on networks of workstations or LANs and WANs. There, the time necessary to setup communication is often relatively large. Thus, it is advantageous to exchange a larger amount of data collectively in one step than to exchange only fractions of the data in many different steps. Otherwise it can happen that the overall execution time is dominated exclusively by the setup time. Then, the time necessary for data exchange and parallel computation is not important any more.

This is where the level- and the point-/domain-oriented methods are different. For the level-oriented method, the number of communication steps is dependent on the number k of levels whereas for the point-/domain-oriented method it only depends on the number of processors P, i.e. it is of the order $\mathcal{O}(\log P)$. However, in the point- and domain-oriented case, the overall amount of data to be exchanged is slightly larger. For practical situations with $n_k \gg P$, this property of the point-/domain-oriented method is a crucial advantage that pays off on certain MIMD computers and especially on workstation networks or LANs and WANs with relatively dominant communication setup. In the following we will explain this difference between the level- and point- or domain-oriented approach in more detail.

An efficient implementation of the level-oriented Gauss-Seidel method for the semidefinite system can follow the multigrid strategy. In general, it consists of the smoothing step, the computation of the residual, the restriction operator and the prolongation operator. For the parallelization of a multigrid method, the domain Ω is usually subdivided into P more or less equally sized subdomains. Then, each subdomain with his grid points on all levels (together with some overlapping border grid points) is associated to its specific processor. Usually array-, tree- or pyramid-type processor topologies are used [9].

Following this domain decomposition principle, the different components of a multigrid method (smoothing, residual computation, restriction, prolongation) can be executed, at least to some extent, in every subdomain independent of each other. This holds especially for grid points that are situated, for each respective level, sufficiently in the interior of each subdomain. However, for grid points that are contained in the local boundary of the subdomains, or that are within one mesh width distance from the local boundary, the situation looks different. In general, for the computation of the multigrid components in these points, data is necessary from adjacent grid points, that can be situated in some adjacent subdomain and is stored on another processor. Therefore, the exchange of data between adjacent processors is necessary. This is usually performed collectively for all border points that belong to two adjacent processors in one communication step. Since the multigrid methods works level by level, this makes clear, that in a parallel multigrid method data has to be exchanged on *each* level of the multigrid hierarchy. For further details, see [8], [9].

Another bottleneck for the parallelization of multigrid methods occurs on the coarser grids. These grids contain relatively few grid points. Thus, only few processors can work here in parallel and the remaining processors have to stay idle. Additionally, in an array-like processor topology, certain adjacent grid points on coarse grids are not longer associated to adjacent processors. Then, communication has to take place sequentially between distant processors and is therefore more costly. Furthermore, processor load on these coarser grids can in general not be distributed equally. Therefore, so-called agglomeration techniques [8] have to be applied.

Altogether, for the parallelization of level-oriented methods, communication is necessary on *every* level. Furthermore, the computations on coarser levels can not employ all available processors, they are usually not well balanced and, additionally, it can be necessary to exchange data between distant processors. Theoretically, it is possible to obtain a parallel complexity of $\mathcal{O}(\log n_k)$ if a sufficiently large number of processors is available. Practically, however, this is often not the case. There, $P \ll n_k$ usually holds, and the overall computation time of a parallel multigrid method is strongly dependent on the number of communication steps and the communication properties of the parallel computing system. It can even happen that the communication setup time dominates both the data exchange time and the computation time and spoils the speed up and the efficiency.

For the simple one-dimensional case with $k = 5$, we see in Figure 2 the hierarchy of grids employed in the multigrid method and its distribution to $P = 3$ processors. Arrows indicate schematically the necessary communication steps.

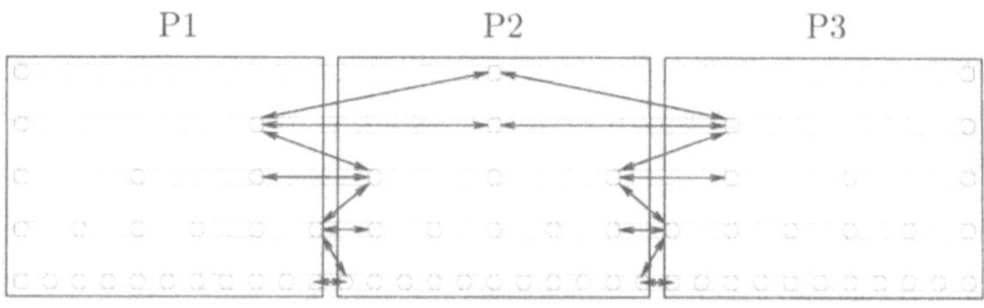

FIGURE 2. Distribution of the level-oriented method to 3 processors.

It is clear that due to the level-wise computation the number of necessary communication steps is of the order $\mathcal{O}(k)$, whereas, at least in our 1D-example, the amount of data to be exchanged in each step is of the order $\mathcal{O}(1)$. Thus, the number of communication steps is dependent on k ond not on P.

Now, we turn to the parallelization properties of the point- and domain-oriented methods. To explain the basic parallelization strategy and the requirements for communication and distributed storage and to show the main difference to the level-oriented methods, we restrict ourselves for reasons of simplicity to the one-dimensional case. In the point-oriented Gauss-Seidel method for the semidefinite system, all unknowns that belong to different levels but to the same grid point are grouped together and treated collectively as block in the iteration procedure.

For the parallelization of this method we apply the "divide and conquer"-principle. In a first step, we assign the center point of the domain together with all its unknowns of w_k^E and further data (like right hand side, certain parts of the residual) that belong to the center point to a processor $P1$. Then, the remaining grid points and all their degrees of freedom with respect to the generating system are split into two mutually independent subsets, that are assigned to two further processors ($P2$, $P3$). Altogether, we obtain the situation as shown schematically in Figure 3. Once again, arrows denote necessary communication steps.

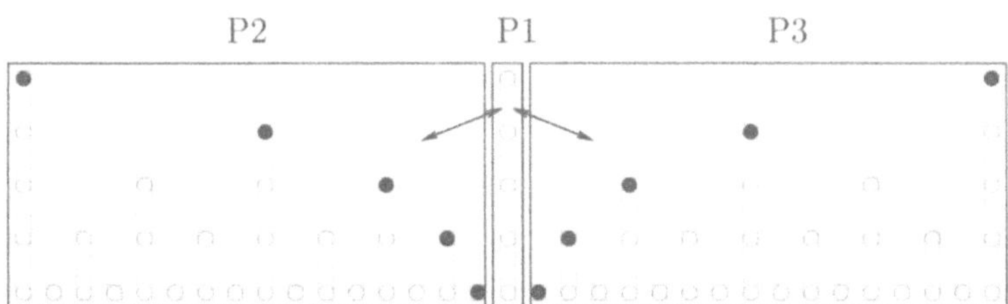

FIGURE 3. Schematical distribution of the point-oriented method to 3 processors.

First, the unknowns that belong to the left and right subsystem of the semidefinite system are relaxed simultaneously on the processors $P2$ and $P3$. Then, necessary data (the associated grid points are marked in Figure 3 by •) is sent to processor $P1$ and the unknowns that belong to the center point can be relaxed or computed exactly. The resulting new values have to be sent to the processors $P1$ and $P2$ and the iteration can continue.

Now, a first difference to the level-oriented method of Figure 2 gets clear: Only the two processors $P2$ and $P3$ can work simultaneously and they have to stay idle if computations are performed on processor $P1$. However, there are only two (simultaneous) communication steps necessary which is independent of k. It can be seen directly that, at least for our simple 1D example with only three processors, the number of communication steps is only of the order $\mathcal{O}(1)$. However, the amount of the data to be exchanged is now of the order $\mathcal{O}(k)$. In contrast to the level-oriented method of Figure 2, the point-oriented approach allows to collect the larger amount of data belonging to different level and to exchange them *collectively* in one communication step.

In the case of many processors, the left and right subproblem of the semidefinite system can be further subdivided in a *recursive* manner. In a natural way, we then obtain a tree-like parallelization structure. Each node of the tree contains the parts of the semidefinite system that are associated to the point block of the respective grid point. The leaves contain the subproblems that belong to the respective subdomains. The point block algorithm sweeps through this tree level by level. Now, the processors of each level work in parallel. This is indicated in Figure 4.

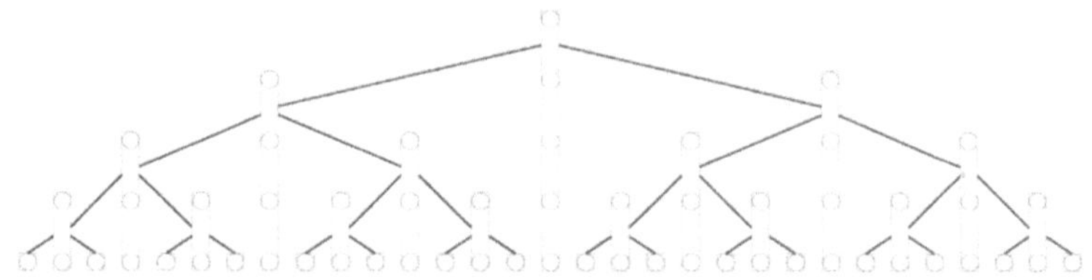

FIGURE 4. Tree-like parallelization structure of the point-oriented method.

To maintain the data supply of the processor nodes over many levels in the case of recursive subdivision, it is not more sufficient to exchange only the data between father- and son-processors, that belong to the points of Figure 3 that are marked by •. For example, for the computation on level l (here $l = 2$), data is necessary that belongs to the points of Figure 5 (below) that are marked by •. However, for the computation on level $l - 1$, data is necessary that belongs to the points of Figure 5 (above) that are marked by •.

Therefore, certain additional values have to be stored and updated in every node of the processor tree to maintain data supply over many levels. The corresponding points on level l are shown in Figure 6. This additional data only doubles the amount of data to be stored in each processor and results in a factor 2 for the size of the data packages to be exchanged between father- and son-processors. The amount of storage as well as the amount of data to be exchanged remains proportional to $k - l + 1$ for each processor on level l.

A short analysis (sum up over the processor tree) then gives a *parallel* complexity of $\mathcal{O}(n_k/P)+\mathcal{O}((\log P-1)(\log n_k-(\log P)/2+1))$ for the number of operations, $\mathcal{O}((\log P-1)(\log n_k-(\log P)/2+1))$ for the amount of data to be exchanged and $\mathcal{O}(\log P)$ for the number of communication steps of the overall algorithm. Theoretically, with $P \approx n_k$, we obtain now a parallel complexity of $\mathcal{O}((\log n_k)^2)$ only, whereas the parallelization of the multigrid method (on a multilevel-type processor topology but not on a binary tree processor topology !) even results in $\mathcal{O}(\log n_k)$. However, remember that the setup time is now not more dependent on $\log n_k$ but

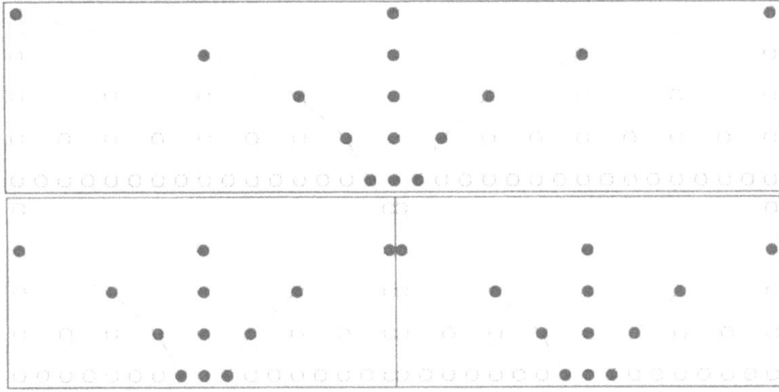

FIGURE 5. Points per processor that correspond to data necessary for the computation.

only on $\log P$. Thus, mainly on networks of workstations as well as on LANs and WANs where the set up time contributes dominantly to the overall execution time, run time advantages can be expected for practical values of P and n_k.

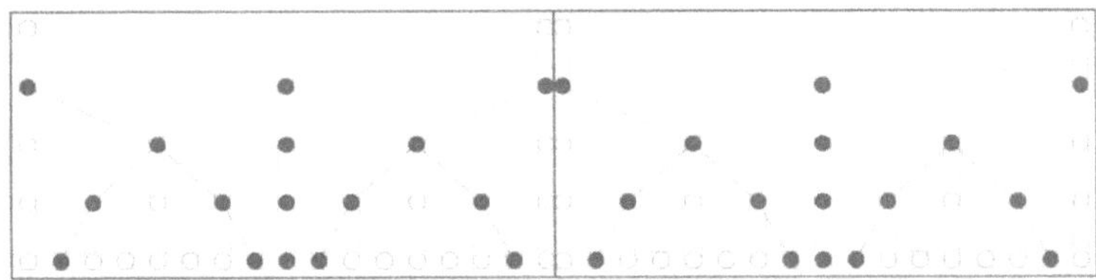

FIGURE 6. Points per processor that correspond to data necessary for the computation and communication.

This "divide and conquer"-approach for the parallelization can be generalized to the two-dimensional case. Then, we decompose the domain Ω along a middle line and assign the corresponding part of the semidefinite system to a first processor. The remaining part of the semidefinite system is split into two independent parts that can be treated in parallel. These two subsystem can be further subdivided in a recursive manner. If this subdivision always takes place in the same direction we obtain a stripe-wise decomposition of the domain, see Figure 7 (left). Furthermore, the direction of the subdivision can be altered with each step. In both cases we obtain a binary tree-type parallelization structure where the subproblems of the semidefinite system that correspond to a node of the tree can be treated on each level in parallel. The alternating subdivision is closely related to the decomposition of the domain into four parts by means of a separating "cross". Then, we obtain a tree-like parallelization structure, where each node possesses four sons, compare

7 (right).

Now, similar to the 1D case (compare Figure 5), data of adjacent points of each level is necessary for the computation of the separator-subproblems and has to be stored on the respective processors. Furthermore, to maintain data supply over many levels of the processor tree, certain additional values have to be stored, updated and exchanged in every node of the processor tree similar to the 1D case (compare Figure 6). A more detailed discussion on this problem is found in [6].

Note at last that the iteration of the one-dimensional multilevel subsystems of the semidefinite system that belong to a separator-line or -cross can be further decomposed in a *second* parallelization step, which is analogous to the parallelization strategy in the one-dimensional case.

FIGURE 7. Stripe- and box-wise nested dissection decompositions and their parallelization structures.

5 Concluding remarks

In this paper we presented different multilevel algorithms based on the generating system approach. We studied level-oriented techniques where Gauss-Seidel methods for the semidefinite system turn out to result just in standard multigrid algorithms. Additionally, we presented new point- and domain-oriented methods. There, block Gauss-Seidel iterations for the semidefinite system are obtained that exhibit a reduction rate independent of the grid size and the number of levels like conventional multigrid methods. Furthermore, these algorithms possess interesting properties with respect to parallelization. Especially the property that the number of communication steps, and thus the setup, is only dependent on P and not on k as for multigrid methods makes these algorithms well suited to networks of workstations. Meanwhile, first parallelization results have been obtained on different MIMD computers (IPSC/860, nCube2, Transputers), virtual shared memory machines (KSR-1), and networks of workstations (HP9000/720 with Ethernet) but this will be reported elsewhere.

In [3], our level- and point-oriented approach is adopted to an extended generating system that additionally contains the basis functions of all grids that result from semi-refinement steps with respect to both coordinate directions. Furthermore, the full and the sparse grid case [13] is considered. For the level-oriented methods, we obtain multigrid algorithms similar to that of Naik and van Rosendale [11] and Mulder [10]. For the point- and domain-oriented approach, analogous point block Gauss-Seidel methods as well as BPX-like preconditioners are derived.

References

[1] James Bramble, Joseph Pasciak, and Jinchao Xu. Parallel multilevel preconditioners. *Math. Comp.*, 31:333–390, 1990.

[2] A. George. *Nested dissection of a regular finite element mesh.* SIAM, J. Numer. Anal. 10, 1973.

[3] M. Griebel, C. Zenger, and S. Zimmer. Multilevel Gauss-Seidel-algorithms for full and sparse grid problems. *Computing*, 50:127–148, 1993.

[4] M. Griebel. *Multilevel algorithms considered as iterative methods on indefinite systems.* TU München, Institut f. Informatik, TUM-I9143; SFB-Report 342/29/91 A, 1991.

[5] M. Griebel. Grid- and point-oriented multilevel algorithms. In W. Hackbusch and G. Wittum, editors, *Incomplete Decomposition (ILU): Theory, Technique and Application, Proceedings of the Eighth GAMM-Seminar, Kiel, 1992, Notes on Numerical Fluid Mechanics*, volume 41, pages 32–46. Vieweg Verlag, Braunschweig, 1992.

[6] M. Griebel. *Punktblock-Multilevelmethoden zur Lösung elliptischer Differentialgleichungen.* Habilitationsschrift, Institut für Informatik, TU München, 1993, to appear in Skripten zur Numerik, Teubner, Stuttgart, 1994.

[7] M. Griebel and S. Zimmer. Multilevel Gauss-Seidel-algorithms for full and sparse grid problems - Part II: Implementational aspects. SFB-Report, TU München, Institut f. Informatik, 1993, in preparation.

[8] R. Hempel and A. Schüller. Experiments with parallel multigrid algorithms using the SUPRENUM communications subroutine library. Arbeitspapiere der GMD 141, GMD, 1988.

[9] O.A. McBryan, P.O. Fredericson, J. Linden, A. Schüller, K. Solchenbach, K. Stüben, C.A. Thole, and U. Trottenberg. Multigrid methods on parallel computers - a survey of recent developments. *Impact of Computing in Science and Engineering*, 3:1–75, 1991.

[10] W. Mulder. A new multigrid approach to convection problems. *CAM Report 88-04*, 1988.

[11] N.H. Naik and J. van Rosendale. The improved robustness of multigrid elliptic solvers based on multiple semicoarsened grids. *ICASE report 51-70*, 1991.

[12] J. Xu. Iterative methods by space decomposition and subspace correction: A unifying approach. *SIAM Review*, 34(4):581–613, 1992.

[13] C. Zenger. *Sparse Grids.* Notes on Numerical Fluid Mechanics 31, W. Hackbusch, ed., Vieweg-Verlag, 1991.

[14] X. Zhang. Multilevel Schwarz methods. *Numerische Mathematik*, 63:521–539, 1992.

9

Large Discretization Step (LDS) Methods For Evolution Equations

Zigo Haras and Shlomo Ta'asan[1]

ABSTRACT A new method for the acceleration of linear time dependent calculations is presented. It solves an extended system of equations on coarse spatial grids using large time steps yielding very accurate solutions. The method employs time stepping on different temporal and spatial scales visiting the finer grids once in many coarse level time steps. Most of the work is performed on the coarse levels in time-space, while the resulting solution is practically the fine grid solution. The proposed method is very general, simple to implement and may be used to accelerate many existing time marching schemes. Numerical examples are given, demonstrating the effectiveness of the method which reduces computational time by more than an order of magnitude.

1 Introduction

Long time simulation of partial differential equations is a highly intensive computational task. While progress has been made in the acceleration of parabolic equations computations [2], little improvement has been achieved for hyperbolic problems.

The solutions of time dependent problems are often smooth and can be spatially approximated on coarse grids. In long-time simulations, this smoothness is hard to exploit due to the fast accumulation of numerical errors. This error necessitates the use of either finer grids or higher order schemes (provided such schemes are available), or a combination of both; resulting in a substantial increase of the computational cost.

In such computations a large scale system of linear equations has to be eval-

[1]This research was made possible in part by funds granted to the second author through a fellowship program sponsored by the Charles H. Revson Foundation and in part by the National Aeronautics and Space Administration under NASA Contract No. NAS1-19480 and NAS1-18605 while the authors were in residence at ICASE, NASA Langley Research Center, Hampton, Va 23681.
Department of Applied Mathematics and Computer Science
The Weizmann Institute of Science, and
Institute for Computer Applications in Science and Engineering

uated or solved for many time steps until a prescribed final time is reached. For implicit schemes some improvement may be achieved by applying a multigrid solver at each time step. However, one is still confined to small time steps and the overall computational cost is very high.

A more advanced multigrid idea, recently investigated in detail [1, 2, 3], applies multigrid in time as well. In that approach, the frozen τ method, some correction terms are added to the coarse grid equations, enabling marching on coarse levels while maintaining the fine grid accuracy. This method has been successfully applied to parabolic equations [3].

The present work suggests a novel approach to long time integration problems. It identifies two grids, the coarse *representation grid* on which the solution appears smooth enough and a finer *computational grid* which is required to achieve a prescribed accuracy with a given time marching scheme at some final time. Our method performs most of the time marching on a grid finer than the *representation grid*, yet significantly coarser than the *computational grid*. It is a generalization of the frozen τ method aimed at achieving a substantially more efficient technique not restricted to parabolic equations. In this approach a set of equations satisfied by the correction term τ are derived and solved on the coarse grid at each time step, resulting in a time varying τ. It visits the *computational grid* or intermediate grids, once in many coarse grid time steps to compute initial conditions for correction terms equations. An important feature of this method is its simplicity and generality which enables acceleration of many existing time marching programs provided they obey a few simple programming conventions. This method, named *Large Discretization Steps*, *LDS* in short, has been successfully applied to linear hyperbolic problems with periodic boundary conditions. These boundary conditions were chosen to investigate the method basic properties and estimate its expected efficiency.

The organization of this paper: In section 2 the LDS-type approximation is introduced and an error bound for a simple case is obtained. Section 3 describes the LDS algorithm and its implementation details. Section 4 presents the numerical results, and Section 5 summarizes the work.

2 Approximation Theorem

The method presented in this paper was motivated by the observation that in many cases the truncation error satisfies approximately the same equation as the solution. Thus, if one could effectively compute the initial conditions for the error term equation, the solution of this equation could be used to increase the accuracy of the approximation scheme. Computing the initial data for the truncation error equation might be a very difficult task; fortunately, there is a simple way to find the initial conditions for the relative truncation error, i.e., the error in a given approximation relative to a more accurate one. This idea can be applied iteratively resulting in a system of equations each correcting the higher order terms of the error.

In this section, this observation is demonstrated for the spatial discretization error and bounds on the global error in this type of approximation are derived in a simple case.

Consider a semi-discretization of a linear initial value problem with constant coefficients

$$\frac{du_n(t)}{dt} - Lu_n(t) = 0 \tag{2.1}$$

with initial conditions $u_n(0) = u_0$, where $u_n(t) = u(n\Delta x, t)$.

Let $\tilde{L}$ be an approximation to L, e.g., a coarse grid representation of the fine grid operator. Define the system

$$\frac{\mathrm{d}}{\mathrm{dt}}\begin{pmatrix} v_{n_0} \\ \vdots \\ v_{n_{m-1}} \\ v_{n_m} \end{pmatrix} - \begin{pmatrix} \tilde{L} & I & 0 & \dots \\ 0 & \tilde{L} & I & \dots \\ & & \ddots & \\ & & & \tilde{L} \end{pmatrix}\begin{pmatrix} v_{n_0} \\ \vdots \\ v_{n_{m-1}} \\ v_{n_m} \end{pmatrix} = \begin{pmatrix} 0 \\ \vdots \\ 0 \\ 0 \end{pmatrix} \tag{2.2}$$

with initial values

$$v_{n_j}(0) = \left(L - \tilde{L}\right)^j u_n(0) \tag{2.3}$$

Henceforth, an approximation of a system of equations by an enlarged system of the form (2.2) will be called *a LDS approximation of degree m.*

Assume that $\tilde{L}$ and L commute. The solution of this linear system of ordinary differential equations is $e^{At}v_n(0)$, where A denotes the above system. Since the matrix A has a block Jordan form, an explicit expression for $v_{n_0}(t)$ is :

$$v_{n_0}(t) \;=\; e^{\tilde{L}t}\sum_{k=0}^{m}\frac{(L-\tilde{L})^k t^k}{k!}u_n(0) = e^{\tilde{L}t}\left(e^{(L-\tilde{L})t} - \frac{(L-\tilde{L})^{m+1}\xi^{m+1}}{(m+1)!}\right)u_n(0)$$

for some $\xi \in [0, t]$.

If $e^{\tilde{L}t}u_n(0)$ are uniformly bounded by some function $K(t)$, then

$$\|u_n(t) - v_{n_0}(t)\| = \frac{\|(L-\tilde{L})^{m+1}\xi^{m+1}e^{\tilde{L}t}u_n(0)\|}{(m+1)!} \leq \frac{\|L-\tilde{L}\|^{m+1}t^{m+1}}{(m+1)!}K(t) \tag{2.4}$$

The commutativity assumption does not hold in general, however, we have obtained similar bounds for the noncommutative case for both the continuous and discrete cases [4]. The bound (2.4) implies that for any fixed final time T the global error goes to zero as the number of equations m goes to infinity and $v_{n_0}(T)$ converges to $u_n(T)$ with convergence rate depending on the magnitude of the relative truncation error. It also suggests that if there exist Fourier components for which $\|\hat{L}(\omega) - \hat{\tilde{L}}(\omega)\|$ is large, the LDS may preform poorly or even fail to work with these schemes.

It should be mentioned that although the LDS transformation allows a polynomial growth of the solution, it maintains the stability properties [5] of the original equation [4].

3 Large Discretization Step (LDS) Methods

This section presents an implementation of the LDS approximation for evolution problems.

Consider a differential equations of the form

$$\begin{aligned} \frac{\partial U(x,t)}{\partial t} - A(x,t,D)U(x,t) = F(x,t) \qquad & (x,t) \in \Omega \times (0,t_0) \\ U(x,0) = U_0 \qquad & x \in \Omega \end{aligned} \tag{3.1}$$

The problem may be multidimensional and possibly a system.

The discretizations considered are of the form

$$U^{n+1} = E(x,t,k,h)U^n + S(x,t,k,h)F^n \tag{3.2}$$

where h, k denotes Δx and Δt, respectively, and $U^n = U_j^n$ approximates $U(jh, nk)$. $E(x,t,k,h)$ is an explicit or implicit two level time marching operator. In the sequel the notation $E^{k,h}$ will be used, omitting the possible dependence on (x,t).

3.1 The LDS Method of General Degree

The bound obtained in the previous section estimates the error in approximating a high accuracy scheme by a system of equations of lower accuracy. In the present work the two discrete operators are the same discretization of a differential operator on two different grids. The LDS method attempts to perform most of the time marching of the extended system of equations on the coarser levels, yet maintaining the fine grid accuracy, by visiting the fine grid once in many coarse grid time steps for error terms computation.

Consider two grids (in space-time), a fine one with spacing (h,k) and a coarse one with (H,K) where $H = \alpha h, K = \alpha k$. Given $U^h(x,0)$ on the (h,k) grid, one is required to calculate the solution up to final time T. Assume appropriate intergrid transfers exist.

The LDS algorithm is composed of two stages : initialization of the correction terms and time marching on the coarse grid for a predetermined number of steps. For presentation simplicity the algorithm will be described in the case $F(x,t) = 0$, it is further assumed that E, S commute, as is the case in many explicit or implicit integration formulas. The general algorithm has the following simple form.

LDS Method of General Degree - d

Initialize $V_{0,d}^{0}$

$N = 0$

While $N \leq \lceil \frac{T}{K} \rceil$ Do

Call Initialize($V_{0,d}^{N}, \ldots, V_{d,d}^{N}$,d)

For i =1,...,Revisit Do

$$\text{Solve } V_{l,d}^{N+1} = E^{H,K}\; V_{l,d}^{N} + V_{l+1,d}^{N}, \qquad \text{l=0,...,d-1}$$

$$V_{d,d}^{N+1} = E^{H,K}\; V_{d,d}^{N}$$

$$\text{Set } N = N + 1$$

End

End

The result presented in the previous section does not suggest how to effectively and efficiently compute the initial values for the correction terms. This can be easily done once observing that if the correction terms were accurately initialized than at the first time steps the LDS solution on the coarse grid should coincide with the fine grid solution.

For presentation simplicity, we first describe the initialization procedure for the case $\frac{H}{h} = 2$.

Initialize($V_{0,d}^{N}, \ldots, V_{d,d}^{N}$,d)

Set $U^{N} = I_{H,K}^{h,k}\, V_{0,d}^{N}$

For i =1,...,d Do

$$\text{Solve } U^{N+\frac{m}{2}} = E^{h,k}\; U^{N+\frac{m-1}{2}}, \qquad \text{m=1,2}$$

$$V_{l,d}^{N+1} = E^{H,K}\; V_{l,d}^{N} + V_{l+1,d}^{N}, \qquad \text{l=0,...,i-2}$$

$$V_{i-1,d}^{N+1} = E^{H,K}\; V_{i-1,d}^{N}$$

$$\text{Set } V_{i,d}^{N} = V_{0,d}^{N+1} - I_{h,k}^{H,K} U^{N+1}$$

$$V_{l,d}^{N+1} = V_{l,d}^{N+1} + V_{i,d}^{N}, \qquad \text{l=0,...,i-1}$$

$$\text{Solve } V_{i,d}^{N+1} = E^{H,K}\; V_{i,d}^{N}$$

$$\text{Set } N = N + 1$$

End

Note that the term $V_{i,d}^{N}$ incorporates in it the $S^{K,H}$ term; this and the commutativity assumption explain the way the LDS integration is performed.

This procedure can be easily adapted for the general case when $\frac{H}{h} = \alpha$. The simplest approach is to perform α time steps on the fine grid for each coarse grid time step and initialize the τ's correspondingly. However, this would render initialization very costly and greatly reduce the LDS efficiency. In case α is a composite number, e.g., $\alpha = 2^l$, a more efficient approach is available, exploiting the LDS high accuracy by employing intermediate grids. In this approach one uses the fine grid to initialize an LDS system of degree m on the grid $H_1 = 2\,h$; since this approximation is very good it may be used to initialize an LDS of degree m on grid $H_2 = 2\,H_1$. This process is repeated until the correction terms on grid $H_l = H$ are initialized.

A more efficient procedure is to initialize the correction terms on all grids simultaneously, i.e., compute an initial condition for a coarse grid equation once enough time marching was performed on the finer grids. Consider, for example, the case $\alpha = 4$ and $m = 2$; the above mentioned procedure would first perform four time steps on the fine grid and two on the H_1 grid to initialize a second degree LDS on that grid, then march with the LDS system four additional steps to initialize the coarse grid system. Since four fine grid time steps are sufficient to initialize the first term in the coarse grid LDS, one may initialize this equation already at this stage and perform only two additional steps on grid H_1 to complete the coarse grid initialization. This way, the minimum number of time steps on the finer grids is performed (which takes a large fraction of the algorithm computation time) resulting in a more accurate initialization.

Note that since initialization is performed by subtracting increasingly closer solutions, each initialized term is a few significant digits less accurate than the previous one. It follows that although the result in Section 2 applies for approximation of any degree m, in practice $m \leq 3$. A simple way to predict the LDS performance is to look at the relative magnitude of the correction terms immediately after initialization. According to the result presented in Section 2, the ratio $\frac{\|\tau_j(x,0)\|}{\|\tau_{j-1}(x,0)\|}$ should be roughly constant. Thus, a large variation in this quantity suggests a large error in the initialization of τ_j, causing the LDS failure. For the above mentioned reason, high precision arithmetic is essential for good results.

3.2 Scheduling

The LDS system of equations is integrated mainly on the coarse grid, thus the accuracy of the correction terms deteriorates at a rate determined by the integration operator on that grid. However, since the magnitude of these terms is significantly smaller than that of the solution, they can be effectively used for several coarse grid time steps. Once a large error had accumulated, the fine grid should be revisited to compute new and more accurate initial conditions for these equations. This work on the fine grids consists a major fraction of the algorithm computational cost. In the sequel an estimate on the number of consecutive time steps one can march on the coarse grid will be derived based on the bound of Section 2.

A simple way to achieve a prescribed accuracy is to maintain the coarse grid error below the fine grid error at all times $t \leq T$. Recall the bound obtained in

Section 2,

$$\|u(x,t) - u^h(x,t)\| \leq |K(t)|\,\|L - L^h\|\,t \tag{3.3}$$

$$\|u^h(x,t) - u^H_{m_0}(x,t)\| \leq |K(t)|\frac{\|L^h - L^H\|^{m+1}t^{m+1}}{(m+1)!} \tag{3.4}$$

In order to achieve the desired accuracy the following inequality must hold

$$|K(t)|\frac{\|L^h - L^H\|^{m+1}t^{m+1}}{(m+1)!} \leq |K(t)|\,\|L - L^h\|\,t \qquad t \leq T \tag{3.5}$$

The operator $L - L^h$ accounts for the local truncation error and its norm satisfies $\|L - L^h\| = Ch^p$. The norm of the relative truncation error can be estimated by $\|L^h - L^H\|^{m+1} \leq C^{m+1}(\alpha^p h^p - h^p)^{m+1}$, where $\frac{H}{h} = \alpha$. Inequality (3.5) holds if,

$$t^m \leq (m+1)!\frac{\|L - L^h\|}{\|L^h - L^H\|^{m+1}} = \frac{(m+1)!\,C\,h^p}{((\alpha^p - 1)Ch^p)^{(m+1)}} \leq \frac{(m+1)!}{(\alpha^p - 1)^m\,C^m\,h^{pm}} \tag{3.6}$$

The Stiriling formula for the factorial function estimates $n! \sim \sqrt{2\pi}\,n^{n+\frac{1}{2}}e^{-n}$. Substituting this in (3.6) yields,

$$t \leq \frac{(m+1)^{\frac{5}{2}}}{e(\alpha^p - 1)\,C\,h^p} \tag{3.7}$$

where e is the Euler constant. Assume that $|K(t)| \leq \|u_0\|$, then the error in the fine grid solution at the final time T satisfies

$$\|u(x,T) - u^h(x,T)\| \leq Ch^p\|u_0\|T = \epsilon\|u_0\| \tag{3.8}$$

for a prescribed relative error tolerance ϵ. Then the time t one can march on the coarse grid without correction satisfies

$$\frac{e\,(\alpha^p - 1)\,\epsilon}{(m+1)^{\frac{5}{2}}} \leq \frac{T}{t} \tag{3.9}$$

The following theorem was thus proved,

Theorem: *For an LDS method of degree m based upon an order p approximation with coarsening ratio* α*, the fine grid should be visited* $\gamma = \frac{e\,(\alpha^p-1)\,\epsilon}{(m+1)^{\frac{5}{2}}}$ *times in order to achieve the fine grid accuracy on the same time interval.*

3.3 Work Considerations

The amount of computational work and the storage requirement in a cycle of LDS of degree m, will be evaluated and compared with the corresponding requirements on the finest grid in the cycle. In order to simplify analysis, it will be assumed that on the fine and coarse grid $\frac{H}{K} = \frac{h}{k}$ and that $H = 2^l h$.

The equation is solved in a d-dimensional space, for $d = 2, 3$. Typically, real world problems occur in 3-dimensional space.

Storage Requirements

The LDS method uses on coarser levels $m+1$ times as many equations as on the finest grid. The size of the spatial grid on any level is 2^d larger than the next coarser one. The storage required to store all the computational grids is given by

$$\left(1+\sum_{j=1}^{l}\frac{m+1}{2^{dj}}\right)S \leq \left(1+\frac{m+1}{2^d-1}\right)S \tag{3.10}$$

where S is the storage requirement for the finest grid.

If memory is at premium, storage may be traded for efficiency by initializing the grid successively rather than simultaneously. This way, at most two successive grids are required simultaneously. Thus, the memory requirement reduces to the size of the two finest grid, given by $(1+\frac{m+1}{2^d})S$.

Efficiency

The cost of a fine grid time step when using the LDS will be estimated. First, we compute the work required to initialize an LDS of degree m. It involves $2m$ steps on the finest grid. On the intermediate grids a total of $2m$ steps is required to initialize the next grid, first, m steps with systems of size $1,\ldots,m$ are required to initialize the LDS on that grid, then additional m time steps with a system of size $m+1$ are required to complete the next coarser level initialization. Thus, the cost of initialization denoted by $I_{l,m}$ is

$$I_{l,m} = 2m + \frac{m(m-1)}{2^{dl+1}} + \left(\frac{m(m-1)}{2} + m(m+1)\right)\sum_{k=1}^{l-1}\frac{1}{2^{dk}} \leq 2m + \frac{m(3m+1)}{2^{d+1}-2} \tag{3.11}$$

During initialization a time equal to $m2^l$ fine grid time step is marched. Denote by N the number of coarse grid time steps marched before revisiting the fine grid. The cost of a fine grid time step in such a cycle is

$$\frac{I_{l,m}+N\,(m+1)\,2^{-dl}}{(N+m)2^l} \leq \frac{2m+\frac{m(3m+1)}{2^{d+1}-2}+\frac{N\,(m+1)}{2^{dl}}}{(N+m)2^l} \tag{3.12}$$

and the efficiency of the LDS cycle is the reciprocal of this quantity.

3.4 Order of Intergrid Transfers

The transfer of the solution among the various grids is a major component of the LDS method. In the initialization stages the solution is first interpolated to the finer grids and after some time stepping on that grids is restricted to the coarsest grid. Appropriate choice of intergrid transfers is essential for obtaining the desired accuracy.

The necessary order of these transfers can be determined by analyzing their effect on the various Fourier components. The pth order interpolation of a $O(1)$ smooth component θ results in an $O(1 - |\theta|^p)$ smooth component on the fine grid and spurious oscillatory components of magnitude $O(|\theta|^p)$, for the frequency harmonics. These components are integrated on the fine grid having dispersive and dissipative errors different from their smooth harmonic. If the equation is of order m, this introduces an error of $O(|\theta|^{p-m})$ in the high frequencies of the solution. The restriction operator couples those components again through the aliasing phenomena. It has two types of errors, fine grid high frequencies may alias with coarse grid low frequencies and fine grid low frequencies may contribute to coarse grid high frequencies. The injection operator transfers all high frequencies to their smooth harmonics and produces no error of the second type. Bearing in mind that the oscillatory modes of the fine grid are spurious components resulting from the interpolation of coarse grid smooth components, this type of error is acceptable. Therefore, if the discretization error is of order q, it is important that the visit to the coarse will not introduce an error larger than the truncation error. Thus, the following relation should hold $p - m \geq q$.

3.5 Relation to Parabolic Multigrid

The LDS was designed to accelerate simulation of linear evolution equations. It may be viewed as a generalization of the frozen τ method [2] for parabolic equations since it can be reduced to that method by computing the initial data for $V_{1,1}$ and subsequently freezing these values.

The smoothness of the change in parabolic equation solutions enables using the same τ for long times. Nevertheless, in order to reduce the number of visits to finer grids required to correct the magnitude of the τ, it was artificially extrapolated based on its change in previous times [3]. It seems that solving a time dependent equation for the τ would be a more appropriate way to achieve this goal. It is hard to predict how this modification will change the overall performance of the parabolic solver, but clearly it will not significantly increase the algorithm cost.

4 Numerical Results

A few numerical examples will be given to demonstrate the potential of our approach. All examples are of linear hyperbolic equations with periodic boundary conditions in two dimensional space. The fine grid had 128×128 points while the coarse grid on which an LDS of degree one is integrated had 32×32 points. The fine grid is visited once in 20 coarse grid time steps, resulting in an efficiency of 14.5. The figures show a cut in the solution at the point u attains its maximum, in the y direction

The first example is the advection equation $u_t = (1 + 0.3\sin(2\pi x))u_x +$

$0.3\,(1+0.3\,\cos(2\pi x))u_y$ discretized second order upwind in space with third order Runge Kutta and CFL=0.3.

The next two are discretization of the linearized shallow water equation

$$\begin{aligned} p_t &= a(x,y)p_x + b(x,y)p_y + c(x,y)\,(u_x + v_y) \\ u_t &= a(x,y)u_x + b(x,y)u_y + c(x,y)p_x \\ v_t &= a(x,y)v_x + b(x,y)v_y + c(x,y)p_y \end{aligned} \tag{4.1}$$

with $a(x,y) = a\,(1. + 0.3\,\cos(2\pi x))$, $b(x,y) = b\,(1. + 0.3\,\sin(2\pi y))$, and $c(x,y) = c\,(1.+0.3\,\sin(2\pi x))$. For both cases $a^2+b^2 > c^2$ and $a^2+b^2 < c^2$ the LDS efficiency can be vividly seen.

5 Summary

A simple and general method for accelerating the long time integration of partial differential equations was introduced. It is highly effective, easily reducing computation time by a factor of 14.

The basic ingredients of the method were analyzed, and it was successfully applied to linear hyperbolic problems with periodic boundary conditions.

The method should be further generalized to treat different discretizations and boundary conditions.

References

[1] A. Brandt, *Multigrid Techniques: Guide with Applications to Fluid Dynamics*, GMD-Studien Nr 85, Bonn 1984.

[2] A. Brandt and J. Greenwald, *Parabolic Multigrid Revisited*, International Series of Numerical Matheamtics, Birkhauser, Verlax, Basel (1991).

[3] J. Greenwald, *Multigrid Methods for Sequences of Problems*, Ph.D. Thesis, Weizmann Institute of Science, 1992.

[4] Z. Haras and S. Ta'asan, *The LDS method for time dependent computations*, In preparation.

[5] R. D. Richtmyer and K. W. Morton, *Difference Methods for Initial Value Problems*, 2nd eds ,(John Wiley & Sons, 1967).

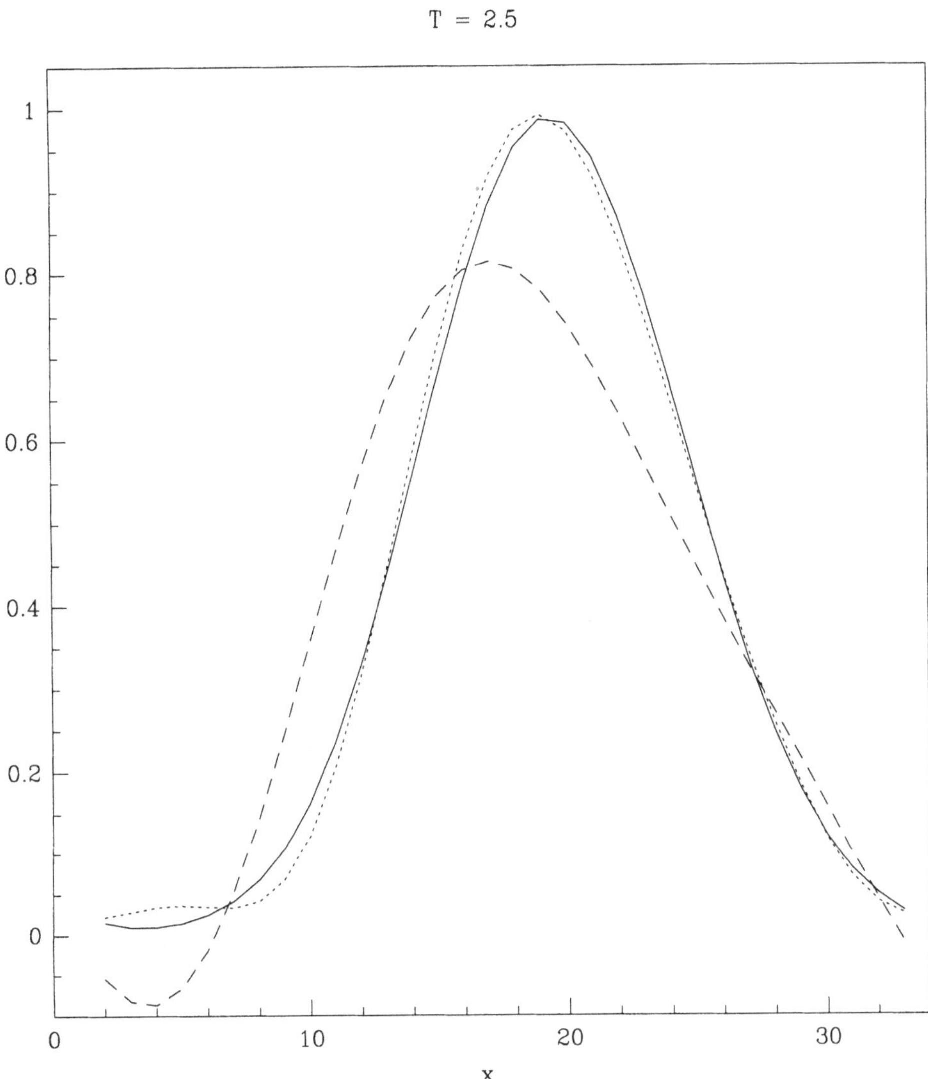

FIGURE 1. Solution of $u_t = (1 + 0.3\sin(2\pi x))u_x + 0.3(1 + 0.3\cos(2\pi x))u_y$, with $u_0 = e^{-20(x^2+y^2)}$. The fine grid solution is drawn with the solid line, the LDS with the dotted line and the coarse grid solution with dashed line.

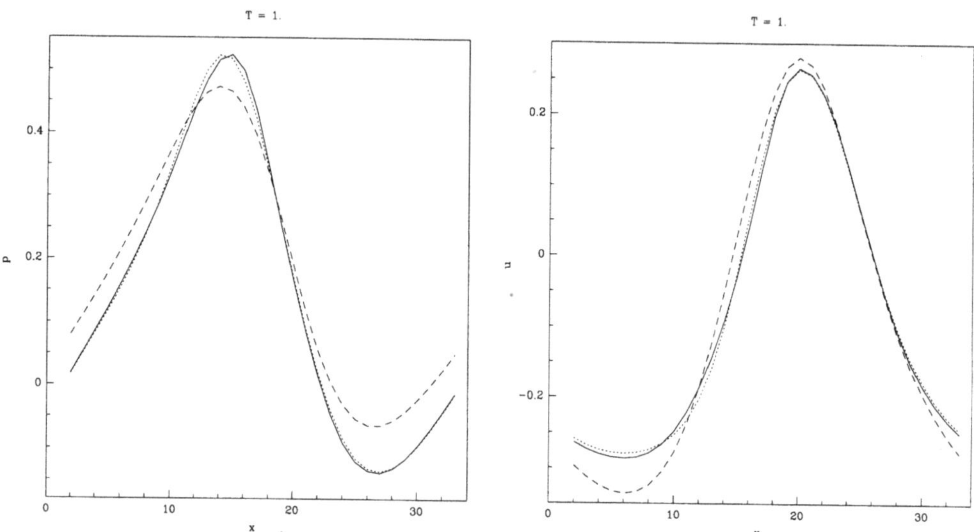

FIGURE 2. The shallow water equation with $a(x) = 0.7\,(1. + 0.3\cos(2\pi x))$, $b(x) = 0.4\,(1. + 0.3\sin(2\pi y))$, $c(x) = 0.3$, with initial conditions $p_0 = e^{-20(x^2+y^2)}$, and $u_0 = 0, v_0 = 0$. The fine grid solution is drawn with the solid line, the LDS with the dotted line and the coarse grid solution with dashed line.

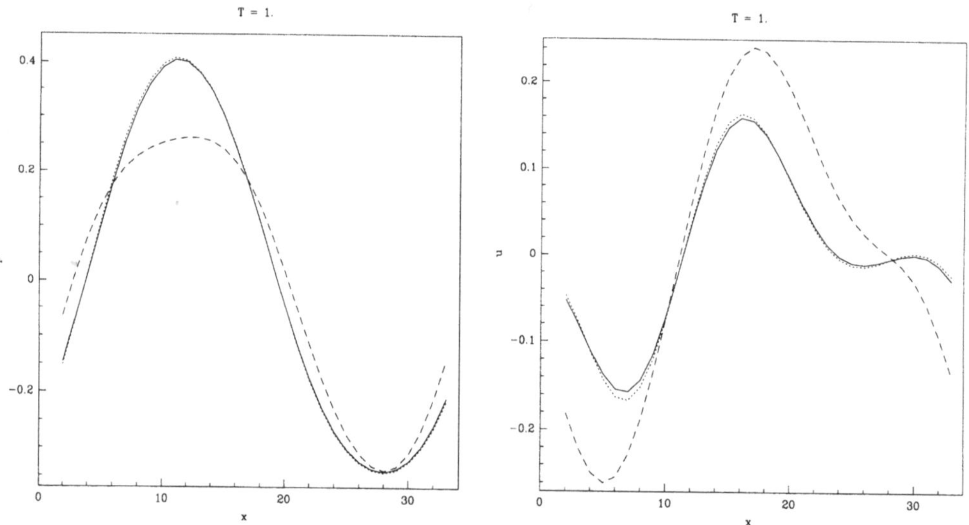

FIGURE 3. The shallow water equation with $a(x) = 0.2\,(1. + 0.3\cos(2\pi x))$, $b(x) = 0.3\,(1. + 0.3\sin(2\pi y))$, $c(x) = 1.0$, with initial conditions $p_0 = e^{-20(x^2+y^2)}$, and $u_0 = 0, v_0 = 0$. The fine grid solution is drawn with the solid line, the LDS with the dotted line and the coarse grid solution with dashed line.

10

A Full Multigrid Method Applied to Turbulent Flow using the SIMPLEC Algorithm Together with a Collocated Arrangement

Peter Johansson and Lars Davidson[1]

ABSTRACT An implementation of a multigrid method in a three-dimensional SIMPLEC code based on a collocated grid arrangement is presented. The multigrid algorithm is FMG-FAS, using a V-cycle described by Brandt [1,2].
The coarse grid is obtained by merging eight fine grid cells in 3D, and four in 2D. Restriction and prolongation of field quantities are carried out by a weighted linear interpolation, and restriction of residuals by a summation. All variables and all equations, including the pressure correction equation, are treated in the same way
.
To stabilize the solution process, a fraction of the multigrid sources is included in the diagonal coefficient a_p, and a damping function is used on negative corrections of the turbulent quantities to prevent them from being negative.
The multigrid method was shown to be relative insensitive to the choice of under-relaxation parameters. Therefore 0.8 or 0.7 is used for all equations, except for the pressure correction equation where 1.5 is used.
Both turbulent and laminar calculations are presented for a 2D backward facing step, a 2D ventilated enclosure, and a 3D ventilated enclosure. The turbulent calculations are made with a two-layer low-Reynolds $k-\epsilon$ model.
Different discretization schemes for the convective schemes are used including the first order hybrid scheme and two higher order schemes (QUICK and a van Leer TVD scheme).

1 Introduction

Except for in some simple situations, calculation of flow problems is always bound to numerical methods. These numerical methods can be based on finite elements, finite volumes, finite differences, etc.

[1]Department of Thermo- and Fluid Dynamics,
Chalmers University of Technology, S-41296 Gothenburg, Sweden

Characteristic of all of these methods is that a high resolution is needed in areas of rapid changes, for example in boundary layers, shocks and recirculation regions. On the other hand, commonly used matrix solvers are dependent on grid density and often the CPU-time is quadratically dependent on the number of nodes. This conflict usually results in the accuracy of the numerical simulation being dictated by limited CPU resources rather than by considerations of the physical flow situation.

Using multigrid, the CPU-time is reduced dramatically. In fact the CPU-time is close to linearly dependent on the number of nodes, which means that higher resolution can be afforded and a more accurate solution achieved.

For incompressible flow, several laminar multigrid implementations have been presented with different variants of FAS or CS. It is much more challenging to adopt FAS to turbulent flow situations, where only a very few efforts have been reported [4,5,6]. The turbulent transport equations (such as the k and ϵ equations) increase the complexity of the equation system in many respects. The k and ϵ equations are nonlinear and source dominated. The value of the turbulent quantities (k and ϵ) must stay positive during every instant of the iteration procedure.

In the present study FAS has been employed with the SIMPLEC algorithm in 2D and 3D turbulent flow using a low-Reynolds $k-\epsilon$ model. An increased convergence rate of a factor 10-100 or even more is obtained. The turbulent multigrid calculations were shown to be stable and the number of iterations was independent of grid density, or even decreased with increasing grid density.

2 The finite volume procedure

2.1 Basic Equations

The conservation equations for incompressible turbulent flow, using the $k-\epsilon$ model, are

$$\frac{\partial}{\partial x_j}(\rho U_j \Phi) = \frac{\partial}{\partial x_j}\left(\Gamma \frac{\partial \Phi}{\partial x_j}\right) + S \tag{1}$$

where Table 1 shows the the different variables and source terms. The turbulence model used is a two-layer $k-\epsilon$ model. In the fully turbulent flow region the standard $k-\epsilon$ model is used, and near walls it is matched at a pre-selected grid line with a one-equation model. In the one-equation region the k equation is solved and the turbulent length scale is prescribed using a mixing length approach [7].

Define a flux vector J_j containing both convection and diffusion as:

$$J_j = \rho U_j \Phi - \Gamma \frac{\partial \Phi}{\partial x_j} \tag{2}$$

Integration over a control volume using Gauss law then yields:

EQUATION	Φ	Γ	S
Continuity	1	0	0
Momentum	U_i	μ_{eff}	$-\frac{\partial p}{\partial x_i}$
Turbulent kinetic energy	k	$\mu + \frac{\mu_t}{\sigma_k}$	$P_k - \rho\epsilon$
Dissipation of k	ϵ	$\mu + \frac{\mu_t}{\sigma_\epsilon}$	$\frac{\epsilon}{k}(C_{\epsilon 1}P_k - C_{\epsilon 2}\rho\epsilon)$
$P_k = \mu_t \frac{\partial U_j}{\partial x_i}\left\{\frac{\partial U_j}{\partial x_i} + \frac{\partial U_i}{\partial x_j}\right\}$		$\mu_{eff} = C_\mu \rho k^2/\epsilon + \mu$	

TABLE 1. The parameters in the general transport equation

$$\int_A n_j J_j dA = \int_V S dV \tag{3}$$

2.2 Numerical method

Representing the flux vector J_j at the cell faces and the sources in the center results in a system of matrix equations that can be written in the form:

$$a_P \Phi_P = \sum_{nb} a_{nb}\Phi_{nb} + s \tag{4}$$

The code CALC-BFC [8], where the multigrid algorithm is implemented, uses the SIMPLEC algorithm of Patankar [9], within a collocated arrangement. To aviod nonphysical oscillations, a third order pressure dissipation is introduced by Rhie and Chow interpolation [10] when the massfluxes are calculated.

The coefficients a_{nb} contain contributions due both to convection and diffusion, and the source terms contain the remaining terms. The convective part is discretized either with the Hybrid-Upwind scheme [9], the quadratic uppstream scheme QUICK [11], or a TVD scheme from van Leer [12] The diffusive part is discretized with central differencing. In SIMPLEC procedure the continuity equation is turned into an equation for a pressure correction Φ_{pp} and the solution procedure is briefly descriebed as:

1. Relax U,V,W-momentum equations
2. Calculate mass fluxes and relax the pressure correction equation.
3. Relax k and ϵ
4. Calculate the turbulent viscosity (using under-relaxation)

"Relax" means first to evaluate the coefficients and source terms in Eq. 4 and then to make a sweep with the TDMA smoother.

3 FAS applied to SIMPLEC

3.1 Description of FAS

The description below is done with two grid levels, but is easily extended to more grid levels using the V-cycle. The FAS algorithm is described here when applied to SIMPLEC. This special multigrid method originates from a laminar multigrid [13], which is similar to the concept presented by Perić *et. al.* [5]. To prepare the multigrid for local mesh refinements some changes were made before the extension to turbulent flow was performed.

For any variable on the fine grid level 2, define the residual r^2 by:

$$a_P^2 \phi_P^2 = \sum_{nb} a_{nb}^2 \phi_{nb}^2 + s^2 + r^2 \tag{5}$$

The representation at a coarse grid by FAS is then:

$$a_P^1 \phi_P^1 = \sum_{nb} a_{nb}^1 \phi_{nb}^1 + s^1 + [\overline{a}_P^1 \overline{\phi}_P^1 - \sum_{nb} \overline{a}_{nb}^1 \overline{\phi}_{nb}^1 - \overline{s}^1 - \overline{r}^1] \tag{6}$$

where $\overline{\phi}^1$ is the restricted field variable and $\overline{r}^1$ is the restricted residual obtained from the fine grid residual r^2. The source term $\overline{s}^1$ and the coefficients $\overline{a}^1$ are calculated at the coarse grid using the restricted field variables, and all overlined terms are held constant under the course of coarse grid iterations.

The overlined quantities are used as an initial guess for a^1, ϕ^1 and s^1 which are changed owing to the restricted residual $\overline{r}^1$ while iterating at the coarse grid. The changes $\delta^1 = \phi^1 - \overline{\phi}^1$ at the coarse grid are then prolongated to correct the approximation ϕ^2 obtained earlier at the fine grid.

The two-dimensional coarse grid control volume is obtained by merging four fine grid cells together, and a three-dimensional coarse grid control volume by merging eight fine grid cells together.

The residuals represent a flux imbalance according to Eq. 3. They are therefore restricted to the coarse grid by a summation of the fine grid residuals that correspond to the fine grid cells that define the coarse grid cell. Restriction and prolongation of the field quantities is made by bilinear interpolation in 2D and trilinear interpolation in 3D.

Since non-uniform grids are used, the interpolation weights are assembled locally. For the restriction these weights are simply defined by the fraction that a fine volume takes of the corresponding coarse one, while for the prolongation the weights are calculated from the distances between the centers of the fine grid volumes and the coarse volumes.

All variables $(U, V, W, P, k, \epsilon)$ are restricted and prolongated equally, and all equations are treated in the same way with just a few exceptions given in the next section.

3.2 Special features of this implementation

1. In the laminar concept [13] the mass fluxes were restricted separately in order to achieve continuity at the coarse grid. At the coarse grid, the mass fluxes were then only corrected by the changes of the velocities. In the present study the pressure correction equation is treated in the same way as the other equations, but since it is a correction equation, $\overline{\Phi_{pp}}$ is equal to zero. No special treatment of mass fluxes is now needed, which simplifies the code, especially in connection with local mesh refinement.

2. In order to prepare the present method for local mesh refinement the pressure is restricted, which produces a problem at the coarse grid with the implicit treatment of the mass fluxes. The problem is related to the Rhie and Chow interpolation, where the coefficient a_P is used to calculate the mass fluxes and the mass fluxes are needed to evaluate a_P. Therefore at a coarse grid, a_P is stored from the last V-cycle and used in the Rhie and Chow interpolation.

3. An extra sweep with the pressure correction equation and calculation of the turbulent vicosity immediately after prolongation has been shown to be efficient in preventing oscillatory behaviour.

4. To stabilize the coarse grid equations special treatment of the multigrid source term s_m

$$s_m = \overline{a}_P^1 \overline{\phi}_P^1 - \sum_{nb} \overline{a}_{nb}^1 \overline{\phi}_{nb}^1 - \overline{s}^1 - \overline{r}^1 \tag{7}$$

is used for the k and ϵ equations. If $s_m > 0$, it is included in the right hand side vector s_Φ^1, while if $s_m < 0$, is it included in the diagonal coefficient a_P^1 via a division by ϕ_P^1

5. The turbulent kinetik energy k and its dissipation ϵ cannot physically be negative. Furthermore during the iterative solution process they must stay positive. If they become negative the turbulent sources would change sign and the turbulent viscosity would be negative, which results in rapid divergence. To prevent the turbulent quantities from becoming negative after prolongation, a damping function on negative corrections, proposed by Lien [6], is used. The positive changes at a coarse grid δ_+^1 are first prolongated to give δ_+^2 and, in the same way, δ_-^2 is obtained. The turbulent quantity ϕ_*^2 at the fine grid is then corrected to ϕ^2 by $\phi^2 = \phi_*^2(\delta_+^1 + \phi_*^2)/(\phi_*^2 - \delta_-^2)$

6. The QUICK scheme can produce negative coefficents, which can destroy the diagonal dominance of the coefficient matrix. Therefore a local under-relaxation is used, as suggested by Hellström [14], so that diagonal dominace is ensured always.

3.3 Special treatment of boundary conditions

The Dirichlet conditions at the inlet at coarse grids are based on global conservation of the mass flux between the grids. Since at a boundary two cells merged together define a coarse boundary cell (in 2D), the velocity is evaluated from the sum of mass fluxes from the two fine cells.

Neumann boundary conditions are applied without any constraints at all coarse grids.

4 Applications

Three cases are presented where both laminar and turbulent calculations are performed using a two-layer $k - \epsilon$ model.

In the laminar calculations the first-order Hybrid-Upwind discretisation scheme is used. For the turbulent calculations two different combinations of discretisations are used. The first combination is as in the laminar case (Hybrid-Upwind for all equations), and the second is QUICK for the velocities together with a bounded TVD scheme of van Leer for the turbulent quantities.

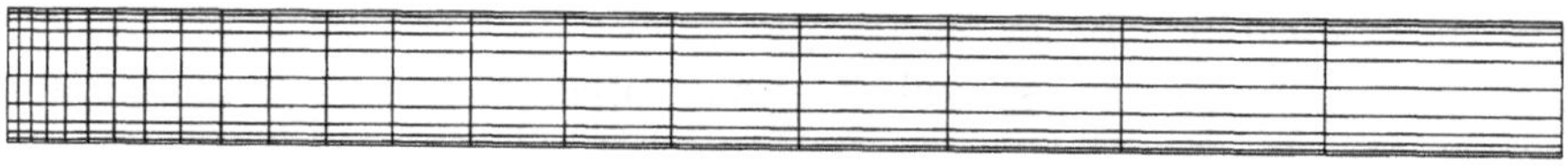

FIGURE 1. 20x10 grid for the backwards facing step

MODEL	LAMINAR		LOW-RE $k - \epsilon$			
RE	100		110 000			
SCHEME	HYBRID		HYBRID		QUICK + VAN-LEER	
	WU	SPEEDUP	WU	SPEEDUP	WU	SPEEDUP
20x10	67	1.0	185	1.0	215	1.0
40x20	58	1.9	108	3.3	130	3.1
80x40	44	6.4	72	10.8	82	∞
160x80	60	15.6	72	23.7	76	∞
320x160	91	37.5	63	100*	73	∞

TABLE 2. Convergence data for backwards facing step

The multigrid calculations were shown to be very stable, and therefore the under-relaxation parameters are set to 1.5 for the pressure correction equation and 0.7 or 0.8 for all other equations.(0.7 was used for the backwards facing step and 0.8 for the other two geometries). Decreasing the under-relaxation parameters to

0.5 decreased the convergence rate by a factor of 2 for both single grid calculations and for multigrid calculations.

Different V-cycles were tested but the convergence rate was not significantly affected and therefore a 2-1-..-1-4-1-..-1 V-cycle was used.

For each case, convergence data of the calculations are shown in tables. Convergence history plots of each geometry are also presented. Superscript * means an estimated result because the single grid is too time consuming to calculate, and ∞ means that the single grid calculation did not converge. The convergence criterion is $maxR = 0.1\%$ where $maxR$ is defined by

$$maxR = \max_{\Phi} \left[\left\{ \sum_{N} \left| a_P \Phi_P - \sum_{nb} a_{nb} \Phi_{nb} - s \right| \right\} / F_\Phi \right] \tag{8}$$

The scaling factor F_Φ is the inlet flux for each equation.

One WU is work equivalent to one SIMPLEC sweep at the finest grid, and is comparable to the calculation of $maxR$.

4.1 2D BACKWARDS FACING STEP

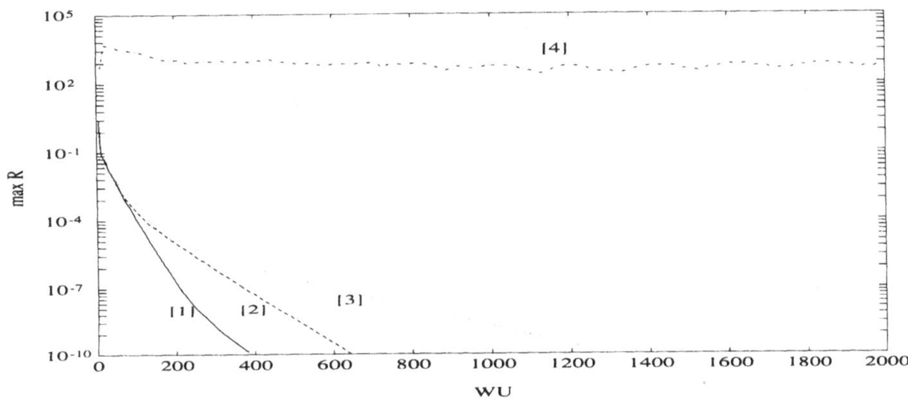

FIGURE 2. Convergence history of the 320x160 calculations for the 2D backwards facing step .[1]: Hybrid,turbulent,5-level FMG. [2]: QUICK,turbulent,5-level FMG. [3]: Hybrid,laminar,5-level FMG. [4]: Hybrid,turbulent,single grid.

The first application is a 2D backwards facing step. The inlet is half of the total height 2H. The total length of the domain is 25H, to be sure that the Neumann condition at the outlet is correct. The coarsest grid is shown in Fig. 1.

For the laminar case, a parabolic velocity profile is used at the inlet and for the turbulent case a 1/7 profile is used. The Reynolds number for the laminar calculation is 100 and 110,000 for the turbulent case.

The single grid calculations for the turbulent case were somewhat unstable at fine grids, mainly because of the high under-relaxation parameters used. To stabilize the single grid calculations, a high viscosity was used during the first ten iterations. That worked nicely for the Hybrid scheme but was insufficient with the QUICK-van Leer combination. Note that the multigrid calculations did not show these instability tendencies.

Convergence data are shown in Table 2. Notice that for the turbulent calculation the number of WU needed for convergence decreases when the mesh gets finer. The low Reynolds treatment demands high resolution near the walls, and although the mesh is expanding (see Fig. 1), 20 nodes were needed in the vertical direction. Therefore the low Reynolds area is defined from the 40x20 grid, where the first grid line is selected to be the one-equation region. A 5-level V-cycle is used for the turbulent case on a 320x160 grid, and the estimated speedup here is around 100, which is significant. Fig. 2 shows the convergence history.

4.2 2D VENTILATED ENCLOSURE

MODEL	LAMINAR		LOW-RE $k-\epsilon$			
RE	100		9000			
SCHEME	HYBRID		HYBRID		QUICK + VAN-LEER	
	WU	SPEEDUP	WU	SPEEDUP	WU	SPEEDUP
10x10	41	1.0	119	1.0	169	1.0
20x20	54	0.9	67	1.7	84	1.6
40x40	100	1.7	51	6.0	59	6.2
80x80	95	5.2	39	52.0	44	44.9
160x160	87	22.6	34	153.6	43	118.1
320x320	134	50*	30	600*	45	450*

TABLE 3. Convergence data for the 2D ventilated enclosure

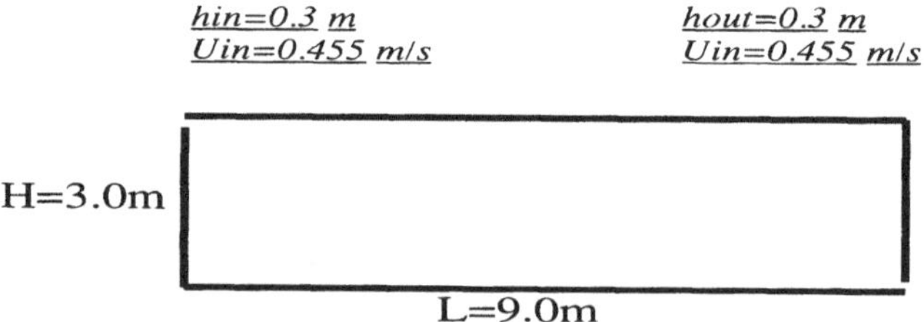

FIGURE 3. The 2D ventilated enclosure

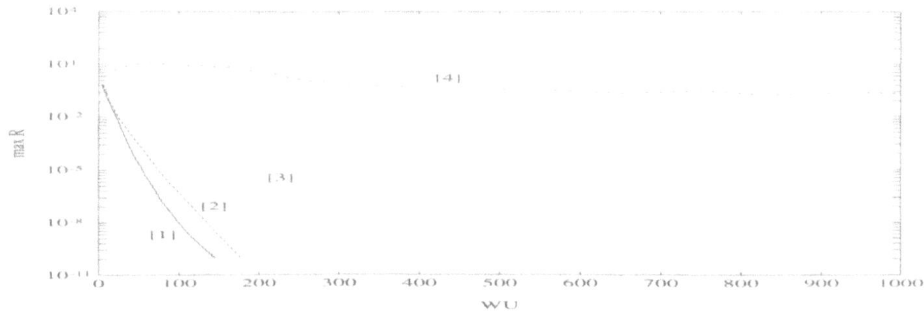

FIGURE 4. Convergence history of the 160x160 calculations for the 2D ventilated enclosure .[1]: Hybrid,turbulent,5-level FMG. [2]: QUICK,turbulent,5-level FMG. [3]: Hybrid,laminar,5-level FMG. [4]: Hybrid,turbulent,single grid.

Next configuration is a two-dimensional model of a ventilated enclosure shown in Fig. 3

The Reynolds number based on the inlet height is 100 for the laminar case and 9000 for the turbulent case. These turbulent calculations proved to be even more robust than for the backwards facing step. This is shown in Table 3 where the number of required WU decreases significantly with the grid density. Here, too, a non-uniform expanding mesh is used in order to be able to have one finite volume of the coarsest grid in the low Reynolds area.

Here the speedup is even more significant, and a 6-level V-cycle is used for the turbulent case on a 320x320 grid, wich converged 600 times faster than the estimated CPU-time of the corresponding single grid calculation. It is worth mentioning that the 320x320 FMG calculation is performed within 30 minutes on a work station (DEC 3000/400). In Figure. 4 the convergence history is shown for the 160x160 grid.

4.3 3D VENTILATED ENCLOSURE

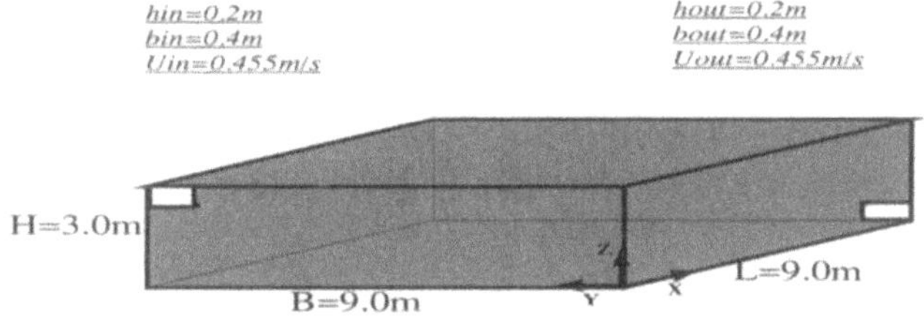

FIGURE 5. The 3D ventilated enclosure

The configuration of the three-dimensional ventilated enclosure is shown in Fig. 5

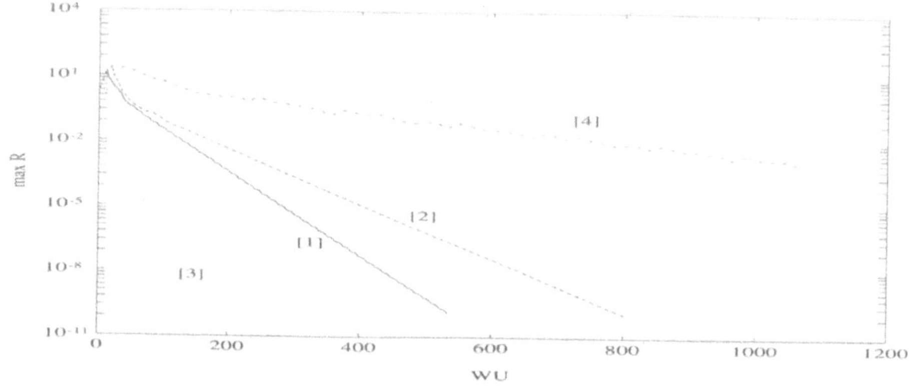

FIGURE 6. Convergence history of the 40x40x40 calculations for the 3D ventilated enclosure. [1]: Hybrid,turbulent,4-level FMG. [2]: QUICK,turbulent,4-level FMG. [3]: Hybrid,laminar,4-level FMG. [4]: Hybrid, turbulent single grid

The Reynolds number based on the inlet hydraulic diameter is 100 for the laminar case and 8200 for the turbulent case. The turbulent flow becomes very complex inside the enclosure, which is shown in Fig. 7.

This complexity also affects the convergence rate, where a typical number of required WU is around 200 while for the other two cases only around 50 were needed. Table 3 shows, however, that the number of required WU is constant or decreasing for increasing grid density, except for the finest laminar cube. Here, too, a non-uniform expanding mesh is used to be able to have one finite volume in the low Reynolds area at the coarsest mesh.

The speedup for this configuration is not very significant since the grid density is not very high. Nevertheless, a speedup factor of 20 at a 4-level V-cycle with the $k - \epsilon$ model on a 80x80x80 grid is considerable. Fig. 6 shows the convergence history for the 40x40x40 grid.

It is interesting to note that when injection was used instead of this prolongation, on the QUICK-van Leer combination, the only effect was that the convergence rate was slowed down by a factor 2. If the weights were set as if the mesh was uniform, the convergence rate was not significantly affected either. On non-orthogonal meshes, where the evaluation of the local weights is too time consuming, the convergence would therefore probably not be much affected much if fixed weights were used.

5 Closure

Some indications from the present investigation are worth pointing out:

1. Multigrid accelleration is highly effective in 2D and 3D laminar and turbulent

MODEL	LAMINAR		LOW-RE $k-\epsilon$			
RE	100		8200			
SCHEME	HYBRID		HYBRID		QUICK + VAN-LEER	
	WU	SPEEDUP	WU	SPEEDUP	WU	SPEEDUP
10x10x10	52	1.0	213	1.0	478	1.0
20x20x20	45	2.8	215	2.0	332	∞
40x40x40	49	9.0	170	4.8	221	∞
80x80x80	116	20*	218	20*	-	-

TABLE 4. Convergence data for 3D ventilated enclosure

flows, with observed speedup factors larger than 100.
2. The CPU-time of the multigrid calculations is linearly dependent of the number of nodes for both laminar and turbulent flows.
3. The effectivness is not greatly affected of grid non-uniformity.
4. Neumann boundary conditions can be handled as well as Dirchlet conditions.

6 References

[1] A. Brandt, Multi-level adaptive solutions to boundary-value problems. *Math. of Comput.*, Vol.31 333-390 (1977)
[2] A. Brandt, *Multigrid techniques: 1984 guide with applications to fluid dynamics.* Computational fluid dynamics lecture notes at von-Karman Institute, (1984)
[3] G. J. Shaw and S. Sivaloganathan, On the smoothing properties of the SIMPLE pressure-correction algorithm, *Int. J. Num. Meth. Fluids*,Vol. 8, 441-461 (1988)
[4] Y. Li, L. Fuchs S. Holmberg, An evaluation of a computer code for predicting indoor airflow and heat transfer., *12th AIVC Conf.* , Ottawa Canada (1991)
[5] M. Perić, M. Rüger and G. Scheuerer, A finite volume multigrid method for calculating turbulent flows., *Proc. 7th Symp. on Turb. Shear Flows*,7.3.1-7.3.6 Stanford, (1989)
[6] F.S. Lien, M.A. Leschziner, Multigrid convergence accelleretion for complex flow including turbulence., *Multigrid methods III* Birkhäuser Verlag. (1991)
[7] H.C. Chen V.C. Patel, Practical near-wall turbulence models for complex flows including separation., *AIAA paper 87-1300*, Honolulu (1987)
[8] L. Davidson, B. Farhanieh, A finite volume code employing colocated vari-

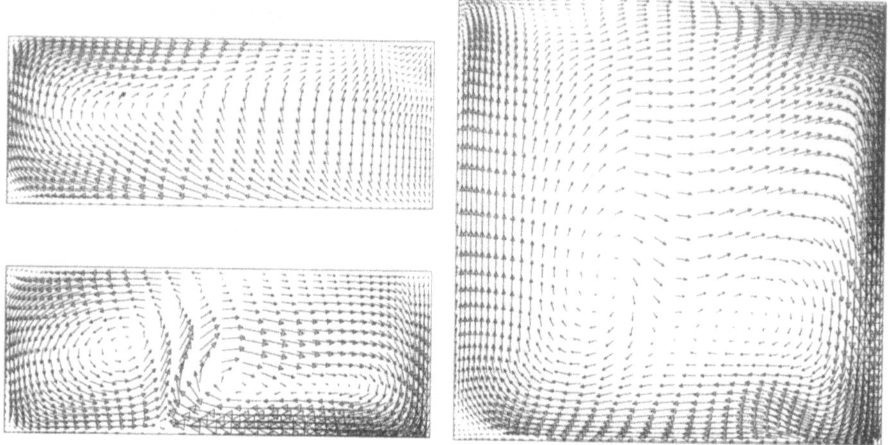

FIGURE 7. Vector plots at x/L=0.5 (upper left), y/B=0.5 (lower left) and z/H=0.5 for the three-dimensional ventilated room

able arrangement and cartesian velocity components of fluid flow and heat transfer in complex three-dimensional geometries. Rept. 91/14 Dep. of Thermo- and Fluid Dyn.,Chalmers Univ. of Techn. (1991)

[9] S.V. PATANKAR, *Numerical heat transfer and fluid flow*, Hemispere Publishing Co., McGraw Hill, (1980)

[10] C.M RHIE, W.L. CHOW, Numerical study of the turbulent flow past an airfoil with trailing edge separation., *AIAA J.* Vol. 21, 1525-1532 (1983)

[11] B.P. LEONARD, A stable and accurate convective modelling procedure based on quadratic upstream interpolation *Comput. Meth. in Appl. Mech and Eng.* Vol. 19, 59-98 (1979)

[12] B. VAN LEER, Towards the ultimate conservation difference scheme. II. Monotonocity and conservation combined in a second-order scheme *J. of Comput. Phys.* Vol. 14, 361-370 1974

[13] P. JOHANSSON, A three-dimensional laminar multigrid method applied to the SIMLEC algorithm, Diploma Thesis, Rept. 92/5 Dep. of Thermo- and Fluid Dyn.,Chalmers Univ. of Techn. (1992)

[14] T. HELLSTRÖM, DAVIDSON L., A multiblock-moving mesh extension to the CALC-BFC code. Rept. 93/3 Dep. of Thermo- and Fluid Dyn.,Chalmers Univ. of Techn. (1993)

11

Multigrid Methods for Mixed Finite Element Discretizations of Variational Inequalities

Tilman Neunhoeffer[1]

ABSTRACT We present a scheme for solving variational inequalities of obstacle type using mixed finite elements for discretization and we show the equivalence to a modified nonconforming method. This is solved using suitable multigrid methods. Numerical results are given for the elastic–plastic torsion of a cylindrical bar and the dam problem.

1 Introduction

We consider stationary variational inequalities of obstacle type:

Problem 1.1 *Find* $\tilde{u} \in C := \{\tilde{v} \in H_0^1(\Omega) : \tilde{v} \geq \Psi$ *a.e. in* $\Omega\}$ *such that*

$$\int_\Omega a \operatorname{grad}\tilde{u} \cdot \operatorname{grad}(\tilde{v} - \tilde{u})\, dx \geq \int_\Omega f\, (\tilde{v} - \tilde{u})\, dx \qquad \forall \tilde{v} \in C \tag{1.1}$$

where Ω denotes a bounded domain in $I\!R^2$, $f \in L^2(\Omega)$, $\Psi \in H^1(\Omega)$ with $\Psi \leq 0$ a.e. on $\Gamma := \partial\Omega$ and a is a symmetric 2×2 matrix-valued function on Ω. We assume that there exists $\alpha > 0$, such that

$$\sum_{i,j} a_{ij}(x)\xi_i\xi_j \geq \alpha|\xi|^2 \qquad \forall x \in \Omega,\ \forall \xi \in I\!R^2, \quad a_{ij} \in L^\infty(\Omega),\ 1 \leq i,j \leq 2. \tag{1.2}$$

It is well known that Problem 1.1 has a unique solution $\tilde{u}$ (cf. e.g. [10]).

For strongly varying or even discontinuous coefficient functions a the discretization by mixed finite element methods is considered as superior to standard conforming techniques. A mixed formulation of Problem 1.1 has been developped by Brezzi, Hager and Raviart [5]. In particular, using the lowest order Raviart–Thomas elements and an appropriate post–processing technique the resulting scheme is equivalent to a nonconforming method of inverse average type

[1]Institut für Informatik, Technische Universität München, D-80290 München, Germany

based on the Crouzeix–Raviart elements augmented by suitable cubic bubble functions. This has been established by Arnold and Brezzi [1] in the unconstrained case where for piecewise constant coefficient functions the components of the discrete solution associated with the nonconforming part and the bubbles totally decouple and thus can be computed independently.

However, as we shall show, such a decoupling does not apply in the case of variational inequalities, since there is an inherent global coupling of the nonconforming part and the bubbles caused by the constraints. Taking care of the constraints by piecewise constant Lagrangian multipliers and using static condensation we end up with a variational inequality in terms of the multipliers. This variational inequality is then solved by an outer–inner iterative scheme with an active set strategy in the outer iterations and suitable multigrid methods for the inner iterations.

The proposed mixed finite element technique is compared with standard conforming methods and its efficiency is illustrated by some numerical results for variational inequalities arising from elastomechanical applications and stationary flow problems in porous media.

2 A mixed formulation for the obstacle problem

The mixed formulation of the obstacle problem can be stated as

Problem 2.1 *Find* $p \in H(\mathrm{div}, \Omega)$ *and* $u \in K := \{v \in L^2(\Omega) : v \geq \Psi \text{ a.e. in } \Omega\}$ *such that*

$$\int_\Omega cp \cdot q \, dx + \int_\Omega u \, \mathrm{div} q \, dx = 0 \qquad \forall q \in H(\mathrm{div}, \Omega), \tag{2.1a}$$

$$\int_\Omega (\mathrm{div} p + f)(v - u) \, dx \leq 0 \qquad \forall v \in K \tag{2.1b}$$

where $c := a^{-1}$ and $H(\mathrm{div}, \Omega) := \{q \in (L^2(\Omega))^2 : \mathrm{div} q \in L^2(\Omega)\}$.

The standard problem 1.1 and the mixed problem 2.1 are related as follows (cf. [5]).

Theorem 2.1 *Suppose that the unique solution* $\tilde{u}$ *of Problem 1.1 is in* $H^2(\Omega)$, *then* $(a \, \mathrm{grad} \tilde{u}, \tilde{u})$ *is the unique solution of Problem 2.1.*

For the discretization of Problem 2.1 we consider a regular triangulation $\mathcal{T}_h$ of Ω which, for simplicity, is supposed to be a polygon. Let $\mathcal{P}^k(T)$ denote the polynomials of degree $\leq k$ on T and

$$RT_{-1}^0(\mathcal{T}_h) := \{ q_h \in (L^2(\Omega))^2 : q_h|_T := \sigma(x) + x\,\tau(x), \quad \tau \in \mathcal{P}^0(T), \sigma \in \mathcal{P}^0(T) \times \mathcal{P}^0(T) \quad \forall\, T \in \mathcal{T}_h\}. \tag{2.2}$$

the space of Raviart–Thomas elements of lowest order [11].
Approximating $H(\mathrm{div}, \Omega)$ by

$$\begin{aligned} RT_0^0(\mathcal{T}_h) &:= RT_{-1}^0(\mathcal{T}_h) \cap H(\mathrm{div}, \Omega) = \\ &= \{q_h \in RT_{-1}^0(\mathcal{T}_h) : q_h \cdot n \text{ is continuous across the interelement boundaries of } \mathcal{T}_h\} \end{aligned} \tag{2.3}$$

and $L^2(\Omega)$ by

$$M_{-1}^0(\mathcal{T}_h) := \{v_h \in L^2(\Omega) : v_h|_T \in \mathcal{P}^0(T) \quad \forall\, T \in \mathcal{T}_h\}, \tag{2.4}$$

we get the finite dimensional counterpart of Problem 2.1:

Problem 2.2 *Find $p_h \in RT_0^0(\mathcal{T}_h)$ and $u_h \in K_h(\mathcal{T}_h) := \{v_h \in M_{-1}^0(\mathcal{T}_h) : v_h \geq \Psi$ a.e. in $\Omega\}$ such that*

$$\int_\Omega c\,p_h \cdot q_h \, dx + \int_\Omega u_h \, \mathrm{div} q_h \, dx = 0 \qquad \forall q_h \in RT_0^0(\mathcal{T}_h), \tag{2.5a}$$

$$\int_\Omega (\mathrm{div} p_h + f)(v_h - u_h) \, dx \leq 0 \qquad \forall v_h \in K_h(\mathcal{T}_h). \tag{2.5b}$$

This problem has a unique solution $(p_h, u_h) \in RT_0^0(\mathcal{T}_h) \times K_h(\mathcal{T}_h)$. A priori estimates for the discretization error can be found in [5].

A common technique for the efficient numerical solution of the saddle point problem 2.2 is hybridization: The continuity constraints on the interelement boundaries are eliminated from the ansatz space $RT_0^0(\mathcal{T}_h)$ and instead are taken care of by appropriate Lagrangian multipliers. For this purpose we denote by $\mathcal{E}_h$ the set of edges of triangles in $\mathcal{T}_h$, $\mathcal{E}_h^\partial := \{e \in \mathcal{E}_h : e \subset \partial\Omega\}$, $\mathcal{E}_h^0 := \mathcal{E}_h \setminus \mathcal{E}_h^\partial$ and we set

$$M_{-1}^0(\mathcal{E}_h^0) := \{\mu_h : \mu_h|_e \in \mathcal{P}^0(e) \quad \forall e \in \mathcal{E}_h^0, \quad \mu_h|_e = 0 \quad \forall e \in \mathcal{E}_h^\partial\}. \tag{2.6}$$

The well known equivalence

Lemma 2.1 (cf. Lemma 1.2 in [1]) *If $q_h \in RT_{-1}^0(\mathcal{T}_h)$, then $q_h \in RT_0^0(\mathcal{T}_h)$ iff*

$$\sum_{T \in \mathcal{T}_h} \int_{\partial T} \mu\, q_h \cdot n_T \, de = 0 \qquad \forall \mu \in M_{-1}^0(\mathcal{E}_h^0) \tag{2.7}$$

tells us how to choose the multipliers. We thus get the extended saddle point problem

Problem 2.3 *Find $\bar{p}_h \in RT^0_{-1}(\mathcal{T}_h)$, $\bar{u}_h \in K_h(\mathcal{T}_h)$, $\lambda_h \in M^0_{-1}(\mathcal{E}^0_h)$ such that*

$$\int_\Omega c\,\bar{p}_h \cdot q_h\,dx + \sum_T \left(\int_T \bar{u}_h\,\mathrm{div} q_h\,dx - \int_{\partial T} \lambda_h\,q_h \cdot n_T\,de \right) = 0 \qquad \forall q_h \in RT^0_{-1}(\mathcal{T}_h), \tag{2.8a}$$

$$\sum_T \int_T \mathrm{div}\bar{p}_h\,(v_h - \bar{u}_h)\,dx + \int_\Omega f\,(v_h - \bar{u}_h)\,dx \le 0 \quad \forall v_h \in K_h(\mathcal{T}_h), \tag{2.8b}$$

$$\sum_T \int_{\partial T} \mu_h\,\bar{p}_h \cdot n_T\,de = 0 \quad \forall \mu_h \in M^0_{-1}(\mathcal{E}^0_h). \tag{2.8c}$$

Problem 2.3 has a unique solution $(\bar{p}_h, \bar{u}_h, \lambda_h)$. Moreover, if (p_h, u_h) is the unique solution of Problem 2.2, then $\bar{p}_h = p_h$ and $\bar{u}_h = u_h$.

Note that the matrix associated with $\int_\Omega c\,p_h \cdot q_h\,dx$ is block diagonal consisting of 3×3 blocks and hence can be inverted easily. Then, elimination of p_h leads to a reduced system with a symmetric, positive definite coefficient matrix. However, we will follow another approach which, in the unrestricted case, has been proposed by Arnold and Brezzi in [1]. Taking into account that the multiplier λ_h represents an approximation of u on the interelement boundaries, by an appropriate post-processing u_h and λ_h can be used to construct a new approximation of u which can be computed as the solution of a variational inequality related to a specific nonconforming discretization. In particular, we define

$$N^1(\mathcal{T}_h) := M^1_{NC}(\mathcal{T}_h) + B^3(\mathcal{T}_h) \tag{2.9}$$

where $M^1_{NC}(\mathcal{T}_h)$ refers to the lowest order nonconforming Crouzeix–Raviart elements

$$\begin{aligned} M^1_{NC}(\mathcal{T}_h) := \{\ & v_h \in M^1_{-1}(\mathcal{T}_h) : v_h|_T \in \mathcal{P}^1(T)\ \ \forall T \in \mathcal{T}_h, \\ & v_h \text{ is continuous at the midpoints of each } e \in \mathcal{E}^0_h \text{ and} \\ & v_h = 0 \text{ at the midpoints of each } e \in \mathcal{E}^\partial_h \} \end{aligned} \tag{2.10}$$

which in (2.9) are augmented by the cubic bubbles

$$B^3(\mathcal{T}_h) := \{ v \in M^3_0(\mathcal{T}_h) : v_h|_T \in \mathcal{P}^3(T)\ \ \forall T \in \mathcal{T}_h,\ \ v|_e = 0\ \ \forall e \in \mathcal{E}_h \}. \tag{2.11}$$

Denoting by P^0_h and Π^0_h the projections onto $M^0_{-1}(\mathcal{T}_h)$ and $M^0_{-1}(\mathcal{E}^0_h)$, respectively, we consider $w_h \in N^1(\mathcal{T}_h)$ such that

$$P^0_h w_h = u_h, \qquad \Pi^0_h w_h = \lambda_h. \tag{2.12}$$

Existence and uniqueness are guaranteed by means of

Lemma 2.2 (cf. Lemma 2.3 in [1]) *For any $v_h \in M^0_{-1}(\mathcal{T}_h)$ and any $\mu_h \in M^0_{-1}(\mathcal{E}^0_h)$ there exists a unique $z_h \in N^1(\mathcal{T}_h)$ such that*

$$P^0_h z_h = v_h, \qquad \Pi^0_h z_h = \mu_h. \tag{2.13}$$

Consequently, we are faced with the following

Problem 2.4 *Find* $p_h \in RT^0_{-1}(\mathcal{T}_h)$, $w_h \in \bar{K}_h(\mathcal{T}_h) := \{z_h \in N^1(\mathcal{T}_h) : P^0_h z_h \geq \Psi\}$ *such that*

$$\int_\Omega c\, p_h \cdot q_h \, dx - \sum_T \int_T \mathrm{grad} w_h \cdot q_h \, dx = 0 \qquad \forall q_h \in RT^0_{-1}(\mathcal{T}_h), \tag{2.14a}$$

$$\sum_T \int_T p_h \cdot \mathrm{grad}(z_h - w_h)\, dx \geq \int_\Omega P^0_h f\, (z_h - w_h)\, dx \quad \forall z_h \in \bar{K}(\mathcal{T}_h). \tag{2.14b}$$

By a simple application of Green's formula we can show:

Theorem 2.2 *Let* (p_h, u_h, λ_h) *be the unique solution of Problem 2.3 and let* w_h *be defined by (2.12), then* (p_h, w_h) *is the unique solution of Problem 2.4.*

Denoting by $P^0_{RT.c}$ the projection onto $RT^0_{-1}(\mathcal{T}_h)$ with respect to the inner product

$$[u, v] := \sum_T \int_T c\, u \cdot v\, dx \tag{2.15}$$

we can write (2.14a) as

$$p_h = P^0_{RT.c}(a\, \mathrm{grad} w_h). \tag{2.16}$$

Substituting (2.16) for p_h in (2.14b), we get

Problem 2.5 *Find* $w_h \in \bar{K}_h(\mathcal{T}_h)$ *such that*

$$\sum_T \int_T P^0_{RT.c}(a\, \mathrm{grad} w_h) \cdot \mathrm{grad}(z_h - w_h)\, dx \geq \int_\Omega (P^0_h f)\, (z_h - w_h)\, dx \quad \forall z_h \in \bar{K}_h(\mathcal{T}_h). \tag{2.17}$$

For elementwise constant a we have a decoupling of (2.17) into the nonconforming and the bubble part. However in contrast to the unrestricted case (cf. [1]) the two parts cannot be solved independently, since there still is a global coupling caused by the constraints.

3 Iterative solution by nonconforming multigrid techniques

We shall now give the algebraic formulation of Problem 2.5 in case of piecewise constant coefficient functions a. Therefore we introduce a set of basis functions

$$\Phi := \{\varphi_1^{NC}, \ldots, \varphi_{p_h}^{NC}, \varphi_1^B, \ldots, \varphi_{q_h}^B\} \tag{3.1}$$

where φ_i^{NC} is a nodal basis function of $M^1_{NC}(\mathcal{T}_h)$ with $\varphi_i^{NC}(m_j) = \delta_{ij}$, $1 \leq i, j \leq p_h$ and φ_i^B is a nodal basis function of $B^3(\mathcal{T}_h)$ with $\varphi_i^B(s_j) = \delta_{ij}$, $1 \leq i, j \leq q_h$.

Here p_h denotes the number of edges of $\mathcal{E}_h^0$ and q_h denotes the number of triangles of $\mathcal{T}_h$, m_j is the midpoint of the edge j and s_j the center of gravity of the triangle j. Then each element of $N^1(\mathcal{T}_h)$ can be represented by a coefficient vector $\underline{z} := (\underline{z}^{NC}, \underline{z}^B)^T \in \mathbb{R}^{p_h+q_h}$ with

$$z_h = \sum_{i=1}^{p_h} z_i^{NC} \varphi_i^{NC} + \sum_{j=1}^{q_h} z_j^B \varphi_j^B. \tag{3.2}$$

In view of the decoupling of the nonconforming and the bubble inequalities, Problem 2.5 has the algebraic form

Problem 3.1 *Find* $\underline{w} := \begin{pmatrix} \underline{w}^{NC} \\ \underline{w}^B \end{pmatrix} \in K$ *such that for all* $\underline{z} := \begin{pmatrix} \underline{z}^{NC} \\ \underline{z}^B \end{pmatrix} \in K$*:*

$$\left\langle \begin{pmatrix} A_{11} & 0 \\ 0 & D \end{pmatrix} \begin{pmatrix} \underline{w}^{NC} \\ \underline{w}^B \end{pmatrix}, \begin{pmatrix} \underline{z}^{NC} - \underline{w}^{NC} \\ \underline{z}^B - \underline{w}^B \end{pmatrix} \right\rangle \geq \left\langle \begin{pmatrix} \underline{b}^{NC} \\ \underline{b}^B \end{pmatrix}, \begin{pmatrix} \underline{z}^{NC} - \underline{w}^{NC} \\ \underline{z}^B - \underline{w}^B \end{pmatrix} \right\rangle \tag{3.3}$$

where

$$K := \left\{ \underline{z} = \begin{pmatrix} \underline{z}^{NC} \\ \underline{z}^B \end{pmatrix} : \underline{z}^{NC} \in \mathbb{R}^{p_h},\ \underline{z}^B \in \mathbb{R}^{q_h},\ P^{NC}\underline{z}^{NC} - P^B\underline{z}^B \geq \underline{\Psi} \right\}. \tag{3.4}$$

A_{11} and D are the matrices corresponding to the bilinear form $\int_\Omega P_{RT.c}^0(a\,\mathrm{grad} w_h) \cdot \mathrm{grad} z_h\, dx$, where D is diagonal and A_{11} is the usual matrix associated with the Crouzeix–Raviart elements. We set $A := \begin{pmatrix} A_{11} & 0 \\ 0 & D \end{pmatrix}$. Further $\underline{b} := (\underline{b}^{NC}, \underline{b}^B)^T$ is the vector associated with the right hand side $\int_\Omega (P_h^0 f) z_h\, dx$, $\underline{\Psi} \in \mathbb{R}^{q_h}$ is the projection of the obstacle function Ψ with $\Psi_i := \int_{T_i} \Psi\, dx$ and $P := (P^{NC}, P^B)$ denotes the projection matrix according to the operator P_h^0.

We would like to solve this system with an active–set–strategy (see [8]) where in an outer iteration a set of active and inactive points is determined and in an inner iteration a linear system reduced on the inactive points is solved. For determining the active and inactive points we need an equal number of inequalities and constraints, but here we have $p_h + q_h$ inequalities and only q_h constraints. That is why we switch to the constrained minimization problem which is equivalent to Problem 3.1:

Problem 3.2 *Find* $\underline{w} \in K$ *such that*

$$J(\underline{w}) = \inf_{\underline{z} \in K} J(\underline{z}) \tag{3.5a}$$

where

$$J(\underline{z}) = \frac{1}{2}\langle A\underline{z}, \underline{z}\rangle - \langle \underline{b}, \underline{z}\rangle. \tag{3.5b}$$

The constraint $P\underline{z} \geq \underline{\Psi}$ can be added to the functional J by a Lagrangian multiplier $\underline{\lambda} \in I\!R^{q_h}$ which can be interpreted as an elementwise constant function $\lambda_h \in M^0_{-1}(\mathcal{T}_h)$. We have $\lambda_h = 0$ for elements where the obstacle for w_h is inactive and $\lambda_h > 0$ for the elements where the obstacle is active.

So we consider the extended problem

Problem 3.3 *Find $\underline{w} \in I\!R^{p_h+q_h}$ and $\underline{\lambda} \in \Lambda := \{\underline{\mu} \in I\!R^{q_h} : \mu_i \geq 0,\ \ 1 \leq i \leq q_h\}$ such that*

$$L(\underline{w}, \underline{\lambda}) = \inf_{\underline{z} \in I\!R^{p_h+q_h}} \sup_{\underline{\mu} \in \Lambda} L(\underline{z}, \underline{\mu}) \tag{3.6a}$$

where

$$L(\underline{z}, \underline{\mu}) := \frac{1}{2}\langle A\underline{z}, \underline{z}\rangle - \langle \underline{b}, \underline{z}\rangle + \langle \underline{\mu}, \underline{\Psi} - P\underline{z}\rangle. \tag{3.6b}$$

The unique solution of Problem 3.3 is characterized by

Problem 3.4 *Find $\underline{w} \in I\!R^{p_h+q_h}$ and $\underline{\lambda} \in \Lambda$ such that*

$$A\underline{w} - P^T\underline{\lambda} \;=\; \underline{b}, \tag{3.7a}$$

$$\langle \underline{\lambda} - \underline{\mu}, \underline{\Psi} - P\underline{w}\rangle \;\geq\; 0 \qquad \forall \underline{\mu} \in \Lambda. \tag{3.7b}$$

Note that (3.7b) is equivalent to the complementarity problem

$$\min(\underline{\lambda}, P\underline{w} - \underline{\Psi}) = 0 \tag{3.8}$$

where the minimum has to be understood componentwise. Eliminating $\underline{w}$ in (3.7b) by means of (3.7a), we end up with a complementarity problem for the multiplier $\underline{\lambda}$:

Problem 3.5 *Find $\underline{\lambda} \in I\!R^d$ such that*

$$\min(\underline{\lambda}, \hat{A}\underline{\lambda} - \hat{\underline{b}}) = 0 \tag{3.9}$$

where $\hat{A} := P^{NC}A_{11}^{-1}(P^{NC})^T + P^BD^{-1}(P^B)^T$ and $\hat{\underline{b}} := \underline{\Psi} - P^{NC}A_{11}^{-1}\underline{b}^{NC} - P^BD^{-1}\underline{b}^B$.

This complementarity problem for the Lagrangian multiplier will be solved by an outer–inner iterative scheme where the outer iteration is based on an active set strategy and the inner iterations are cg–iterations with the action of A_{11}^{-1} being replaced by multigrid iterations for nonconforming finite elements as proposed by Braess and Verfürth [3]. For the motivation of the active set strategy we remark that (3.9) is equivalent to the constrained minimization problem

$$\hat{J}(\underline{\lambda}) = \inf_{\underline{\mu} \in \Lambda} \hat{J}(\underline{\mu}) \tag{3.10}$$

where $\hat{J}(\underline{\mu}) := \frac{1}{2}\,\underline{\mu}^T\hat{A}\underline{\mu} - \underline{\mu}^T\hat{\underline{b}}$. Then, given an iterate $\underline{\lambda}^{(\nu)}$ and proceeding in descent direction $-\nabla\hat{J}(\underline{\lambda}^{(\nu)})$ we have

$$\underline{\lambda}^{(\nu)} - \nabla\hat{J}(\underline{\lambda}^{(\nu)}) \in \mathrm{int}\Lambda$$

$$\Longleftrightarrow$$

$$\lambda_i^{(\nu)} - (\hat{A}\underline{\lambda}^{(\nu)} - \underline{\hat{b}})_i > 0, \quad i \in I_h := \{1, 2, \ldots, q_h\}. \tag{3.11}$$

Consequently, an element is said to be active if (3.11) is violated, i.e.

$$\lambda_i^{(\nu)} \leq (\hat{A}\underline{\lambda}^{(\nu)} - \underline{\hat{b}})_i. \tag{3.12}$$

We denote by $I_h^{(2)}(\underline{\lambda}^{(\nu)})$ the set of all active elements and refer to its complement $I_h^{(1)}(\underline{\lambda}^{(\nu)}) := I_h \setminus I_h^{(2)}(\underline{\lambda}^{(\nu)})$ as the set of inactive elements. The new iterate is then determined by

$$\lambda_i^{(\nu+1)} = 0, \quad i \in I_h^{(2)}(\underline{\lambda}^{(\nu)})$$

with respect to the active elements while the components associated with inactive elements are computed as the solution of the reduced linear system

$$(\hat{A}\underline{\lambda}^{(\nu+1)})_i = \underline{\hat{b}}, \quad i \in I_h^{(1)}(\underline{\lambda}^{(\nu)}). \tag{3.13}$$

The outer iteration will be stopped if $I_h^{(1)}(\underline{\lambda}^{(\nu+1)}) = I_h^{(1)}(\underline{\lambda}^{(\nu)})$.

The reduced systems will be solved by means of the following cg–iterations constituting the inner iterations.

Inner cg-iterations:

Step 1: Let $\underline{\mu}^{(0)}$ be given by

$$\mu_i^{(0)} := \begin{cases} \lambda_i^{(\nu)} & \text{if } \ i \in I_h^{(1)}(\underline{\lambda}^{(\nu)}) \\ 0 & \text{if } \ i \in I_h^{(2)}(\underline{\lambda}^{(\nu)}) \end{cases},$$

compute $\underline{\eta}^{(0)}$ as the solution of

$$A_{11}\underline{\eta}^{(0)} = \underline{b}^{NC} + (P^{NC})^T\underline{\mu}^{(0)} \tag{3.14}$$

and set

$$\begin{aligned} r_i^{(0)} &:= \begin{cases} \Psi_i - (P^{NC}\underline{\eta}^{(0)})_i - (P^B D^{-1}(P^B)^T\underline{\mu}^{(0)})_i - \\ \qquad\qquad -(P^B D^{-1}\underline{b}^B)_i \quad \text{if } \ i \in I_h^{(1)}(\underline{\lambda}^{(\nu)}) \\ 0 \ \text{ if } \ i \in I_h^{(2)}(\underline{\lambda}^{(\nu)}) \end{cases} \\ \underline{p}^{(0)} &:= \underline{r}^{(0)}. \end{aligned}$$

Step 2: For $m = 0, 1, 2, \ldots$

Compute $\underline{\eta}^{(m+1)}$ as the solution of

$$A_{11}\underline{\eta}^{(m+1)} = (P^{NC})^T\underline{p}^{(m)} \tag{3.15}$$

$$
\begin{aligned}
z_i^{(m)} &:= \begin{cases} (P^{NC}\underline{\eta}^{(m+1)})_i + (P^B D^{-1}(P^B)^T \underline{p}^{(m)})_i & \text{if} \quad i \in I_h^{(1)}(\underline{\lambda}^{(\nu)}) \\ 0 \quad \text{if} \quad i \in I_h^{(2)}(\underline{\lambda}^{(\nu)}) \end{cases} \\
\alpha &:= < \underline{r}^{(m)}, \underline{r}^{(m)} > / < \underline{z}^{(m)}, \underline{p}^{(m)} > \\
\underline{\mu}^{(m+1)} &:= \underline{\mu}^{(m)} + \alpha\, \underline{p}^{(m)} \\
\underline{r}^{(m+1)} &:= \underline{r}^{(m)} - \alpha\, \underline{z}^{(m)} \\
\beta &:= < \underline{r}^{(m+1)}, \underline{r}^{(m+1)} > / < \underline{r}^{(m)}, \underline{r}^{(m)} > \\
\underline{p}^{(m+1)} &:= \underline{r}^{(m+1)} - \beta\, \underline{p}^{(m)}
\end{aligned}
$$

Step 3: Set $\underline{\lambda}^{(\nu+1)} := \underline{\mu}^{(m+1)}$

The linear systems (3.14) and (3.15) are solved iteratively by the multigrid algorithm for nonconforming finite elements suggested by Braess and Verfürth [3].

Having determined the multiplier $\underline{\lambda}$, we can compute $\underline{w}$ by (3.7a). For edges e_i where $\lambda_{i_1} > 0$ or $\lambda_{i_2} > 0$ for the two adjacent elements T_{i_1} and T_{i_2} of e_i, we say that the obstacle is active and we set $w_i^{NC} = 1/2\,(\Psi_{i_1} + \Psi_{i_2})$. For the inactive edges $i \in I^{(1)}(\underline{\lambda}) := \{1 \le i \le p_h : \lambda_{i_1} = \lambda_{i_2} = 0\}$ we have $((P^{NC})^T \underline{\lambda})_i = 0$. So we only must solve the linear system

$$(A_{11}\underline{w}^{NC})_i = b_i^{NC} \tag{3.16}$$

reduced on the inactive edges $i \in I^{(1)}(\underline{\lambda})$. This is done again by the multigrid algorithm of Braess and Verfürth. For the bubble part, we set

$$w_i^B = (D_{ii})^{-1} b_i^B \quad \text{if} \quad \lambda_i = 0$$

and we determine w_i^B, such that $w_h = \Psi$ in the center of gravity of T_i if $\lambda_i > 0$. We see that actually we do not need the multiplier. We only need the active and inactive elements.

4 Numerical Results

In our computations [2] we used 2 pre– and 2 postsmoothing steps in the nonconforming multigrid–method (V–cycle) to solve the linear systems where A_{11} appears. Normally we stopped the iteration, when the difference between two iterates was lower than $\varepsilon_{NC} = 10^{-5}$, but at latest after 10 multigrid cycles.

We restrict ourselve on examples with piecewise constant coefficient matrices, which are diagonal with same diagonal elements. So they can be replaced by scalars.

The results of the proposed mixed algorithm are compared with the results of a conforming technique where a preconditioner is used in the inner iterations (see [7],[9]). It is part of an adaptive scheme but we only used uniform refinements.

[2] All computations have been performed on a SUN SPARCstation 2

4.1 The Elastic–Plastic Torsion of a Cylindrical Bar

First we consider the torsion of an elastic–plastic cylinder of cross–section $\Omega := [0,1]\times[0,1]$ consisting of two different materials. In the inner part $\Omega_2 := [1/3, 2/3]\times[1/3, 2/3]$ holds $a = a_2$ and in the outer part $\Omega_1 := \Omega \setminus \Omega_2$ we have $a = a_1$. The matrix a describes the elastic behaviour of the material. As shown in [6], the stress potential u_C for a positive twist angle per unit length C is the solution of the variational inequality

Problem 4.1 *Find* $u_C \in K := \{v \in H_0^1(\Omega) : v(x) \leq \operatorname{dist}(x, \partial\Omega)$ *a.e. in* $\Omega\}$ *such that*

$$\int_\Omega a \operatorname{grad} u_C \cdot \operatorname{grad}(v - u_C)\, dx \geq 2C \int_\Omega (v - u_C)\, dx \qquad \forall v \in K. \tag{4.1}$$

where $\operatorname{dist}(x, \partial\Omega)$ denotes the distance between a point $x \in \Omega$ and the boundary $\partial\Omega$. In the region where the obstacle is active, the bar is plastic, in the other part it remains elastic.

We have chosen $c = 3\, a_1/a_2$ and we have used a hierarchy $(\mathcal{T}_l)_{l=0}^4$ of 4 triangulations starting from an initial coarse triangulation consisting of 18 elements. Both for the mixed and the conforming method Table 1 contains the number of inner iterations per outer iteration which are necessary to get a satisfactory convergence in the outer iterations. We see that for increasing ratios $a_1 : a_2$ the number of inner iterations for the conforming method increases more than for the mixed method.

ratio $a_1 : a_2$	1:1	10:1	50:1	100:1	500:1	1000:1
mixed method	3	4	6	9	11	12
conforming method	4	6	13	16	21	24

Table 1: Elastic–plastic torsion: Number of inner iterations

4.2 The Dam Problem

As a second example we consider a dam of porous material with cross–section $\Omega := [0,1] \times [0,1]$. We suppose the dam to consist of two vertical layers of two different materials. In the left part $\Omega_1 := [0, 1/2] \times [0,1]$ we have a permeability coefficient $k = k_1$, in the right part $\Omega_2 := \Omega \setminus \Omega_1$ we have $k = k_2$. The dam separates two water reservoirs of different levels $y_1 = 0.8$ on the left hand side and $y_2 = 0.2$ on the right hand side. The pressure can be computed by using the Baiocchi–Transformation

$$w(x,y) := \int_y^1 p(x,t)\, dt \tag{4.2}$$

as the solution of the following variational inequality.

Problem 4.2 *Find* $w \in K := \{v \in H^1(\Omega) : v \geq 0 \ in\ \Omega,\ \ v|_{\partial\Omega} = g\}$ *such that*

$$\int_\Omega k \,\mathrm{grad} w \cdot \mathrm{grad}(v-w)\, dx\, dy \geq -\int_\Omega k\,(v-w)\, dx\, dy \qquad \forall v \in K. \tag{4.3}$$

with appropriate inhomogeneous Dirichlet boundary conditions g. (For details see [2].) In the part where the obstacle for w is active the dam is dry, the other part represents the wet region.

Here we have used a hierarchy $(\mathcal{T}_l)_{l=0}^4$ of 4 triangulations starting from an initial coarse triangulation $\mathcal{T}_0$ consisting of 8 elements. Table 2 illustrates the advantage of the mixed method for high ratios as Table 1.

ratio $k_1 : k_2$	1:1	10:1	50:1	100:1	500:1	1000:1
mixed method	3	4	4	4	5	6
conforming method	3	6	10	11	15	15

Table 2: Dam problem: Number of inner iterations

References

[1] D.N. Arnold, F. Brezzi: *Mixed and Nonconforming Finite Element Methods: Implementation, Postprocessing and Error Estimates*, M²AN 19 (1985), 7–32

[2] C. Baiocchi, V. Comincioli, L. Guerri, G. Volpi: *Free Boundary Problems in the Theory of Fluid Flow through Porous Media: A Numerical Approach*, Estratto da Calcolo 10 (1973), 1–85

[3] D. Braess, R. Verfürth: *Multigrid Methods for Nonconforming Finite Element Methods*, SIAM J. Numer. Anal. 27 (1990), 979–986

[4] F. Brezzi, M. Fortin: *Mixed and Hybrid Finite Element Methods*, Springer, Berlin (1991)

[5] F. Brezzi, W.W. Hager, P.A. Raviart: *Error Estimates for the Finite Element Solution of Variational Inequalities, Part II: Mixed Methods*, Numer. Math. 31 (1978), 1–16

[6] H. Brezis, M. Sibony: *Equivalence de deux inéquations variationelles et applications*, Arch. Rat. Mech. Anal. 41 (1971), 254–265

[7] B. Erdmann, M. Frei, R.H.W. Hoppe, R. Kornhuber, U. Wiest: *Adaptive Finite Element Methods for Variational Inequalities*, to appear in East–West J. Num. Math. (1993)

[8] R.H.W. Hoppe: *Multigrid Algorithms for Variational inequalities*, SIAM J. Numer. Anal. 24 (1987), 1046–1065

[9] R.W.H. Hoppe, R. Kornhuber: *Adaptive Multilevel–Methods for Obstacle Problems*, to appear in SIAM J. Numer. Anal. (1993)

[10] D. Kinderlehrer, G. Stampacchia: *An Introduction to Variational Inequalities and their Applications*, Academic Press, New York (1980)

[11] P.A. Raviart, J.M. Thomas: *A Mixed Finite Element Method for Second Order Elliptic Problems*, in: Mathematical Aspects of the Finite Element Method, Lecture Notes in Mathemathics 606, Springer, Berlin (1977)

FIGURE 1. Elastic–plastic torsion, ratio 1000:1, $c = 3000$
The shaded area represents the plastic zone.

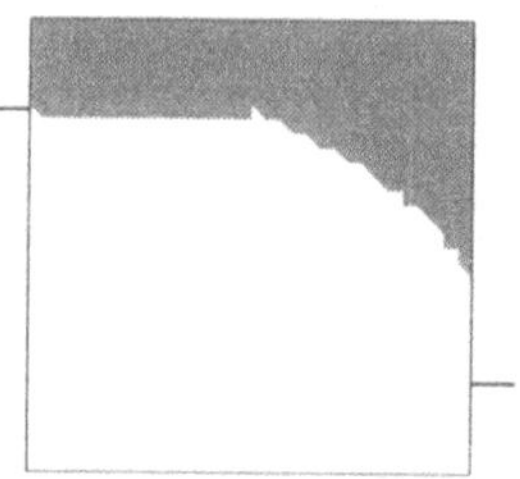

FIGURE 2. Dam problem, ratio 100:1
The shaded area represents the dry area.

12

Multigrid with Matrix–dependent Transfer Operators for Convection–diffusion Problems

Arnold Reusken[1]

1 Introduction

At several places (e.g. [1,6,8,9,13,14]) it is claimed that one should use matrix–dependent prolongations and restrictions when solving interface problems or convection–diffusion problems using multigrid. Recently, in [10], a theoretical analysis has been presented which yields a further justification of this claim. In [10] it is proved that for 1D convection- diffusion problems the use of suitable matrix–dependent transfer operators results in a multigrid method which is robust w.r.t. variation in the amount of convection, even if one uses damped Jacobi for smoothing. In section 2 we briefly discuss the approach used in [10] and give some important results from [10].
The main subject of this paper is a generalization of the 1D approach in [10] (cf. Section 2), resulting in a new multigrid method for 2D convection- diffusion problems. This 2D method is based on the following. A given matrix on a "fine" grid is modified using a suitable *lumping* procedure (e.g. a 9-point star is reduced to a 5-point star by using an approximation based on linear interpolation), then for this modified operator the Schur complement w.r.t. the coarse grid is local and can be computed with low costs. In a preprocessing phase, starting with a given matrix on the finest grid, this approach is applied recursively and (in a natural way) results in matrix- dependent prolongations and restrictions and coarse–grid operators which satisfy the Galerkin condition. Also, in this framework, a modification of the standard multigrid method is suggested: a prescribed additional correction, complementary to the coarse grid correction, is introduced.

[1]Department of Mathematics and Computing Science, Eindhoven University of Technology, P.O. Box 513, 5600 MB Eindhoven, the Netherlands. e-mail: wsanar@win.tue.nl

2 Matrix–dependent transfer operators for 1D convection–diffusion problems

In this section we briefly discuss the approach used in [10] and we give some results from [10]. In section 3 we then generalize this approach, resulting in a method for 2D problems.

We consider a second order linear elliptic two–point boundary value problem. For convenience we use a sequence of *uniform* grids, although our analysis still holds if we only assume that the grids are quasi–uniform. Let $h := 1/(N+1)$ be the mesh size parameter and $x_i := ih,\ i = 0, 1, \ldots, N+1$. We use the notation $U_h := \mathbb{R}^N$; $i \in U_h$ corresponds to the grid point x_i. For discretization we use a three–point difference scheme; the resulting operator $L_h : U_h \to U_h$ can be represented with a difference star $[-a_{h,i} \quad b_{h,i} \quad -c_{h,i}]$ in the grid point x_i. For notational convenience we drop the h index in the difference star, and thus (with $a_1 = c_N = 0$):

$$(2.1) \qquad [L_h]_i = [-a_i \quad b_i \quad -c_i] \qquad i = 1, 2, \ldots, N\ .$$

We make the following stability assumption concerning the discretization method:

$$(\text{A1}) \qquad L_h \text{ is a weakly diagonally dominant M} - \text{matrix}\ .$$

We refer to [7] for numerical discretization methods that satisfy (A1). We take an arbitrary "fine" grid with mesh size h and a corresponding "coarse" grid with mesh size $H = 2h$. The coarse grid space is denoted by $U_H = \mathbb{R}^{N_H}$ with $N_H := \frac{1}{2}(N-1)$.

We now discuss the components for a two–grid method.

Smoothing operator. For smoothing we use a (damped) Jacobi or Gauss–Seidel method.

Grid transfers. We use the matrix–dependent prolongation $p : U_H \to U_h$ and restriction $r : U_h \to U_H$ as given in [8, 13]. In stencil notation we have:

$$(2.2) \qquad [p]_i = [c_{2i-1}/b_{2i-1} \quad 1 \quad a_{2i+1}/b_{2i+1}]\ , \quad i = 1, 2, \ldots, N_H\ ,$$

$$(2.3) \qquad [r]_i = \tfrac{1}{2}[a_{2i}/b_{2i-1} \quad 1 \quad c_{2i}/b_{2i+1}]\ , \quad i = 1, 2, \ldots, N_H\ .$$

Coarse–grid operator. We use the Galerkin approach:

$$(2.4) \qquad L_H := rL_hp\ .$$

For an analysis of the two–grid method we introduce a red–black ordering of the fine–grid nodes. We define the permutation matrix $Q : U_h \to U_h$ by

$$(2.5) \qquad (Qu)_i = u_{2i-1} \quad 1 \le i \le N_H + 1, \quad (Qu)_{i+N_H+1} = u_{2i} \quad 1 \le i \le N_H\ .$$

We use the notation:

$$(2.6) \qquad \hat{p} := Qp \;,\; \hat{r} := rQ^T \;,\; \hat{L}_h := QL_hQ^T =: \begin{bmatrix} A_{11} & -A_{12} \\ -A_{21} & A_{22} \end{bmatrix} .$$

An easy calculation (cf. [10]) yields the following
LEMMA 2.1. The following holds:

$$(2.7a) \qquad \hat{p} = \begin{bmatrix} A_{11}^{-1}A_{12} \\ I \end{bmatrix} \;,\; \hat{r} = \frac{1}{2}[A_{21}A_{11}^{-1} \;\; I] \;,$$

$$(2.7b) \qquad L_H = \frac{1}{2}(A_{22} - A_{21}A_{11}^{-1}A_{12}) \;,$$

$$(2.7c) \qquad [L_H]_i = \frac{1}{2}\Big[-\frac{a_{2i-1}}{b_{2i-1}}\, a_{2i} \quad -\frac{c_{2i-1}}{b_{2i-1}}\, a_{2i} + b_{2i} - \frac{a_{2i+1}}{b_{2i+1}}\, c_{2i} \quad -\frac{c_{2i+1}}{b_{2i+1}}\, c_{2i} \Big] .$$

REMARK 2.2. From (2.7b) we see that, apart from a scaling factor, L_H equals the Schur complement of $\hat{L}_h$. As a consequence L_H is an M-matrix. Using (2.7c) one can verify that L_H is weakly diagonally dominant. It follows that the Galerkin approach with matrix–dependent grid transfer operators yields stable three–point operators on all coarser grid.

Let $Q_0 : U_h \to U_h$ be a diagonal matrix with $(Q_0)_{i,i} = 1$ if i is odd and $(Q_0)_{i,i} = 0$ is i is even; also $D_h := \mathrm{diag}(L_h)$.
The proof of the following lemma is included because it is fundamental for the generalization to the 2D case in Section 3.
LEMMA 2.3. The following holds:

$$L_h^{-1} - pL_H^{-1}r = Q_0D_h^{-1} .$$

PROOF. Let $\mathcal{S} := A_{22} - A_{21}A_{11}^{-1}A_{12}$. Using lemma 2.1 we get

$$L_h^{-1} - pL_H^{-1}r = Q^T(\hat{L}_h^{-1} - \hat{p}L_H^{-1}\hat{r})Q = Q^T(\hat{L}_h^{-1} - \begin{bmatrix} A_{11}^{-1}A_{12} \\ I \end{bmatrix} \mathcal{S}^{-1}[A_{21}A_{11}^{-1} \;\; I])Q .$$

Note that

$$(2.8) \qquad \hat{L}_h^{-1} = \begin{bmatrix} A_{11}^{-1} + A_{11}^{-1}A_{12}\mathcal{S}^{-1}A_{21}A_{11}^{-1} & A_{11}^{-1}A_{12}\mathcal{S}^{-1} \\ \mathcal{S}^{-1}A_{21}A_{11}^{-1} & \mathcal{S}^{-1} \end{bmatrix}$$

$$= \begin{bmatrix} A_{11}^{-1}A_{12} \\ I \end{bmatrix} \mathcal{S}^{-1}[A_{21}A_{11}^{-1} \;\; I] + \begin{bmatrix} A_{11}^{-1} & \emptyset \\ \emptyset & \emptyset \end{bmatrix} .$$

Hence $L_h^{-1} - pL_H^{-1}r = Q^T \begin{bmatrix} A_{11}^{-1} & \emptyset \\ \emptyset & \emptyset \end{bmatrix} Q = Q_0D_h^{-1}$. □

Lemma 2.3 yields the following approximation property (where $\|\cdot\|_\infty$ can be replaced by another norm):

$$(2.9) \qquad \|(L_h^{-1} - pL_H^{-1}r)D_h\|_\infty = 1$$

which clearly shows robustness. Numerical experiments in [10] indicate that if one uses standard prolongation (linear interpolation) and standard weighted restriction a robust approximation property (as in (2.9)) does not hold.
In [10] the approximation property in (2.9) is combined with an analysis of the smoothing property.

3 Generalization to a multigrid method for 2D convection–diffusion problems

In this section we derive a multigrid method for 2D convection– diffusion problems based on the approach of Section 2.
We consider a second order elliptic BVP on $\Omega = [0,1]^2$ and use uniform grids for discretization. In this section we remain close to the red–black structure of Section 2, i.e. we use red–black coarsening $h \to \sqrt{2}h$.

3.1 Lumping method

We consider grids as indicated in Figure 1.
Spaces of grid functions are denoted by $G(\Omega_h), G(\Omega_H)$, etc. For ease we assume Dirichlet boundary conditions, and thus the grids only contain points in the interior of the unit square.
Clearly on Ω_h we have a red–black partitioning of the nodes: $\Omega_h = (\Omega_h \backslash \Omega_H) \cup \Omega_H$. The points in $(\Omega_h \backslash \Omega_H)$ have label 1 and the points in Ω_H (coarse grid) have label 2. Let there be given a discretization method resulting in a nonsingular operator $A_h: \ G(\Omega_h) \to G(\Omega_h)$. We use the standard nodal basis with red–black ordering of the nodes. Then A_h is represented in block–matrix form as

$$(3.1) \qquad A_h = \begin{bmatrix} A_{11} & -A_{12} \\ -A_{21} & A_{22} \end{bmatrix}.$$

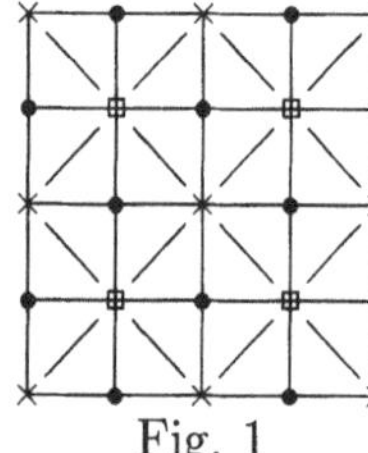

$\Omega_h : \ \{\times\} \cup \{\square\} \cup \{\bullet\}$
$\Omega_H := \Omega_{\sqrt{2}h} : \ \{\times\} \cup \{\square\}$
$\Omega_{2h} : \ \{\times\}$

Fig. 1

The fundamental step in the proof of Lemma 2.3 is based on the factorization

$$(3.2) \qquad A_h^{-1} = \begin{bmatrix} A_{11}^{-1}A_{12} \\ I \end{bmatrix} S^{-1}[A_{21}A_{11}^{-1} \quad I] + \begin{bmatrix} A_{11}^{-1} & \emptyset \\ \emptyset & \emptyset \end{bmatrix},$$

with $\mathcal{S} = A_{22} - A_{21}A_{11}^{-1}A_{12}$.

If A_h corresponds to a 5-point stencil $\begin{bmatrix} & * & \\ * & * & * \\ & * & \end{bmatrix}$ then A_{11} is diagonal and the grid transfer operators

$$(3.3) \qquad p = \begin{bmatrix} A_{11}^{-1}A_{12} \\ I \end{bmatrix}, \quad r = [A_{21}A_{11}^{-1} \;\; I]$$

are local. The coarse grid operator $A_H := rA_hp$ is also local but corresponds to a 9-point stencil and thus we cannot repeat the same procedure to go to the next coarser grid: A red–black partitioning yields a matrix $(A_H)_{11}$ which corresponds to a 5-point stencil and thus p and r as in (3.3) are not suitable anymore.
In order to be able to repeat the same procedure we introduce a *lumping* strategy on a given grid Ω_h as in Fig. 1 which maps a 9-point difference star in a point of $\Omega_h \backslash \Omega_H$ on a 5- point star in the same point. The lumping method we use is as follows. Assume we have the following 9-point star in a grid point of $\Omega_h \backslash \Omega_H$:

$$(3.4) \qquad \begin{bmatrix} \alpha_{NW} & \alpha_N & \alpha_{NE} \\ \alpha_W & \alpha_M & \alpha_E \\ \alpha_{SW} & \alpha_S & \alpha_{SE} \end{bmatrix}.$$

We replace this stencil by:

$$(3.5a) \qquad \begin{bmatrix} 0 & \beta_N & 0 \\ \beta_W & \beta_M & \beta_E \\ 0 & \beta_S & 0 \end{bmatrix},$$

$$(3.5b) \qquad \begin{aligned} &\beta_N = \alpha_N + \alpha_{NW} + \alpha_{NE}, \quad \beta_W = \alpha_W + \alpha_{NW} + \alpha_{SW}, \\ &\beta_M = \alpha_M - (\alpha_{NW} + \alpha_{NE} + \alpha_{SW} + \alpha_{SE}), \quad \beta_E = \alpha_E + \alpha_{NE} + \alpha_{SE}, \\ &\beta_S = \alpha_S + \alpha_{SW} + \alpha_{SE}. \end{aligned}$$

Obvious modifications are used close to the boundary. The same procedure can be used if we have a 9-point stencil on a square rotated grid (as Ω_H in Fig. 1). The lumping procedure is based on a linear interpolation approximation: in the given equation (with star as in (3.4)) the unknown u_{NW} is replaced by $-u_M + u_N + u_W$ which is just the value in NW of the plane through u_M, u_N, u_W.
In matrix form, the lumping procedure yields a matrix

$$(3.6) \qquad \tilde{A}_h = \begin{bmatrix} \tilde{A}_{11} & -\tilde{A}_{12} \\ -A_{21} & A_{22} \end{bmatrix}$$

with $\tilde{A}_{11}$ diagonal and $\tilde{A}_{12}$ corresponding to a 4-point stencil. The prolongation and restriction

$$(3.7) \qquad p = \begin{bmatrix} \tilde{A}_{11}^{-1}\tilde{A}_{12} \\ I \end{bmatrix}, \quad r = [A_{21}\tilde{A}_{11}^{-1} \;\; I]$$

are local and easy to compute.

REMARK 3.1. A lumping procedure combined with Schur complement computation is used in [2] too. However, in [2] only SPD problems are considered. Also the lumping in [2] (adding off–diagonal coefficients to the diagonal) is different from ours; numerical experiments have shown that the use of this lumping procedure in many cases does not yield a convergent multigrid iteration for convection–diffusion problems.

3.2 PREPROCESSING PHASE

Let there be given a sequence of square grids as in Fig.1 $\Omega_1 \subset \Omega_2 \subset \ldots \subset \Omega_\ell$ (mesh size ratio $\sqrt{2}$), and let there be given a discretization operator on Ω_ℓ which corresponds to a 9-point (or 5-point) stencil. We construct A_k, $1 \le k < \ell$ and $p_k : \Omega_{k-1} \to \Omega_k$, $r_k : \Omega_k \to \Omega_{k-1}$, $2 \le k \le \ell$ as follows:

$$
\begin{array}{llll}
A_\ell \overset{\text{lumping (if necessary)}}{\longrightarrow} & \tilde{A}_\ell \overset{\text{as in (3.7)}}{\longrightarrow} p_\ell, r_\ell & & \\
& \big\downarrow \text{Schur complement} & & \\
(3.8) \qquad & A_{\ell-1} \overset{\text{lumping}}{\longrightarrow} & \tilde{A}_{\ell-1} \overset{\text{as in (3.7)}}{\longrightarrow} p_{\ell-1}, r_{\ell-1} & \\
& & \big\downarrow \text{Schur complement} & \\
& & A_{\ell-1} \to \text{etc.}\ . &
\end{array}
$$

It is easy to check that the Schur complement (A_{k-1}) of an operator with a 5-point stencil in the red (label 1) points has a 9-point stencil, and thus the lumping procedure of subsection 3.1 can be applied to A_{k-1}. In the multigrid algorithm below we use p_k, r_k $(2 \le k \le \ell)$, A_k and the diagonal block $(\tilde{A}_k)_{11}$ $(1 \le k \le \ell)$. Note that $(\tilde{A}_k)_{11}$ is diagonal.

A Galerkin property which can be used for constructing the Schur complement is given in the following lemma (cf. (2.7), (2.9b))

LEMMA 3.2. Let $B := \begin{bmatrix} B_{11} & -B_{12} \\ -B_{21} & B_{22} \end{bmatrix}$, $p := \begin{bmatrix} B_{11}^{-1}B_{12} \\ I \end{bmatrix}$,

$$r := [B_{21}B_{11}^{-1} \quad I]\ , \quad r_{inj} := [0 \quad I]\ .$$

Then for the Schur complement $B_{22} - B_{21}B_{11}^{-1}B_{12}$ of B the following holds:

$$B_{22} - B_{21}B_{11}^{-1}B_{12} = rBp = r_{inj}Bp\ . \tag{3.9}$$

COROLLARY 3.3. For computing the Schur complement A_{k-1} of $\tilde{A}_k$ one can use (3.9). For x_i, x_j in the coarse (H) grid Ω_{k-1} we have

$$(A_{k-1})_{ij} = < A_{k-1}e_j^H, e_i^H > = < r_{inj}\tilde{A}_k p_k e_j^H, e_i^H > = < p_k e_j^H, \tilde{A}_k^T e_i^h > .$$

In Theorem 3.4 we present a stability result for the coarse grid matrices A_k $(1 \le k < \ell)$. A point $x_j \neq x_i$, with $x_j, x_i \in \Omega_k$, is called a "closest neighbour of x_i" if $\mathrm{dist}(x_j, x_i) = \min\{\mathrm{dist}(x_m, x_i) \mid x_m \in \Omega_k,\ x_m \neq x_i\}$.

THEOREM 3.4. We assume that the 9-point operator $A_\ell = (a_{ij}^{(\ell)})$ has the following properties:

(3.10a) "sign property" : $a_{ij}^{(\ell)} \le 0$ for all $i \neq j$, $a_{ii}^{(\ell)} > 0$ for all i .

(3.10b) "diagonal dominance" : $\sum_{j \neq i} |a_{ij}^{(\ell)}| \le a_{ii}^{(\ell)}$ for all i, with inequality for at least one i .

(3.10c) "connected graph property" : for every i : $a_{ij}^{(\ell)} \neq 0$ for all j for which $x_j \in \Omega_\ell$ is a closest neighbour of $x_i \in \Omega_\ell$.

Then for all k with $1 \le k < \ell$ $A_k = (a_{ij}^{(k)})$ has the properties (3.10a-c) with ℓ replaced by k.

PROOF. Assume that A_k (with a 9-point stencil) is such that (3.10a-c) are satisfied with ℓ replaced by k. A_{k-1} results from A_k as follows: A_k–lumping $\rightarrow$ $\tilde{A}_k$–Schur complement $\rightarrow$ A_{k-1}. We first consider the lumping procedure. In $x_i \in \Omega_k \backslash \Omega_{k-1}$ we have stars for A_k, $\tilde{A}_k$ as in (3.4), (3.5) respectively; coefficients corresponding to boundary points ($x_j \in \partial\Omega$) are taken zero. Due to (3.10a,c) we have $\alpha_P < 0$ for all $P \in \{N, W, E, S\} \backslash \partial\Omega$, $\alpha_P \le 0$ for all $P \in \{NW, NE, SW, SE\}$, $\alpha_M > 0$. Using this and (3.10b) it follows that $\beta_P < 0$ for all $P \in \{N, W, E, S\} \backslash \partial\Omega$ and $\beta_M > 0$. Also note that $\beta_M - \sum_{P \neq M} \beta_P = \alpha_M + \sum_{P \neq M} \alpha_P$ holds. Hence for $\tilde{A}_k$ the sign property, diagonal dominance and the connected graph property hold.
We now consider the step in which the Schur complement is formed. With $\tilde{A}_k := \begin{bmatrix} \tilde{A}_{11} & -\tilde{A}_{12} \\ -A_{21} & A_{22} \end{bmatrix}$ (cf. (3.6)) we have $A_{k-1} = A_{22} - A_{21}\tilde{A}_{11}^{-1}\tilde{A}_{12}$. Using Corollary 3.3 it is easy to check that the connected graph property holds for A_{k-1}. Note that $\tilde{A}_k$ is an irreducibly diagonally dominant matrix with the sign property (3.10a), and thus $\tilde{A}_k$ is an M-matrix. Hence A_{k-1} (Schur complement of $\tilde{A}_k$) is an M-matrix and thus the sign property holds for A_{k-1}. We use the notation $e := (1, 1, \ldots, 1)^T$; the length of this vector varies but it is clear from the context what it should be. Note that: $\tilde{A}_{11}e - \tilde{A}_{12}e \ge 0$, $\tilde{A}_{11}^{-1} \ge 0$, $\tilde{A}_{11}^{-1}\tilde{A}_{12}e \le e$, $A_{21} \ge 0$, $-A_{21}\tilde{A}_{11}^{-1}\tilde{A}_{12}e \ge -A_{21}e$ and $-A_{21}e + A_{22}e \ge 0$. The last two inequalities yield $A_{k-1}e = (A_{22} - A_{21}\tilde{A}_{11}^{-1}\tilde{A}_{12})e \ge 0$. Because A_{k-1} is regular, $(A_{k-1}e)_i > 0$ holds for at least one i. Now it follows, using the sign property for A_{k-1}, that diagonal dominance holds for A_{k-1}. □

3.3 Two- and multigrid method

We introduce the following two–grid method on level k for solving $A_k x_k = b_k$ $(1 \leq k \leq \ell)$.

(3.11) Procedure $TGM_k(x_k, b_k)$;
Begin if $k = 1$ then $x_k := A_k^{-1} b_k$ else
Begin
(3.11a) for $i := 1$ to ν do $x_k := \mathcal{S}_k(x_k, b_k)$; (* smoothing *)
(3.11b) $d_k := A_k x_k - b_k$;
(3.11c) $z_k := A_{k-1}^{-1} r_k d_k$;
(3.11d) $y_k := \begin{bmatrix} (\tilde{A}_k)_{11}^{-1} & \emptyset \\ \emptyset & \emptyset \end{bmatrix} d_k$;
(3.11e) $x_k := x_k - p_k z_k - y_k$;
end;
end;

Note that the correction in (3.11d) is not used in standard multigrid methods. The direct solver in (3.11c) can be replaced by a recursive call, which then yields a multigrid algorithm.
The error iteration matrix of the linear smoothing method $\mathcal{S}_k$ is denoted by S_k, and the iteration matrix of the two-grid method is denoted by M_k.

For the error iteration the following holds:

$$M_k = (I - \tilde{A}_k^{-1} A_k) S_k^{\nu} \ . \tag{3.12}$$

From (3.12) we see that the two–grid convergence is determined by the combined effect of lumping $(I - \tilde{A}_k^{-1} A_k)$ and smoothing (S_k^{ν}). Note that the coarse grid operator and transfer operators do not occur in the error iteration matrix. This is due to the fact that the coarse–grid correction + additional correction (y_k in (3.11d)) yield a direct solver for the defect equation $\tilde{A}_k w_k = d_k$.
For a two–grid convergence analysis one needs suitable tools for analyzing the effect of the lumping procedure. This is a subject of current research.

Remark 3.5. We state two convergence results which can be proved. These proofs, however, will not be given in this paper.
1. Let A_ℓ be the standard 5-point discretization of the Poisson equation. Lumping in *all* points of Ω_k + Schur complement computation yields coarse–grid matrices A_k $(k < \ell)$ with star

$$c_k \begin{bmatrix} -1 & -2 & -1 \\ -2 & 12 & -2 \\ -1 & -2 & -1 \end{bmatrix} . \tag{3.13}$$

Fourier analysis (on $\Omega = [0,1]^2$; Dirichlet BC) can be used to prove the following:

- in (3.12) we take $\nu = 0$; then $M_\ell = 0$, $\|M_k\|_2 \leq \frac{1}{2}$ $1 \leq k < \ell$.

- in (3.12) we take damped Jacobi with damping $\theta = 0.75$; then $M_\ell = 0$, $\|M_k\|_2 \leq \frac{1}{2}(2\nu)^{2\nu}/(2\nu+1)^{2\nu+1}$ $(1 \leq k < \ell)$ (note: $\nu = 1 \rightarrow 2/27$).

The latter bound, and numerical results (cf. below) show that our method is competitive with the many fast Poisson equation solvers that already exist.
2. We neglect boundary conditions and consider an infinite grid with a constant difference star

$$[A_\ell]_i = \frac{\varepsilon}{h_\ell^2}\begin{bmatrix} 0 & -1 & 0 \\ -1 & 4 & -1 \\ 0 & -1 & 0 \end{bmatrix} + \frac{1}{h_\ell}\begin{bmatrix} 0 & 0 & 0 \\ -1 & 2 & 0 \\ 0 & -1 & 0 \end{bmatrix}$$

(i.e. $-\varepsilon\Delta u + u_x + u_y$ with full upwind discretization).
Analysis of lumping in *all* points of Ω_k + Schur complement computation yields explicit recursion formulas for the coefficients in the star of A_k $(k < \ell)$. Using Fourier analysis one can prove:

$$\|I - \tilde{A}_k^{-1}A_k\|_2 \leq 0.8 \quad (1 \leq k < \ell)\ .$$

This shows robustness: even with $\nu = 0$ (cf. (3.11)) we have two–grid contraction numbers ≤ 0.8 for all k with $1 \leq k \leq \ell$ (i.e. independent of h_k) and all $\varepsilon > 0$.

REMARK 3.6. We note that for the Poisson equation the bound for the two–grid contraction number of our method (cf. Remark 3.5.1) is the same as for the two–grid methods in [11,3,4]. In these papers red–black coarsening and a matrix–dependent prolongation are used too. However, there are significant differences. This is already clear from the form of the two–grid iteration matrix as in (3.12). There is no lumping procedure in [11,3,4] and in [11,3,4] the coarse–grid matrices are all derived from (standard) discretization of the differential operator, whereas in our method we have other coarse–grid matrices (cf. (3.13)) based on the Galerkin condition.

3.4 NUMERICAL EXPERIMENTS

We consider the following class of convection–diffusion problems ($\varepsilon > 0$):

$$(3.14) \qquad \begin{cases} -\varepsilon\Delta u + a(x,y)u_x + b(x,y)u_y = 0 & \text{in } \Omega =]0,1[^2 \\ u = g & \text{on } \partial\Omega\ . \end{cases}$$

The meshes we use are as in Fig. 1, with mesh size $h_k := (\frac{1}{2})^{\frac{1}{2}k+1}$, $k = 1, 2, \ldots, \ell$. For the preprocessing phase we only need a discretization on the finest grid Ω_ℓ. For Δ we use the standard 5-point difference star and the convection part is discretized using standard full upwind differences (cf. e.g. [13]). Note that this results in a 5-point M-matrix A_ℓ.
In the preprocessing phase coarse–grid matrices A_k $(1 \leq k < \ell)$, which are 9-point M-matrices, prolongations and restrictions and diagonal matrices $(\tilde{A}_k)_{11}^{-1}$ are constructed. The multigrid algorithm of Section 3.3 can now be applied; the

only components which have to be specified are: $\mathcal{S}$ (smoothing operator), ν (# smoothings), γ (# recursive calls).
In the experiments below we always take a four–direction Gauss–Seidel smoothing. The following sequence of directions is used in all experiments: "for $i=1$ to n do for $j=1$ to n, for $i=n$ downto 1 do for $j=n$ downto 1, for $j=1$ to n do for $i=1$ to n, for $j=n$ downto 1 do for $i=n$ downto 1", where (i,j) corresponds to the grid point (ih, jh). So we do not adjust the ordering of the grid points to the direction of the flow.

We measure arithmetic costs per iteration in terms of the unit D_ℓ: the arithmetic costs for one defect calculation on the finest grid. Note that for a standard multigrid V-cycle ($h \to \sqrt{2}h$ coarsening) with $\nu = 2$ smoothings the costs are roughly $8\mathrm{D}_\ell$.

Numerical results for the following algorithms are presented:

Alg. 1: $\nu = 1,\ \gamma = 0$ (only Gauss–Seidel). Costs $\approx 4\mathrm{D}_\ell$.
Alg. 2: $\nu = 0,\ \gamma = 1$ ($\mathcal{S}$ not used). Costs $\approx 5\mathrm{D}_\ell$.
Alg. 3: $\nu = \nu(k)$, with $\nu(k) = 0$ if $k = \ell$, $\nu(k) = 1$ if $k \le \ell - 1$, $\gamma = 1$. Costs $\approx 13\mathrm{D}_\ell$.

W.r.t. Algorithm 3 we note that we do not use smoothing on level ℓ because $I - \tilde{A}_\ell^{-1} A_\ell = 0$.

In all the experiments we take a fixed arbitrary starting vector $x^{(0)}$. If the error after the m-th iteration is denoted by $e^{(m)}$, we use as a measure for the error reduction:

$$\delta_E := [\|e^{(10)}\|_2 / \|e^{(0)}\|_2]^{1/10} \text{ or } \delta_D := [\|A_\ell e^{(10)}\|_2 / \|A_\ell e^{(0)}\|_2]^{1/10} .$$

Experiment 3.1. (as in [12,14]). We take:

$$a(x,y) = (2y-1)(1-x^2)\ , \quad b(x,y) = 2xy(y-1)\ ;$$
$$g(x,y) = \sin(\pi x) + \sin(13\pi x) + \sin(\pi y) + \sin(13\pi y)\ .$$

In Table 1 we show values of δ_D.

	algorithm 1		algorithm 2		algorithm 3	
ε	$h_\ell = 1/64$	$h_\ell = 1/128$	$h_\ell = 1/64$	$h_\ell = 1/128$	$h_\ell = 1/64$	$h_\ell = 1/128$
10^0	0.76	0.78	0.25	0.26	0.0031	0.0035
10^{-2}	0.84	0.80	0.43	0.45	0.029	0.042
10^{-4}	0.46	0.90	0.85	0.91	0.0016	0.013

Table 1.

EXPERIMENT 3.2. (rotating flow in part of the domain, cf.[5]). We take

$$\begin{cases} a(x,y) = \sin(\pi(y-\frac{1}{2}))\cos(\pi(x-\frac{1}{2})) \\ b(x,y) = -\cos(\pi(y-\frac{1}{2}))\sin(\pi(x-\frac{1}{2})) \end{cases} \quad \text{if } (x-\tfrac{1}{2})^2 + (y-\tfrac{1}{2})^2 < 1/16\,,$$

$a,\ b$ zero otherwise ; $g(x,y) = 0$.

In Table 2 we show values of δ_E.

	algorithm 1		algorithm 2		algorithm 3	
ε	$h_\ell = 1/64$	$h_\ell = 1/128$	$h_\ell = 1/64$	$h_\ell = 1/128$	$h_\ell = 1/64$	$h_\ell = 1/128$
10^{-2}	0.97	0.99	0.29	0.35	0.019	0.015
10^{-4}	0.96	0.99	0.47	0.55	0.10	0.26
10^{-6}	0.96	0.98	0.48	0.56	0.11	0.25

Table 2.

EXPERIMENT 3.3. (random 5-point M-matrix). We use a random number generator, which generates random numbers in $[0,1]$, to fill all off–diagonal places of 5-point stars in all interior grid points. The diagonal coefficients are taken such that the sum in each star is zero. Clearly this results in a 5-point M-matrix.
We take the right hand side and boundary values equal to zero.
In Table 3 we show values of δ_E.

algorithm 1		algorithm 2		algorithm 3	
$h_\ell = 1/64$	$h_\ell = 1/128$	$h_\ell = 1/64$	$h_\ell = 1/128$	$h_\ell = 1/64$	$h_\ell = 1/128$
0.98	0.99	0.36	0.39	0.11	0.20

Table 3

REMARK 3.7. From the experiments above one may infer the following. Algorithm 1 shows results as expected. In all experiments algorithm 2, in which only the diagonal matrix $(\tilde{A}_k)_{11}^{-1}$ (cf. (3.11d,e)) is used for smoothing, converges; this indicates that our approach w.r.t. the coarse grid correction is satisfactory. In certain cases, (cf. Exp. 3.2, 3.3) algorithm 2 which has low costs per iteration and is easy parallizable has an "acceptable" convergence rate. In cases where the convergence rate of algorithm 2 is rather low, (cf. Exp. 3.1) the use of four–direction Gauss–Seidel on level $k \le \ell - 1$ (i.e. algorithm 3) yields a significant improvement.
Clearly Experiment 3.3 is not related to a pde problem. This experiment is done to give some further indication of the robustness of algorithm 2.

REMARK 3.8. If the block matrix A_h in (3.1) corresponds to a grid decomposition in which standard h-$2h$ coarsening is used, then the approach of Section 3.2, 3.3 can be applied, provided we use a suitable lumping strategy. Numerical experiments have shown the same robustness as for the case with red-black coarsening.

ACKNOWLEDGEMENT

The author wishes to thank Prof.dr. Gabriel Wittum for fruitful discussions.

References

[1] R.E. Alcouffe, A. Brandt, J.E. Dendy Jr., J.W. Painter, *The multi-grid method for the diffusion equation with strongly discontinuous coefficients*, SIAM J. Sci. Stat. Comput. 2: 430-454 (1981).

[2] O. Axelsson, V. Eijkhout, *The nested recursive two- level factorization method for nine-point difference matrices*, SIAM J. Sci. Stat. Comput. 12: 1373-1400 (1991).

[3] D. Braess, *The contraction number of a multigrid method for solving the Poisson equation*, Numer. Math. 37: 387-404 (1981).

[4] D. Braess, *The convergence rate of a multigrid method with Gauss–Seidel relaxation for the Poisson equation*, Math. Comp. 42: 505-519 (1984).

[5] A. Brandt, I. Yavneh, *Accelerated multigrid convergence and high–Reynolds recirculating flows*, SIAM J. Sci. Comput. 14: 607- 626 (1993).

[6] J.E. Dendy Jr., *Black box multigrid for nonsymmetric problems*, Appl. Math. Comp. 13: 261-283 (1983).

[7] E.P. Doolan, J.J.H. Miller, W.H.A. Schilders, *Uniform numerical methods for problems with initial and boundary layers*, Boole Press, Dublin, 1980.

[8] W. Hackbusch, *Multi–grid Methods and Applications*, Springer, Berlin, 1985.

[9] P.W. Hemker, R. Kettler, P. Wesseling, P.M. de Zeeuw, *Multigrid methods: development of fast solvers*, Appl. Math. Comp. 13: 311-326 (1983).

[10] A. Reusken, *Multigrid with matrix–dependent transfer operators for a singular perturbation problem*, Computing 50, 199-211 (1993).

[11] M. Ries, U. Trottenberg, G. Winter, *A note on MGR methods*, Linear Algebra Appl. 49: 1-26 (1983).

[12] J.W. Ruge, K. Stüben, *Algebraic multigrid*, In Multigrid Methods (S.F. McCormick, ed.), SIAM, Philadelphia 1987.

[13] P. Wesseling, *An Introduction to Multigrid Methods*, Wiley, Chichester, 1992.

[14] P.M. de Zeeuw, *Matrix–dependent prolongations and restrictions in a blackbox multigrid solver*, J. Comput. Appl. Math. 33: 1- 27 (1990).

13

Multilevel, Extrapolation, and Sparse Grid Methods

U. Rüde[1]

ABSTRACT Multigrid Methods are asymptotically optimal solvers for discretized partial differential equations (PDE). For the optimal solution of PDEs, however, the quality of the discretization is of the same importance as the speed of the algebraic solution process. Especially for high accuracy requirements, high order discretizations become increasingly attractive. We describe higher order techniques, like *extrapolation* and *sparse grid combination* that are particularly interesting in the context of multilevel algorithms, because they are based on discretizing the problems on grids with different mesh sizes. Classical *Richardson extrapolation* can be extended and generalized in many ways. One generalization is to consider the mesh widths in the different coordinate directions as distinct parameters. This leads to the so-called *multivariate extrapolation* and the *combination technique.*

1 Introduction

Multigrid methods are generally considered to be among the fastest solvers for discretized elliptic boundary value problems. The efficiency of multigrid is derived from the use of different discretizations with different mesh widths such that fine mesh approximations are corrected recursively by approximations on coarser levels. The same *hierarchical structure* that leads to fast solvers can also be used to support the discretization process itself. Such techniques, where several distinct meshes participate to define a combined discrete solution, are generally called *extrapolation methods.*

The basic idea of *Richardson* extrapolation for elliptic equations (see Marchuk and Shaidurov [9]) is to take a linear combination of approximations on different grids. If $u^h(x)$ denotes the discrete solution at point x in a uniform mesh with spacing h, the combined solution

$$u_1^h(x) = 4/3u^{h/2}(x) - 1/3u^h(x) \tag{1}$$

is chosen such that the first term of an h^2-*error expansion* is eliminated. Thus the

[1]Institut für Informatik, Technische Universität, D-80290 München, Germany, e-mail: ruede@informatik.tu-muenchen.de

extrapolation process is based on the existence of an asymptotic error expansion of the form

$$u^h(x) = u^*(x) + h^2 e_2(x) + \cdots + h^{2m} e_{2m}(x) + h^{2m+2} R_{2m+2}(x,h), \qquad (2)$$

where u^* denotes the exact solution and where we assume $R_{2m+2}(x) \leq C(x)$ is bounded, and where $C(x)$ and $e_{2k}(x), 1 \leq k \leq m$ are independent of h.

The discrete solution $u^h(x)$ may be a grid- or finite element function. In any case, the spaces where the linear combination (1) is formed must be chosen carefully. In (1) we have implicitly assumed that this is the coarse mesh with spacing h that is naturally embedded in the fine mesh, so that the fine mesh solution may be transferred to the coarse mesh with *injection.*

The existence of asymptotic expansions (2) can be assumed only in cases when the solution is sufficiently regular and when the discretizations are uniform. In elliptic problems the smoothness of the solution may be disturbed at reentrant corners or where the data is non-smooth. The form of the domain or the need for local refinement may make the use of uniform meshes difficult. However, even in these more complicated cases, the existence of generalized expansions with fractional powers of h can be shown under assumptions, like *local* uniformity of the meshes, see Blum, Lin, and Rannacher [2, 3, 1]. The *local smoothness* of the solution is a basic characteristic of many elliptic problems, so that extrapolation can be used locally, even when the global solution is non-smooth.

Extrapolation is a natural supplement of multigrid-like methods and has been investigated in this context in several papers, see Brandt [4], Hackbusch [8], Rüde [11], and Schaffer [15].

With this background, several interesting new extrapolation-based approaches have been developed within the past few years, including the *sparse grid combination technique* and *multivariate extrapolation.*

In this paper we will focus on *explicit extrapolation* methods that are based on the (linear) combination of solutions on different grids. *Implicit extrapolation* methods, in contrast, obtain higher order by applying the extrapolation idea on quantities like the truncation error or the numerical approximation of the functional. Such methods are discussed in Rüde [12]. Some methods of this type do not need uniform meshes and are therefore especially attractive in an adaptive refinement setting. An analysis is given in Rüde [13].

2 Combination extrapolation

Extrapolation exploits the asymptotic behavior of a solution depending on a small parameter. For most extrapolation techniques it has been assumed that there is a single parameter to use — the mesh size h. In the case of partial differential equations, however, the location x as well as the mesh spacing h may be considered as vector valued quantities.

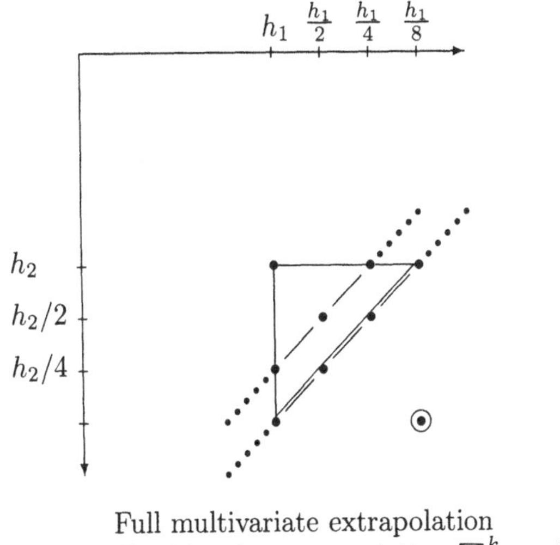

Full multivariate extrapolation
Combination extrapolation $\sum_{j=0}^{k} - \sum_{j}^{k-1}$, see (4)
⊙ Full grid $u^{h2^{-k},h2^{-k}}$

FIGURE 1. Schematic representation of multivariate extrapolation techniques

Let us now assume that the solution domain is the unit square $[0,1]^2$, and that it is possible to compute numerical approximations on grids with mesh spacing $h_1 = 1/N_1$ in the x_1-direction and $h_2 = 1/N_2$ in the x_2-direction. Denoting these approximations by $u^{h_1,h_2}(x_1,x_2)$, we assume that the error satisfies a *splitting* condition

$$u^{h_1,h_2}(x_1,x_2) = u^*(x_1,x_2) + e_1^{h_1}(x_1,x_2) + e_2^{h_2}(x_1,x_2) + R^{h_1,h_2}(x_1,y_1), \tag{3}$$

where the dominating terms are assumed to be e_1 and e_2. The level k combination is now defined as

$$\begin{aligned} u_k^{h_1,h_2} &\stackrel{\text{def}}{=} \sum_{j=0}^{k} u^{2^{-j}h_1,2^{j-k}h_2} - \sum_{j=0}^{k-1} u^{2^{-j}h_1,2^{j-k+1}h_2} = \\ &u^* + e_1^{h_1} + e_2^{h_2} + \sum_{j=0}^{k} R^{2^{-j}h_1,2^{j-k}h_2} - \sum_{j=0}^{k-1} R^{2^{-j}h_1,2^{j-k+1}h_2} \end{aligned} \tag{4}$$

so that the dominating terms are the same as for the solution $u^{2^{-k}h_1,2^{-k}h_2}$. This special variant of extrapolation does not need any explicit knowledge of the form of the error terms e_1 and e_2. It is only required that each of the leading error terms depends on only one of the mesh parameters. For an illustration of the combination extrapolation process, see Figure 1. The diagonal lines in the scheme of solutions indicate the approximations contributing to the combination.

The resulting grids are depicted in Figure 2 for different k. Note that, according to the above consideration, the combined solution on all these grid should provide similar accuracies.

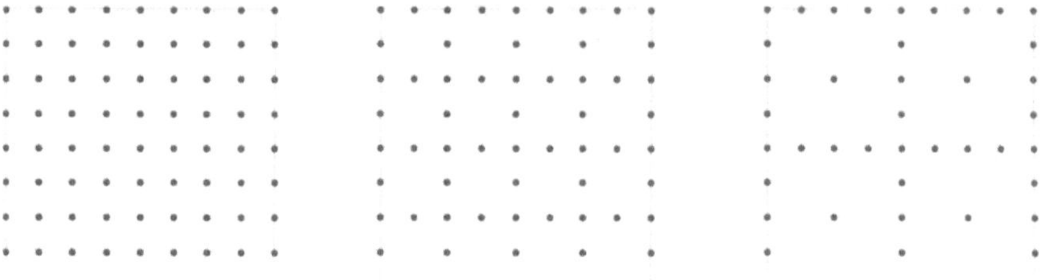

FIGURE 2. Grids for $u^{1/8,1/8}, u_1^{1/4,1/4}, u_2^{1/2,1/2}$

The extreme case of $u_k^{1/2,1/2}$ has been introduced by Griebel, Schneider, and Zenger [7] as the *combination technique.* The grid resulting in this case is called a *sparse grid*, see $u_2^{1/8,1/8}$ in Figure 2 for a simple example. If we assume that e_1 and e_2 are still dominating over the sum of $2k-1$ remainder terms, as they now appear in (4), we can compute a solution with the accuracy of a grid with N^2 nodes by combining the solution of $2k-1 = O(\log_2(N))$ grids, each with approximately only N or $N/2$ nodes.

This turns out to be not quite correct, because here even for smooth solutions the remainder terms are of the same order as the smallest $e_1^{h_1}$ and $e_2^{h_2}$. Recent theoretical results confirm earlier heuristical considerations and practical evidence (see further below) that the error in the combination technique is now dominated by the remainder terms. Under certain smoothness assumptions it can be shown to be of the order $O(h^2|\log(h)|)$, as compared to the full grid accuracy of $O(h^2)$ under the same assumptions. For more details based on finite difference and Fourier techniques consult Bungartz, Griebel, Röschke, and Zenger [5]. A proof based on Sobolev space techniques for finite element discretizations is given by Pflaum [10]. Both papers are based on the recursion formula

$$u_{k-1}^{h/2,h/2} = u_k^{h,h} + \sum_{j=1}^{k} H^{2^{-j}h,2^{j-k}h}, \tag{5}$$

where

$$H^{h_1,h_2} \stackrel{\text{def}}{=} u^{2h_1,2h_2} - u^{2h_1,h_2} - u^{h_1,2h_2} + u^{h_1,h_2},$$

and on finding bounds of the form

$$\|H^{h_1,h_2}\| \le Ch_1^2h_2^2.$$

Based on this estimate, the difference of two consecutive terms in (5) is bounded by

$$\|u_{k-1}^{h/2,h/2} - u_k^{h,h}\| \le Ck2^{-2k}h^4.$$

Thus

$$\|u^{2^{-k}h,2^{-k}h} - u_k^{h,h}\| \le C\sum_{j=1}^{k} j2^{-2j}(2^{j-k}h)^4 = Ch^{-4}(\frac{4k}{3}2^{-2k} - \frac{4}{9}2^{-2k} + \frac{4}{9}2^{-4k}),$$

so that we obtain for the special case of the combination technique

$$\|u^{2^{-k},2^{-k}} - u_{k-1}^{1/2,1/2}\| \leq C\frac{2^{-2k}k}{3} - \frac{4\,2^{-2k}}{9} + \frac{4\,2^{-4k}}{9}.$$

If we define $h = 2^{-k}$ this becomes the sparse grid error estimate

$$\|u^{h,h} - u_{k-1}^{1/2,1/2}\| \leq \bar{C}h^2|\log_2 h|.$$

The combined solution $u_k^{h_1,h_2}$ would conventionally be defined on the *intersection* of all grids participating in the extrapolation process. Here the common intersection of all grids participating in the computation of $u_k^{h,h}$ is the grid of $u^{h,h}$ itself, so that the combination solution $u_k^{1/2,1/2}$ would be defined on just a single point within the solution domain. Using bilinear interpolation for each single component function, we can extend the the domain of definition to the *union* of all participating grids. This is possible, because bilinear interpolation can be shown to be compatible with the error splitting (3). In (4) it is implicitly assumed that a suitable interpolation has been applied to all terms.

Assuming that the computation of u^{h_1,h_2} requires $O(h_1^{-1} \cdot h_2^{-1}) = O(N_1 \cdot N_2)$ operations, the total cost for finding $u_k^{h,h}$ is $O(k2^{-k}N^2)$, where $h = 1/N$, as compared to $O(N^2)$ for the direct computation of $u^{h,h}$. Note that for $k \geq 2$ the computation of $u_k^{h,h}$ is significantly cheaper than finding $u^{2^{-k}h,2^{-k}h}$. Furthermore, as with any explicit extrapolation technique, the different solutions can be computed independently in parallel. Note, however, that the collection of the results is not trivial in a parallel environment, and for an optimal strategy we must form the interpolants and their combination in a tree-like algorithmic scheme. Furthermore, the parallelization effect is even more profitable for higher dimensions, see Griebel, Huber, Rüde, Störtkuhl, and Zenger [6].

The efficient computation of the solutions contributing in a combination extrapolation requires the solution of highly nonisotropic problems. This requires modified multigrid algorithms using line relaxation or semi-coarsening.

The following example illustrates the features of combination extrapolation. We study the solution of the homogeneous Dirichlet problem of Poisson's equation

$$\begin{array}{rcll} -\Delta u & = & f(x,y) & \text{in } \Omega, \\ u & = & 0 & \text{on } \partial\Omega \end{array} \tag{6}$$

discretized by 5-point differences with the data chosen such that the true solution is

$$u(x,y) = \sin(\omega\pi x)\sin(\omega\pi y) \quad \text{for } \omega = 1,2,3,\cdots. \tag{7}$$

This function is an eigenfunction of the discrete Laplace operator, so that the errors can be computed analytically without pollution by interpolation effects.

In Table 1 we compare the efficiency of various combination extrapolation methods directly. For the smoothest solution ($\omega = 1$) the results for $u_k^{1/2,1/2}$ provide the highest accuracy relative to the total number of unknowns.

TABLE 1. Combination extrapolation

$\omega = 1$							
$h_1 = h_2$		$\hat{u}_0$	$\hat{u}_1$	$\hat{u}_2$	$\hat{u}_3$	$\hat{u}_4$	$\hat{u}_5$
1/2	Work	1	7	29	95	273	723
	Error	1.18e-2	1.96e-3	2.80e-4	1.94e-5	-7.70e-6	-5.06e-6
1/4	Work	9	51	181	535	1433	
	Error	2.68e-3	6.16e-4	1.41e-4	3.21e-5	7.26e-6	
1/8	Work	49	259	869	2471		
	Error	6.56e-4	1.60e-4	3.93e-5	9.46e-6		
1/16	Work	225	1155	3781			
	Error	1.63e-4	4.05e-5	1.01e-5			
1/32	Work	961	4867				
	Error	4.07e-5	1.01e-5				
1/64	Work	3969					
	Error	1.01e-5					
$\omega = 2$							
1/2	Work	1	7	29	95	273	723
	Error	1.85e-2	-2.24e-3	-1.31e-3	-5.18e-4	-1.78e-4	-5.70e-5
1/4	Work	9	51	181	535	1433	
	Error	2.95e-3	4.91e-4	7.01e-5	4.86e-6	-1.93e-6	
1/8	Work	49	259	869	2471		
	Error	6.71e-4	1.54e-4	3.53e-5	8.04e-6		
1/16	Work	225	1155	3781			
	Error	1.64e-4	4.02e-5	9.84e-6			
1/32	Work	961	4867				
	Error	4.07e-5	1.01e-5				
1/64	Work	3969					
	Error	1.01e-5					

For $\omega = 2$, that is a less smooth solution, $u_3^{1/4,1/4}$ and $u_2^{1/8,1/8}$ are of better or the similar efficiency as $u_5^{1/2,1/2}$. This seems to get more prominent for $k \to \infty$. For example, $u_{10}^{1/2,1/2}$ provides an accuracy of 1.17e-7 with total system size of 53277 unknowns, which is less efficient than to compute $u_6^{1/8,1/8}$ giving an accuracy of 8.89e-8 with total system size 33933. At present, it is therefore unclear, whether and under which circumstances the combination technique should be used in its original form with $u_k^{1/2,1/2}$ or whether restricted forms using $u_k^{h_1,h_2}$ are more efficient.

3 Multivariate extrapolation

The asymptotic expansion (2) can be generalized to a multivariate form, e.g.

$$u^{h_1,h_2} = \begin{array}{lllllllll} u^* & + & h_1^2 e_{20} & + & h_1^4 e_{40} & + & h_1^6 e_{60} & + & h_1^8 e_{80} + \cdots \\ h_2^2 e_{20} & + & h_1^2 h_2^2 e_{22} & + & h_1^4 h_2^2 e_{24} & + & h_1^6 h_2^2 e_{26} & + & \cdots \\ h_2^4 e_{04} & + & h_1^2 h_2^4 e_{24} & + & h_1^4 h_2^4 e_{44} & + & \cdots & & \\ h_2^6 e_{06} & + & h_1^2 h_2^6 e_{26} & + & \cdots & & & & \\ h_2^8 e_{08} & + & \cdots & & & & & & \end{array}, \tag{8}$$

where all terms up to a total order of 8 are shown. We note that the error terms correspond to a Table of approximate solutions arranged like

$$\begin{array}{lllll} u^{h_1,h_2} & u^{h_1/2,h_2} & u^{h_1/4,h_2} & u^{h_1/8,h_2} & \cdots \\ u^{h_1,h_2/2} & u^{h_1/2,h_2/2} & u^{h_1/4,h_2/2} & \cdots & \\ u^{h_1,h_2/4} & u^{h_1/2,h_2/4} & \cdots & & \\ u^{h_1,h_2/8} & \cdots & & & \end{array}, \tag{9}$$

so that we can use a combination of approximations $u^{h_1 2^{-i}, h_2 2^{-j}}$ for $0 \le i$, $0 \le j$ and $i + j \le p$ to eliminate the error terms $h_1^{2i} h_2^{2j}$ for $2i + 2j \le 2p$ (see Figure 1).

The extrapolation coefficients can be computed by symbolic algebra techniques. A Maple[2] program for this purpose is shown in Figure 3. Referring to the arrangement of terms as in (9) (see also Figure 1) we find the extrapolation coefficients for order $k = 1, 2$ and 3 as follows:

$$k = 1: \quad \begin{array}{rr} -5/3 & 4/3 \\ 4/3 & \end{array} \tag{10}$$

$$k = 2: \quad \begin{array}{rrr} 37/45 & -100/45 & 64/45 \\ -100/45 & 80/45 & \\ 64/45 & & \end{array} \tag{11}$$

$$k = 3: \quad \begin{array}{rrrr} -485/2835 & 3108/2835 & -6720/2835 & 4096/2835 \\ 3108/2835 & -8400/2835 & 5376/2835 & \\ -6720/2835 & 5376/2835 & & \\ 4096/2835 & & & \end{array} \tag{12}$$

We repeat the experiments of Table 1 for multivariate extrapolation. $\bar{u}_k^{h_1,h_2}$ denotes the corresponding result, where $\bar{u}_0^{h_1,h_2} \equiv u^{h_1,h_2}$ is the result without extrapolation.

---

[2]Maple V, Copyright (c) 1981-1990 by the University of Waterloo, is a symbolic computer algebra system

```
T:=[ 1,
      h1^2, h2^2,
      h1^4, h1^2*h2^2, h2^4,
      h1^6, h1^4*h2^2, h1^2*h2^4, h2^6 ];
S:=[ [h1=8, h2=8],
      [h1=4, h2=8], [h1=8, h2=4],
      [h1=2, h2=8], [h1=4, h2=4], [h1=8, h2=2],
      [h1=1, h2=8], [h1=2, h2=4], [h1=4, h2=2],
                                  [h1=8, h2=1] ];
n:=nops(T);
X:=['x.i'$i=1..n];
B:=[1,'0'$i=2..n];
EQS:={
      'convert(
            ['subs(S[i],T[j])*X[i]'$i=1..n],
            `+`)
  =
     B[j]'$j=1..n
};
X:=convert(X,set);
solve(EQS,X);
```

FIGURE 3. Maple Program to calculate the extrapolation coefficients in (10-12)

A comparison of Table 1 with Table 2 shows that for these smooth solutions extrapolation provides significantly more accurate results than the combination technique.

We now consider more oscillatory functions and compare the accuracy of the full grid solution $u^{1/64,64}$, the combination technique $\hat{u}_5^{1/2,1/2}$ and a multivariate extrapolation $\bar{u}_1^{1/8,1/8}$. Figure 4 shows the error in a logarithmic scale versus the frequency parameter. The error of the full grid solution is almost independent of ω. Of course, this is also the most expensive of the methods. For all other solutions the error grows with the frequency. The error for the multivariate extrapolation is smaller than for the combination method, though $\bar{u}_1^{1/8,1/8}$ is significantly cheaper to compute than $\hat{u}_5^{1/2,1/2}$.

The most efficient method is conventional Richardson extrapolation which is even more accurate than multivariate extrapolation for a comparable computational cost. Note that for more extrapolation steps and more than two dimensions, the multivariate extrapolation becomes more attractive in terms of work and parallelization possibilities. Also note, that with multivariate extrapolation it is possible to accommodate different expansions in the mesh parameters, as they may occur

TABLE 2. Multivariate extrapolation

$\omega = 1$					
$h_1 = h_2$		$\bar{u}_0$	$\bar{u}_1$	$\bar{u}_2$	$\bar{u}_3$
1/2	Work	1	7	30	102
	Error	1.18e-2	1.13e-3	4.64e-5	1.02e-6
1/4	Work	9	51	190	586
	Error	2.68e-3	7.32e-5	6.25e-7	3.50e-9
1/8	Work	49	259	918	2730
	Error	6.56e-4	4.43e-6	9.41e-9	1.32e-11
1/16	Work	225	1155	4006	11722
	Error	1.63e-4	2.75e-7	1.45e-10	5.11e-14
1/32	Work	961	4867	16710	48522
	Error	4.07e-5	1.171e-8	2.27e-12	1.99e-16
1/64	Work	3969	19971	68230	197386
	Error	1.01e-5	1.07e-9	3.54e-14	7.79e-19

$\omega = 2$					
1/2	Work	1	7	30	102
	Error	1.85e-2	-9.19e-3	1.39e-3	-1.16e-4
1/4	Work	9	51	190	586
	Error	2.95e-3	-3.32e-4	1.16e-5	-2.56e-7
1/8	Work	49	259	918	2730
	Error	6.71e-4	-1.83e-5	1.56e-7	-8.74e-10
1/16	Work	225	1155	4006	11722
	Error	1.64e-4	-1.11e-6	2.34e-9	-3.30e-12
1/32	Work	961	4867	16710	48522
	Error	4.07e-5	-6.88e-8	3.64e-11	-1.27e-14
1/64	Work	3969	19971	68230	197386
	Error	1.01e-6a	-4.29e-9	5.68e-12	4.98e-17

for time-dependent equations, when the time step size is one of the parameters. For the given situation, the combination technique seems to be generally inferior to extrapolation schemes.

4 Extrapolation and Full Multigrid Algorithms

Extrapolation raises the approximation accuracy beyond the truncation error of a single solution by itself, so that, in contrast to conventional multigrid techniques, it does not suffice to compute each problem to an accuracy matching its own truncation error. Each subproblem must be solved with the accuracy that is expected for the extrapolated result.

Typically, the initial guess for a multigrid solution is computed by nested iteration. When the objective is a solution with truncation error accuracy, then

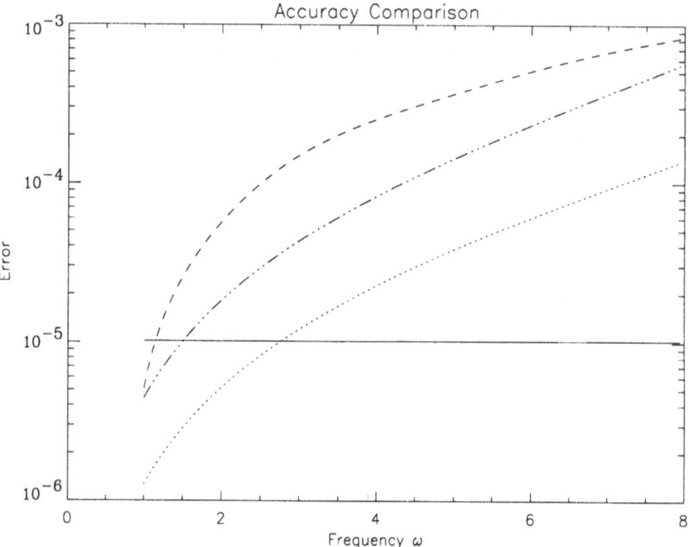

FIGURE 4. Accuracy of various methods depending on frequency: solid line = full grid solution ($u^{1/64,1/64}$), dashed line = combination method ($\hat{u}_5^{1/2,1/2}$), dashed-dotted line = multivariate extrapolation ($\bar{u}_1^{1/8,1/8}$), dotted line = Richardson extrapolation $4/3u^{1/16,1/16} - 1/3u^{1/8,1/8}$

the initial guess can be obtained by interpolation, so that a fixed, small number of correction cycles is sufficient to compute the solution on each grid. For higher accuracy, as required with extrapolation, the analysis in Hackbusch [8] and Schüller and Lin [16] shows that the number of cycles with this strategy depends logarithmically on h.

For an extrapolation method, it does therefore not suffice to compute an approximation with h^2 accuracy as initial guess, but we must provide sufficient accuracy so that the terms $h_1^p h_2^q$ are represented correctly up to the order of the extrapolation process. This cannot be accomplished easily by a nested iteration for each single subproblem, but we must use additional information. For more conventional extrapolation techniques this has been discussed by Hackbusch [8] and Schüller and Lin [16]. The optimal starting values for the iteration on each grid can be obtained by a modified extrapolation that is designed not to produce the best approximation to the differential equation, but to the discrete solution on the new grid. Additionally, the initial guesses must be computed by sufficiently high order interpolation.

For the combination extrapolation, suitable starting values in the above sense can be obtained by using a combination extrapolation of the form

$$u^{h_1/2,h_2/2} \approx u^{h_1/2,h_2} + u^{h_1,h_2/2} - u^{h_1,h_2},$$

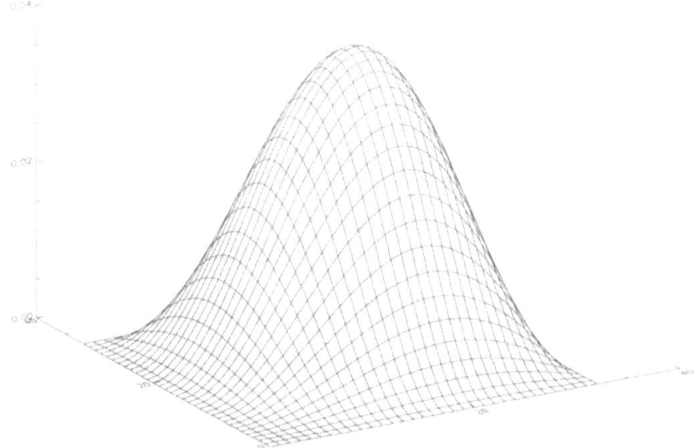

FIGURE 5. Test solution

wherever the necessary terms are available. At the borders of the extrapolation scheme (see Figure 1) this is not possible and a prediction of $u^{h_1/2,h_2}$ from u^{h_1,h_2} and u^{2h_1,h_2} requires the explicit knowledge of the form of $e_1^{h_1}$ (analogous for $u^{h_1,h_2/2}$).

Analogous techniques must be used for multivariate extrapolation, where the initial guesses must be computed from the existing approximations by suitable linear combinations plus interpolation. Note that requires the exchange of information from all existing solutions, so that a parallel implementation requires a global communication step, that should be arranged in a hierarchical form.

5 Numerical experiment

We now study solutions of the discretization of (6) with bilinear finite elements and data such that the solution becomes

$$u(x,y) = x\,(1-x)\cos(\frac{\pi x}{2})y\,(1-y)\cos(\frac{\pi y}{2}). \tag{13}$$

This solution is depicted in Figure 5. In Figure 6 we display the performance of our *explicit* extrapolation methods. We plot the accuracy versus the total number of unknowns involved in the solution process (neglecting the extra work required to obtain higher accuracy). The slope of the lines reflects the order, so that the two methods of 4th order and 6th order have parallel lines, respectively. The plot of the standard method clearly shows the $O(h^2)$ behavior by a factor 4 reduction

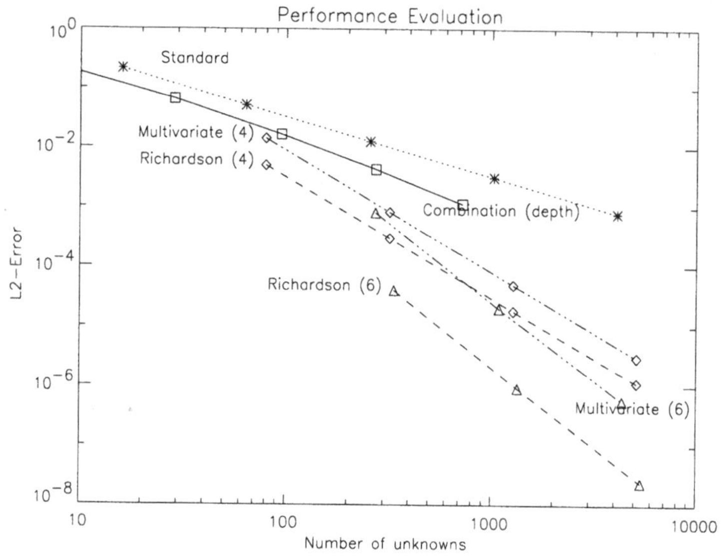

FIGURE 6. Performance of explicit extrapolation methods for the smooth model problem.

when the number of unknowns is increased by a factor of 4.

The combination solution produces a curved line, sloping upward, reflecting the fact that for comparatively small numbers of unknowns the logarithmic factors in complexity and accuracy are still essential. The graph for the combination technique is obtained from a L_2-error norm evaluated on the smallest regular grid including the combined grid. For larger number of unknows the combination method should asymptotically provide close to order 4 (with respect to the number of unknowns) and therefore the graph should asymptotically tend to the same slope as the 4th order methods. The graph also shows that for this example straightforward Richardson extrapolation is most efficient. The potentially more effective multivariate extrapolation variants are not yet significantly cheaper, but their errors (though of the same order) have larger constants. This depends very much on the details of the problem as well as the details of the evaluation.

Clearly, the multivariate extrapolation methods are very attractive when we are concerned about parallelization. The same is true for the combination technique. The smoothness can only be efficiently exploited by high order methods.

6 Conclusions

In this paper we have discussed several extrapolation based PDE solvers, including the sparse grid combination technique and multivariate extrapolation methods.

For the model cases considered, extrapolation seems to be the most efficient choice, with multivariate extrapolation being an interesting alternative in a parallel environment. For simple model problems in two dimensions it seems that various extrapolation techniques present a more efficient solution method than the combination technique.

Future work will include the analysis of more complicated problems and will explore the various combinations of sparse grid and extrapolation methods. A more comprehensive study must include the sparse grid finite element technique and other implicit extrapolation methods (see Rüde [12]) and must address the combination of these methods with local mesh refinement. Suitable techniques are discussed in Rüde [14].

References

[1] H. Blum. Asymptotic error expansion and defect correction in the finite element method. Habilitationsschrift, Universität Heidelberg, 1991.

[2] H. Blum, Q. Lin, and R. Rannacher. Asymptotic error expansions and Richardson extrapolation for linear finite elements. *Numer. Math.*, 49:11–37, 1986.

[3] H. Blum and R. Rannacher. Extrapolation techniques for reducing the pollution effect of reentrant corners in the finite element method. *Numer. Math.*, 52:539–564, 1988.

[4] A. Brandt. Multigrid techniques: 1984 guide with applications to fluid dynamics. *GMD Studien*, 85, 1984.

[5] H. Bungartz, M. Griebel, D. Röschke, and C. Zenger. A proof of convergence for the combination technique for the Laplace equation using tools of symbolic computation. SFB Bericht 342/4/93 A, Institut für Informatik, TU München, April 1993.

[6] M. Griebel, W. Huber, U. Rüde, and T. Störtkuhl. The combination technique for parallel sparse-grid-preconditioning or -solution of PDEs on workstation networks. In L. Bougé, M. Cosnard, Y. Robert, and D. Trystram, editors, *Parallel Processing: CONPAR 92 - VAPP V*, volume 634 of *Lecture Notes in Computer Science*, pages 217–228. Springer Verlag, 1992. Proceedings of the Second Joint International Conference on Vector and Parallel Processing, Lyon, France, September 1–4, 1992.

[7] M. Griebel, M. Schneider, and C. Zenger. A combination technique for the solution of sparse grid problems. SFB Bericht 342/19/90, Institut für Informatik, TU München, October 1990. Also available in: Proceedings of the

International Symposium on Iterative Methods in Linear Algebra, Bruxelles, April 2-4, 1991.

[8] W. Hackbusch. *Multigrid Methods and Applications.* Springer Verlag, Berlin, 1985.

[9] G. Marchuk and V. Shaidurov. *Difference Methods and their Extrapolations.* Springer, New York, 1983.

[10] C. Pflaum. Konvergenz der Kombinationstechnik in der Energie- und L^2-Norm. Manuscript, 1993.

[11] U. Rüde. Multiple tau-extrapolation for multigrid methods. Bericht I-8701, Institut für Informatik, TU München, January 1987.

[12] U. Rüde. Extrapolation and related techniques for solving elliptic equations. Bericht I-9135, Institut für Informatik, TU München, September 1991.

[13] U. Rüde. Extrapolation techniques for constructing higher order finite element methods. Bericht I-9304, Institut für Informatik, TU München, 1993.

[14] U. Rüde. *Mathematical and computational techniques for multilevel adaptive methods*, volume 13 of *Frontiers in Applied Mathematics.* SIAM, Philadelphia, 1993.

[15] S. Schaffer. Higher order multigrid methods. *Math. Comp.*, 43:89–115, 1984.

[16] A. Schüller and Q. Lin. Efficient high order algorithms for elliptic boundary value problems combining full multigrid techniques and extrapolation methods. Arbeitspapiere der GMD 192, Gesellschaft für Mathematik und Datenverarbeitung, December 1985.

14

Robust Multi-grid with 7-point ILU Smoothing

Rob Stevenson[1]

1 Introduction

This paper deals with MG applied to discretized anisotropic boundary value problems. A model problem is given by

$$\begin{cases} -(\epsilon\partial_1^2 + \partial_2^2)u &= f \quad \text{on } \Omega = (0,1)^2 \\ u &= 0 \quad \text{on } \partial\Omega \end{cases},$$

discretized using linear finite elements corresponding to a regular triangulation of Ω. Besides this problem, we consider the problems that arise when the unit square is replaced by a domain having a re-entrant corner (less-regular problems) or when the differential operator is replaced by a rotated operator so that possibly the grid is no longer oriented with the direction of the anisotropy. We are interested in the question of whether the MGM is robust, that is, whether it converges uniformly not only in the "level" l, but also in ϵ. We concentrate on 7-point ILU smoothers which are known to give often good results in practice.

The paper is organized as follows: In section 2, some general MG convergence theory is treated. In [10], sufficient conditions on the MGM were given for robustness of the W-cycle applied to anisotropic problems on convex domains and on domains with re-entrant corners. The same was done in [9] for V-cycle MG applied to anisotropic problems on convex domains. Both papers follow and in some aspects extend the general MG theory developed by Hackbusch ([1]), which was extended by Wittum in [11, 12]. In the present paper, we show robustness of the W- and V-cycle under the same conditions as in [10] and [9], but now using the general convergence theory of Mandel, McCormick and Bank ([5]). In some sense the new results are stronger (robustness of the W-cycle for less regular problems with only one smoothing step) and the arguments are much simpler. This is particularly true for less-regular problems, since unlike Hackbusch's analysis of the W-cycle, the estimate needed for the smoother is independent of the regularity

[1]Utrecht University. From August 1, 1993: Department of Mathematics and Computer Science, Eindhoven University of Technology, P.O. Box 513, 5600 MB Eindhoven, The Netherlands. Email: stevenso@win.tue.nl .

of the boundary value problem. Essential for our results is a new estimate of the so-called "smoothing factor" ([5]) for smoothers having negative eigenvalues.

In sections 3 and 4, we apply the theory of section 2 to our model (rotated) anisotropic problems. On level l, let $A_l(\epsilon) = W_l(\epsilon) - R_l(\epsilon)$ be the 7-point ILU decomposition of the stiffness matrix, h_l be the mesh-width and let $\phi \in [0, \pi)$ be the angle of rotation. Using an estimate from [10] concerning the so-called "approximation property" ([1]), the essential conditions for robustness turn out to be

(a). $\sup_l h_l^2 \|R_l(\epsilon)\|_\infty = \begin{cases} \mathcal{O}(\epsilon) & \text{if } \phi \in \{0, \frac{\pi}{2}, \frac{3\pi}{4}\} \\ \mathcal{O}(\epsilon^2) & \text{if } \phi \notin \{0, \frac{\pi}{2}, \frac{3\pi}{4}\} \end{cases}$

(b). $\sup_{l,\epsilon} \rho(W_l(\epsilon)^{-1} A_l(\epsilon)) < 2$.

In section 3, we consider $\phi \in \{0, \frac{\pi}{2}, \frac{3\pi}{4}\}$ which means that the grid is oriented with the direction of the anisotropy. We show that $\sup_l h_l^2 \|R_l(\epsilon)\|_\infty \sim \epsilon$. Even when we consider only smoothers that are convergent, condition (b) is not trivial. For example, Jacobi iteration applied to our model problem is convergent for each h_l and ϵ. However, even for fixed ϵ, the spectral radius of the preconditioned system tends to 2 if $h_l \downarrow 0$. Until now, for 7-point ILU, (b) was proved only for a kind of modified decomposition. Numerical experiments show that modification is also necessary for obtaining robustness if $\phi \in \{\frac{\pi}{2}, \frac{3\pi}{4}\}$. For $\phi = 0$ however, it appears that unmodified ILU generally yields a more efficient MGM than modified ILU. We will give an upper bound for the spectral radius of the preconditioned system with 7-point ILU for a general 7-point symmetric M-matrix with constant coefficients. Using this bound it will follow that (b) is also satisfied for unmodified ILU if $\phi = 0$.

The case $\phi \notin \{0, \frac{\pi}{2}, \frac{3\pi}{4}\}$ is treated in section 4. For these angles and ϵ small, $A_l(\epsilon)$ is not an M-matrix. The 7-point decomposition is not exact for $\epsilon = 0$ and for this reason we are not able to give a complete proof of robustness. Yet, numerical experiments with modified ILU do indicate robustness if $\phi \in (0, \frac{\pi}{2}) \cup (\frac{3\pi}{4}, \pi)$. As a partial explanation, we prove that the "asymptotic rest" (the rest "away from the boundary") is of the order ϵ^2 (cf. (a)) if $\phi \in (0, \frac{\pi}{2}) \cup (\frac{3\pi}{4}, \pi)$. This is quite a surprising result since for $\phi \in \{0, \frac{\pi}{2}, \frac{3\pi}{4}\}$, the asymptotic rest was only of order ϵ. Using this difference in order, we will be able to clarify some discrepancies between the local mode smoothing factors and the actual MG contraction numbers.

2 General multi-grid convergence theory

We assume sequences of linear operators

$$A_l : \mathbf{C}^{n_l} \to \mathbf{C}^{n_l},\ p : \mathbf{C}^{n_l} \to \mathbf{C}^{n_{l+1}} \text{ and } r : \mathbf{C}^{n_{l+1}} \to \mathbf{C}^{n_l} \qquad l \in \mathbf{N}_0 = \{0, 1, 2, \ldots\}$$

with $l \mapsto n_l$ increasing, which are the sequences of stiffness matrices, prolongations and restrictions respectively. We equip $\mathbf{C}^{n_l}$ with the scaled Euclidian scalar product defined by

$$(u_l, v_l) = \frac{1}{n_l} \sum_{i=1}^{n_l} (u_l)_i \overline{(v_l)_i}$$

and norm $\|\cdot\| = (\cdot,\cdot)^{\frac{1}{2}}$. Adjoints relative to $(\cdot,\cdot)$ will be denoted by $\ ^*$.

Basic assumptions that we make throughout the paper are

- $A_l = A_l^* > 0$, $r = p^*$ and $A_{l-1} = rA_lp$.

Because of the first assumption, we may define the energy scalar product by

$$(u_l, v_l)_E = (A_l u_l, v_l)$$

and norm $\|\cdot\|_E = (\cdot,\cdot)_E^{\frac{1}{2}}$. Adjoints relative to $(\cdot,\cdot)_E$ will be denoted by $\ ^H$.

To solve $A_l u_l = f_l$ on some level l, we apply the induced MGM with ν_1 pre-smoothing steps of type $u_l \leftarrow u_l + (W_l^{(1)})^{-1}(f_l - A_l u_l)$, ν_2 post-smoothing steps of type $u_l \leftarrow u_l + (W_l^{(2)*})^{-1}(f_l - A_l u_l)$ and γ recursive calls on each coarser grid. We assume that the equations on level 0 are solved sufficiently accurately. Let M_l' be the error amplification operator of the resulting MGM on level l. We put $K_l = I_l - pA_{l-1}^{-1}rA_l$, $S_l^{(i)} = I_l - (W_l^{(i)})^{-1}A_l$ and $R_l^{(i)} = W_l^{(i)} - A_l$ $(i \in \{1,2\})$. We consider only smoothers that satisfy

- $\|S_l^{(i)}\|_{E\leftarrow E} < 1$, for each $l \in \mathbf{N}_0$ $(i \in \{1,2\})$.

We recall the MG convergence theorem from [5]:

Theorem 2.1 *Let $\gamma = 1$ (V-cycle) or $\gamma = 2$ (W-cycle) and let $\alpha \in (0,1]$ with $\alpha = 1$ if $\gamma = 1$. For $i \in \{1,2\}$, put*

$$\beta_i := \begin{cases} \left(\rho(K_l A_l^{-\alpha})^{\frac{1}{\alpha}} \rho(A_l[(I_l - S_l^{\nu_i}(S_l^{\nu_i})^H)^{-1} - I_l])\right)^{-1} & \text{if } \nu_i > 0 \\ 0 & \text{if } \nu_i = 0 \end{cases} .$$

Define the bijection $\Psi_\alpha : (0,1] \to [0,\infty)$ by

$$\Psi_\alpha(x) = \begin{cases} \frac{1}{x} - 1 & \text{if } \alpha \geq \frac{1}{1+x} \\ (1-x)(1+\frac{1}{x})^{\frac{1}{\alpha}}\alpha(1-\alpha)^{\frac{1}{\alpha}-1} & \text{if } \alpha \leq \frac{1}{1+x} \end{cases},$$

then

$$\|M_l'\|_{E\leftarrow E} \leq \left(\Psi_\alpha^{-1}(\beta_1)\Psi_\alpha^{-1}(\beta_2)\right)^{\frac{1}{2}}. \tag{1}$$

Note that Ψ_α^{-1} is decreasing and so the rhs of (1) is less than 1 uniform in l if the β_i^{-1}'s are bounded uniform in l and $\nu_1 + \nu_2 > 0$. For each $y \geq 0$, $\alpha \mapsto \Psi_\alpha^{-1}(y)$ is decreasing. For $\alpha \geq \frac{y+1}{y+2}$, in particular for $\alpha = 1$, it holds that $\Psi_\alpha^{-1}(y) = \frac{1}{1+y}$.

Proof. [5, lemmas 4.3, 4.4, 5.1 and theorems 5.1, 5.2] □

We give an upper bound for $\rho(A_l[(I_l - S_l^\nu(S_l^\nu)^H)^{-1} - I_l])$ for symmetric smoothers ($S_l = S_l^H$) that possibly have negative eigenvalues ($\rho(W_l^{-1}A_l) > 1$). In contrast to what was suggested in [5, remark 4.4], such a bound has been missing until now. Applications are given by the ILU smoothers.

Theorem 2.2 *Let $\nu \geq 1$, $W_l = W_l^*$, so that $S_l = S_l^H$, and $\rho(W_l^{-1}A_l) \leq \rho < 2$. Then*

$$\rho(A_l[(I_l - S_l^{2\nu})^{-1} - I_l]) \leq \rho(R_l S_l) \max\left\{\frac{1}{2\nu}, \frac{\rho(1-\rho)^{2\nu-2}}{1-(1-\rho)^{2\nu}}\right\}.$$

Proof. Let the operator S_l^+ be such that $S_l^+ S_l$ is the projection on Im S_l orthogonal w.r.t. $(\cdot,\cdot)_E$ and $(S_l^+)^H = S_l^+$. Then by Ker $[(I_l - S_l^{2\nu})^{-1} - I_l] =$ Ker S_l and $\sigma(S_l) \subset [1-\rho, 1]$, we have

$$\begin{aligned}
&\rho(A_l[(I_l - S_l^{2\nu})^{-1} - I_l]) \\
&\quad = \rho(S_l(I_l - S_l)^{-\frac{1}{2}} A_l (I_l - S_l)^{-\frac{1}{2}} S_l (S_l^+)^2 (I_l - S_l)[(I_l - S_l^{2\nu})^{-1} - I_l]) \\
&\quad \leq \|S_l(I_l - S_l)^{-\frac{1}{2}} A_l (I_l - S_l)^{-\frac{1}{2}} S_l\|_{E\leftarrow E} \|(S_l^+)^2 (I_l - S_l)[(I_l - S_l^{2\nu})^{-1} - I_l]\|_{E\leftarrow E} \\
&\quad \leq \rho(R_l S_l) \max_{\lambda \in [1-\rho, 1]} |\tfrac{1-\lambda}{\lambda^2}((1-\lambda^{2\nu})^{-1} - 1)|.
\end{aligned}$$

Elementary analysis shows that the maximum is obtained in one of the end points of the interval, which gives the result. □

Theorem 2.2 shows that the negative influence of eigenvalues of $W_l^{-1}A_l$ close to 2 on the MG convergence speed disappears when more smoothing steps are applied. If we had estimated only $\rho(A_l[(I_l - S_l^2)^{-1} - I_l])$ and then had used $\rho(A_l[(I_l - S_l^{2\nu})^{-1} - I_l]) \leq \frac{1}{\nu}\rho(A_l[(I_l - S_l^2)^{-1} - I_l])$ ([5, lemmas 4.2, 4.4]), then this effect would not have become visible.

Using theorem 2.2, we can estimate $\beta^{-1} = \beta_i^{-1}$ from theorem 2.1 by

$$\beta^{-1} \leq (\sup_l \rho(K_l A_l^{-\alpha})\rho(A_l^\alpha))^{\frac{1}{\alpha}} \sup_l \frac{\rho(R_l S_l)}{\rho(A_l)} \max\left\{\frac{1}{2\nu}, \frac{\rho(1-\rho)^{2\nu-2}}{1-(1-\rho)^{2\nu}}\right\}, \tag{2}$$

where $\rho = \sup_l \rho(W_l^{-1}A_l)$. When we insert the rhs of (2) with $\alpha = 1$ into the rhs of (1), we find exactly the V-cycle bound of [9, (3.7) with $s = 2$].

If W_l is constructed by making a (weakly) regular splitting of A_l, e.g. an ILU decomposition of an M-matrix A_l, then $\|S_l\|_\infty \leq 1$ if the row sums of A_l are non-negative ([8, proposition 3.3]). In that case, we can estimate $\rho(R_l S_l) \leq \|R_l\|_\infty$.

When theorem 2.2 is applied, $\sup_l \rho(W_l^{-1}A_l) < 2$ is a condition for obtaining a bound on $\|M_l'\|_{E\leftarrow E}$ which is less than 1 uniform in l. That this condition is also necessary for obtaining an uniformly converging MGM becomes plausible by the results of numerical tests and the following heuristic argument: Let e_l be such that $W_l^{-1}A_l e_l = \lambda e_l$ with λ close to 2. Then $S_l e_l = \mu e_l$ with $\mu = 1 - \lambda$ close to -1. So

e_l is hardly reduced by the smoother. Moreover, because of $W_l^{-1}A_le_l \approx 2e_l$, the error e_l is not "almost in the kernel of A_l"; in other words, e_l is not "smooth" and therefore it is hardly reduced by the coarse grid correction either.

To prove robustness for anisotropic problems, in theorem 2.1 we need β_i^{-1}'s that are bounded uniform in l ánd ϵ. If we consider (2), then the expected increase of $(\sup_l \rho(K_l(\epsilon)A_l(\epsilon)^{-\alpha})\rho(A_l(\epsilon)^{\alpha}))^{\frac{1}{\alpha}}$ when $\epsilon \downarrow 0$ due to the loss of ellipticity might be compensated by a corresponding decrease of $\sup_l \frac{\rho(R_l(\epsilon)S_l(\epsilon))}{\rho(A_l(\epsilon))}$. Indeed, if $R_l(0) = 0$, we can expect that $\lim_{\epsilon\downarrow 0} \sup_l \frac{\rho(R_l(\epsilon)S_l(\epsilon))}{\rho(A_l(\epsilon))} = 0$. Assuming that $\|S_l(\epsilon)\|_\infty \leq 1$, we obtain from theorems 2.1 and 2.2 the following sufficient conditions for robustness

- $\sup_{l,\epsilon} \rho(K_l(\epsilon)A_l(\epsilon)^{-\alpha})^{\frac{1}{\alpha}}\|R_l(\epsilon)\|_\infty < \infty$
- $\sup_{l,\epsilon} \rho(W_l(\epsilon)^{-1}A_l(\epsilon)) < 2.$

In sections 3 and 4, we will discuss these conditions for our model problems.

Finally in this section, we show that the factor $\rho(R_lS_l)$ from theorem 2.2 may be replaced by $\rho(R_l)$ if $\rho(W_l^{-1}A_l) \leq 1$. Applications are given by the SGS-smoothers and the modified ILU smoothers (ILU_ω), characterized by a rest R_l satisfying $(R_l)_{ii} = \omega\sum_{j\neq i}|(R_l)_{ij}|$, if $\omega \geq 1$.

Theorem 2.3 *Let $\nu \geq 1$, $W_l = W_l^*$ and $\rho(W_l^{-1}A_l) \leq 1$. Then*

$$\rho(A_l[(I_l - S_l^{2\nu})^{-1} - I_l]) \leq \frac{\rho(R_l)}{2\nu}.$$

Proof. [5, lemma 4.4, theorem 4.4 and remark 4.4] or

$$\begin{aligned}&\rho(A_l[(I_l - S_l^{2\nu})^{-1} - I_l]) \\ &= \rho(S_l^{\frac{1}{2}}(I_l - S_l)^{-\frac{1}{2}}A_l(I_l - S_l)^{-\frac{1}{2}}S_l^{\frac{1}{2}}S_l^{+}(I_l - S_l)[(I_l - S_l^{2\nu})^{-1} - I_l]) \\ &\leq \rho(R_l)\max_{\lambda\in[0,1]}|\tfrac{1-\lambda}{\lambda}((1-\lambda^{2\nu})^{-1} - 1)|.\end{aligned}$$

□

3 7-point ILU smoothing

We consider a symmetric M-matrix A with constant coefficients that can be described by a 7-point difference stencil on a finite two-dimensional grid:

$$A \triangleq \begin{bmatrix} \cdot & -as & -asw \\ -ae & \delta & -ae \\ -asw & -as & \cdot \end{bmatrix}. \tag{3}$$

The 7-point ILU decomposition $A = W - R$ w.r.t. an east-to-west, south-to-north lexicographical ordering of the grid points is given by $W = (D-L)D^{-1}(D-L^T)$,

$$L \triangleq \begin{bmatrix} \cdot & \cdot & \cdot \\ \cdot & \cdot & \mathit{le}_{ij} \\ \mathit{lsw}_{ij} & \mathit{as} & \cdot \end{bmatrix},\ D \triangleq \begin{bmatrix} \cdot & \cdot & \cdot \\ \cdot & d_{ij} & \cdot \\ \cdot & \cdot & \cdot \end{bmatrix},\ R \triangleq \begin{bmatrix} & \cdot & \cdot & \cdot & \cdot & r_{i+1j+2} \\ & \cdot & \cdot & \cdot & \cdot & \cdot \\ r_{ij} & \cdot & \cdot & \cdot & & \cdot \end{bmatrix},$$

where

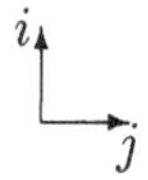

$$d_{ij} = \delta - \frac{\mathit{le}_{ij}^2}{d_{ij+1}} - \frac{\mathit{as}^2}{d_{i-1j}} - \frac{\mathit{lsw}_{ij}^2}{d_{i-1j-1}},$$

$$\mathit{le}_{ij} = \mathit{ae} + \frac{\mathit{as}}{d_{i-1j}}\mathit{lsw}_{ij+1},\ \mathit{lsw}_{ij} = \mathit{asw} + \frac{\mathit{as}}{d_{i-1j}}\mathit{le}_{i-1j-1},\ r_{ij} = \frac{\mathit{le}_{i-1j-2}\mathit{lsw}_{ij}}{d_{i-1j-1}} \tag{4}$$

skipping terms containing indices (p,q) that correspond to grid points outside the domain.

In theorem 3.2 we estimate $\|R\|_\infty$ and $\rho(W^{-1}A)$. Our main application is given by our (rotated) anisotropic model problems described precisely in the following example.

Example 3.1 Let $A = A_l(\epsilon)$ be the stiffness matrix resulting from the application of the linear finite element method to the rotated anisotropic boundary value problem

$$\begin{cases} \left(-(\epsilon c^2 + s^2)\partial_1^2 - 2(\epsilon - 1)sc\partial_2\partial_1 - (\epsilon s^2 + c^2)\partial_2^2\right) u &= f \quad \text{on } \Omega \\ u &= 0 \quad \text{on } \partial\Omega \end{cases} \quad (\epsilon \in (0,1]),$$

where $c = \cos(\phi)$, $s = \sin(\phi)$, $\phi \in [0,\pi)$, w.r.t. a regular triangulation of Ω assuming that $\partial\Omega$ coincides with grid lines. Then $A_l(\epsilon)$ is of the form (3) with

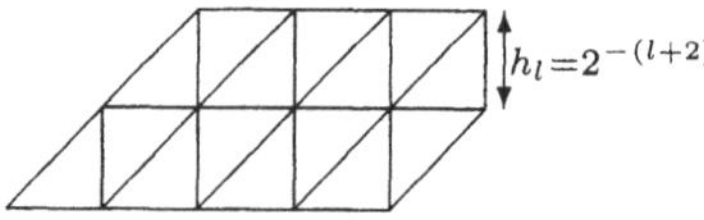

$\mathit{ae} = h_l^{-2}(s(c+s) + \epsilon c(c-s))$, $\mathit{as} = h_l^{-2}(c(c+s) + \epsilon s(s-c))$, $\mathit{asw} = h_l^{-2}(\epsilon - 1)sc$ and $\delta = 2(\mathit{ae} + \mathit{as} + \mathit{asw})$.

Note that $A_l(\epsilon)$ is an M-matrix for all $\epsilon \in (0,1]$ if and only if $\phi \in \{0, \frac{\pi}{2}, \frac{3}{4}\pi\}$.

We assume canonical (7-point) prolongations and restrictions. Then from [10, theorems 2.1 and 2.4] we know that for $\phi \in \{0, \frac{\pi}{2}, \frac{3}{4}\pi\}$,

$$\left(\sup_l \rho(K_l(\epsilon)A_l(\epsilon)^{-\alpha})\rho(A_l(\epsilon)^{\alpha})\right)^{\frac{1}{\alpha}} = \mathcal{O}(\epsilon^{-1})\ (\epsilon \downarrow 0) \tag{5}$$

with $\alpha = 1$ if Ω is convex and $\alpha < 1$ otherwise (for a discussion about the value of α we refer to [10, remark 2.5]). For $\phi \notin \{0, \frac{\pi}{2}, \frac{3}{4}\pi\}$, (5) is not valid, but instead only $\mathcal{O}(\epsilon^{-2})$ holds ([10, remark 2.3]).

Theorem 3.2 *Let* $\mathit{as} > 0$, $\mathit{asw}, \mathit{ae} \geq 0$, $\mathit{asw} + \mathit{ae} > 0$ *and* $\delta = 2(\mathit{ae} + \mathit{as} + \mathit{asw})$. *Let* $\zeta = \overline{\zeta}$ *be the largest root of*

$$f(\zeta) := \zeta^3 - 2(\mathit{as} + \mathit{ae} + \mathit{asw})\zeta^2 + (\mathit{ae}^2 + \mathit{asw}^2 - 4\mathit{as}^2)\zeta + 4\mathit{as}(\mathit{ae}\,\mathit{asw} + 2\mathit{as}(\mathit{asw} + \mathit{ae} + \mathit{as})) = 0.$$

Put $r = \dfrac{ae\,asw\,\overline{\zeta} + as(asw^2 + ae^2)}{\overline{\zeta}^2 - 4as^2}$ *and* $q = 2\dfrac{ae\,asw + asw\,as + ae\,as}{ae + asw + 4as}$. *Then*

(a). $r_{ij} \leq r$ *(and* $r_{ij} \uparrow r$ *"away from the boundary") and so* $\|R\|_\infty \leq 2r$

(b). $\rho(W^{-1}A) \leq \frac{q}{q-r}$.

(Note that q *and* r *are homogeneous functions of* A *of order* 1 *and thus* $\frac{q}{q-r}$ *is homogeneous of order* 0. *So e.g. multiplication of* A *by* h_l^{-2} *makes* r *and* q *a factor* h_l^{-2} *larger but does not change* $\frac{q}{q-r}$.*) We have* $(\frac{2ae\,asw + as\,ae + as\,asw}{2ae + 2asw + 4as} <) 2r < q$ *and thus* $\|R\|_\infty \leq q$ *(explicit bound) and* $\rho(W^{-1}A)(\leq \frac{q}{q-r}) < 2$.

Proof. Since A is an M-matrix, it is well known (cf. [6]) that

$$d_{ij} \geq \overline{d}, \quad le_{ij} \leq le, \quad lsw_{ij} \leq lsw, \quad r_{ij} \leq \frac{le\,lsw}{\overline{d}}$$

(and convergence to these bounds "away from the boundary"), where $d = \overline{d}$ is the largest root less than δ of the system

$$\left\{d = \delta - \tfrac{le^2}{d} - \tfrac{as^2}{d} - \tfrac{lsw^2}{d},\ (1 - \tfrac{as^2}{d^2})le = ae + \tfrac{as\,asw}{d},\ (1 - \tfrac{as^2}{d^2})lsw = asw + \tfrac{as\,ae}{d}\right\} \quad (6)$$

$$\iff d^6 - 2(as+ae+asw)d^5 + (ae^2+asw^2-as^2)d^4 + 4as(asw\,ae+as(as+ae+asw))d^3$$
$$+(ae^2+asw^2-as^2)as^2d^2 - 2(as+ae+asw)as^4d + as^6 = 0.$$

By noting that d is a solution of (6) if and only if $\frac{as^2}{d}$ is a solution, (6) can be reduced to the equation $f(\zeta) = 0$, where $\zeta = d + \frac{as^2}{d}$. It can be verified that $f(\delta)$ $(= 2(ae + asw)(ae^2 + as\,ae + asw\,as + asw^2)) > 0$, $f(2as + ae + asw)$ $(= -2(ae + asw)(ae\,asw + asw\,as + as\,ae)) < 0$, $f(0) > 0$ and $\lim_{\zeta\to-\infty} f(\zeta) = -\infty$. We conclude that all solutions of $f(\zeta) = 0$, and thus all solutions d of (6), are less than δ. So the solution $d = \overline{d}$ is the largest solution of (6) and $\overline{d} = \frac{\overline{\zeta}+\sqrt{\overline{\zeta}^2-4as^2}}{2}$ where $\zeta = \overline{\zeta} \in (2as + ae + asw, \delta)$ is the largest solution of $f(\zeta) = 0$.

We can now write

$$\frac{le\,lsw}{\overline{d}} = \frac{ae\,asw\left(\overline{d} + \frac{as^2}{\overline{d}}\right) + as(asw^2 + ae^2)}{\left(\overline{d} - \frac{as^2}{\overline{d}}\right)^2} = \frac{ae\,asw\,\overline{\zeta} + as(asw^2 + ae^2)}{\overline{\zeta}^2 - 4as^2} \ (= r) \quad (7)$$

which shows (a). The expression r is a decreasing function of $\overline{\zeta} > 0$. Substituting the bounds that we found for $\overline{\zeta}$ yields the inequalities $\frac{2ae\,asw+as\,ae+as\,asw}{2ae+2asw+4as} < 2r < q$.

It remains to show that $\rho(W^{-1}A) \leq \rho := \frac{q}{q-r}$. Because of $A = A^* > 0$ and $W = W^*$, this inequality is equivalent to $A + \frac{\rho}{\rho-1}R \geq 0$ ([8, lemma 2.9]). We have

$$R = \frac{1}{2}(R + R) \ \geq \ \frac{1}{2}(R - \mathrm{diag}\{(\mathrm{rowsum}R)_{ij}\})$$

$$\hat{=} \ \frac{1}{2}\begin{bmatrix} \cdot & \cdot & \cdot & \cdot & r_{i+1j+2} \\ \cdot & \cdot & -(r_{ij}+r_{i+1j+2}) & \cdot & \cdot \\ r_{ij} & \cdot & \cdot & \cdot & \cdot \end{bmatrix}. \qquad (8)$$

Although it is not the case for R, in (8) we have an operator that is a decreasing function of the coefficients r_{pq} (use Gershgorin's circle theorem). Since $r_{pq} \leq r$ and $\frac{\rho}{\rho-1}r = q$, we thus have

$$A + \frac{\rho}{\rho-1}R \geq \begin{bmatrix} \cdot & \cdot & -as & -asw & \frac{1}{2}q \\ \cdot & -ae & \delta - q & -ae & \cdot \\ \frac{1}{2}q & -asw & -as & \cdot & \cdot \end{bmatrix}.$$

We compare this operator in the rhs with the semi-positive definite operator

$$(\sigma - \tilde{L})\sigma^{-1}(\sigma - \tilde{L}^T) \hat{=} \begin{bmatrix} \cdot & \cdot & -as & -y + \frac{as\,x}{\sigma} & \frac{xy}{\sigma} \\ \cdot & -x + \frac{as\,y}{\sigma} & \sigma + \frac{x^2+y^2+as^2}{\sigma} & -x + \frac{as\,y}{\sigma} & \cdot \\ \frac{xy}{\sigma} & -y + \frac{as\,x}{\sigma} & -as & \cdot & \cdot \end{bmatrix},$$

where $\tilde{L} \hat{=} \begin{bmatrix} \cdot & \cdot & \cdot \\ \cdot & \cdot & x \\ y & as & \cdot \end{bmatrix}$, $\sigma = y+x+as$ and x, y are real. Both operators have zero row sums ($(\sigma - \tilde{L})$ has zero row sums). The system of equations $\begin{cases} -y + \frac{as\,x}{\sigma} = -asw \\ -x + \frac{as\,y}{\sigma} = -ae \end{cases}$ has two (one, if $as = 0$) solutions $(x, y) \in \mathbf{R}^2$. It can be verified that these solutions satisfy the remarkable equality $\frac{xy}{\sigma} = \frac{1}{2}q$ which completes the proof. □

We will now consider $A = A_l(\epsilon)$ from example 3.1 for $\phi \in \{0, \frac{\pi}{2}, \frac{3\pi}{4}\}$. Theorem 3.2(b) shows that for each ϵ, $\sup_l \rho(W_l(\epsilon)^{-1}A_l(\epsilon)) < 2$. Therefore, from theorem 2.1 it follows that for *fixed* ϵ the MGM with 7-point ILU smoothing converges uniformly in l.

For each $\phi \in \{0, \frac{\pi}{2}, \frac{3\pi}{4}\}$, the inequalities $\frac{2ae\,asw + as\,ae + as\,asw}{2ae + 2asw + 4as} < 2r < q$ show that $\sup_l h_l^2 \|R_l(\epsilon)\|_\infty \sim \epsilon$. (For $\phi = 0$ this is a remarkable result since for fixed l, $h_l^2\|R_l(\epsilon)\|_\infty = \mathcal{O}(\epsilon^2)$ as can be deduced from the recursions defining the decomposition). Since by (5), $\sup_l h_l^{-2}\rho(K_l(\epsilon)A_l(\epsilon)^{-\alpha})^{\frac{1}{\alpha}} = \mathcal{O}(\epsilon^{-1})$, robustness will follow whenever $\sup_{l,\epsilon} \rho(W_l(\epsilon)^{-1}A_l(\epsilon)) < 2$.

For $\phi = \frac{3\pi}{4}$, we obtain $q = h_l^{-2}\frac{4\epsilon}{1+9\epsilon}$ and $r = h_l^{-2}\frac{\frac{1}{2}\epsilon(1-\epsilon)h_l^2\overline{\zeta} + \epsilon(\frac{1}{4} - \epsilon + \frac{5}{4}\epsilon^2)}{(h_l^2\overline{\zeta})^2 - 4\epsilon^2}$ in theorem 3.2. For $\epsilon = 0$, the equation $f(\zeta) = 0$ reduces to $h_l^2\zeta(h_l^2\zeta - \frac{1}{2})^2 = 0$. Therefore, it holds that $\lim_{\epsilon \downarrow 0} h_l^2\overline{\zeta} = \frac{1}{2}$ and so the upper bound $\frac{q}{q-r}$ for $\sup_l \rho(W_l(\epsilon)^{-1}A_l(\epsilon))$ from theorem 3.2 tends to 2 if $\epsilon \downarrow 0$.

We have computed $\|M_l'(\epsilon)\|_{E \leftarrow E}$ numerically with the Lanczos method using symmetry of $M_l'(\epsilon)$ w.r.t. $(\cdot,\cdot)_E$. The results are given in table 1. With the same number j of MG evaluations, the Lanczos method gives a much more accurate approximation of $\|M_l'(\epsilon)\|_{E \leftarrow E}$ than the averaged reduction factor $\sqrt[j]{\frac{\|r_j\|}{\|r_0\|}}$ where r_i

$\epsilon \setminus h_l$	$\frac{1}{32}$	$\frac{1}{64}$	$\frac{1}{128}$	$\frac{1}{256}$	$\frac{1}{512}$	$\frac{1}{1024}$
1	.028(.039)	.032(.044)	.034(.047)	.034(.047)	.035(.047)	–
.1	.058(.059)	.062(.064)	.064(.066)	.064(.067)	.064(.067)	–
.01	.185(.087)	.230(.109)	.243(.119)	.245(.120)	.247(.121)	–
.001	.106(.036)	.317(.091)	.510(.130)	.576(.145)	.592(.156)	.595(.159)
.0001	.004(.001)	.046(.016)	.263(.067)	.585(.117)	.761(.149)	.811(.168)

TABLE 1: $\|M_l'\|_{E \leftarrow E}$ for $\Omega = (0,1)^2$, 7-point ILU smoothing, $\gamma = 1$, $\nu_1 = \nu_2 = 1$ and $\phi = \frac{3\pi}{4}$. In parentheses the results for 7-point ILU$_\omega$ smoothing with $\omega = \frac{1}{2}$.

denotes the residual after step i. Table 1 shows unsatisfactory contraction numbers for small ϵ and h_l and it strongly indicates that indeed this MGM is not robust. This would also mean that in this case the upper bound $\frac{q}{q-r}$ is sharp when $\epsilon \downarrow 0$.

In order to obtain a robust method, modified ILU (ILU$_\omega$) was proposed by Wittum ([12]), Oertel and Stüben ([7]) and Khalil ([4]). From [8], we recall the following theorem:

Theorem 3.3 ([8, theorem 3.2]) *Let $A = W_\omega - R_\omega$ be a symmetric ILU$_\omega$ decomposition of a symmetric M-matrix A. Then for $\omega \geq 0$,*

$$(a).\ \|R_\omega\|_\infty \leq (1+\omega)\|R_0\|_\infty \quad \text{and} \quad (b).\ \rho(W_\omega^{-1}A) \leq \max\{1, \tfrac{2}{1+\omega}\}.$$

So for the modified decomposition the eigenvalues of the preconditioned system stay away from 2, whereas the rest is only at most a factor $1+\omega$ larger than with the unmodified decomposition.

Since we already knew that for $\phi \in \{0, \frac{\pi}{2}, \frac{3\pi}{4}\}$ the rest of unmodified ILU is of order ϵ, we conclude that for these angles MG with 7-point ILU$_\omega$ smoothing with $\omega > 0$ is robust. In table 1, numerically computed contraction numbers $\|M_l'(\epsilon)\|_{E \leftarrow E}$ are given for $\phi = \frac{3\pi}{4}$ and $\omega = \frac{1}{2}$. We took $\omega = \frac{1}{2}$ because at least for $\phi = \frac{\pi}{2}$, this value yields the optimum so-called "local mode smoothing factor" if $\epsilon \downarrow 0$ ([7]). However, the contraction numbers do not vary much as a function of ω as long as ω stays away from 0.

For $\phi = \frac{\pi}{2}$, it holds that $q = h_l^{-2}\frac{2\epsilon}{1+4\epsilon}$ and $r = h_l^{-2}\frac{\epsilon}{(h_l^2\zeta)^2 - 4\epsilon^2}$ in theorem 3.2. For $\epsilon = 0$, the equation $f(\zeta) = 0$ reduces to $h_l^2\zeta(h_l^2\zeta - 1)^2 = 0$. So $\lim_{\epsilon\downarrow 0} h_l^2\zeta = 1$, and again the upper bound $\frac{q}{q-r}$ for $\sup_l \rho(W_l(\epsilon)^{-1}A_l(\epsilon))$ tends to 2 if $\epsilon \downarrow 0$. Numerical computations show contraction numbers for unmodified and modified 7-point ILU which are similar to those found for $\phi = \frac{3\pi}{4}$. So also for $\phi = \frac{\pi}{2}$, we may conclude that the upper bound $\frac{q}{q-r}$ is sharp when $\epsilon \downarrow 0$ and that a robust method is only obtained with modified ILU.

Finally, we discuss the case $\phi = 0$. We obtain $q = h_l^{-2}\frac{2\epsilon}{4+\epsilon}$ and $r = h_l^{-2}\frac{\epsilon^2}{(h_l^2\zeta)^2 - 4}$. For $\epsilon = 0$, $f(\zeta) = 0$ reduces to $(h_l^2\zeta + 2)(h_l^2\zeta - 2)^2 = 0$ so that $\lim_{\epsilon\downarrow 0} h_l^2\zeta = 2$, which gives no information as to whether or not $\frac{q}{q-r}$ is less than 2 uniform in ϵ.

Solving $f(\zeta)=0$ explicitly yields $h_l^2\overline{\zeta}=2+(1+\frac{1}{2}\sqrt{2})\epsilon-\frac{1}{16}(1+\sqrt{2})\epsilon^2+\mathcal{O}(\epsilon^3)$ and so

$$\frac{q}{q-r}=\sqrt{2}-\frac{1}{8}(\sqrt{2}-1)\epsilon+\mathcal{O}(\epsilon^2).$$

We conclude that for the unrotated anisotropic equation the MGM with unmodified 7-point ILU smoothing is robust. In table 2 numerically computed contraction numbers are given for unmodified and modified ILU with $\omega=\frac{1}{2}$. A comparison of

$\epsilon \setminus h_l$	$\frac{1}{32}$	$\frac{1}{64}$	$\frac{1}{128}$	$\frac{1}{256}$	$\frac{1}{512}$	$\frac{1}{1024}$
1	.028(.039)	.032(.044)	.034(.047)	.034(.047)	.035(.047)	–
.1	.023(.021)	.026(.028)	.027(.030)	.027(.031)	.027(.032)	–
.01	.006(.004)	.020(.014)	.027(.023)	.028(.028)	.029(.029)	–
.001	.000(.000)	.001(.001)	.011(.006)	.023(.018)	.028(.025)	.029(.028)
.0001	.000(.000)	.000(.000)	.000(.000)	.003(.002)	.016(.010)	.025(.021)

TABLE 2: As table 1 with now $\phi=0$.

these results shows that for small ϵ modified ILU gives only slightly better results, whereas for $\epsilon\approx 1$ unmodified ILU yields a faster converging method.

To see whether the bound $\frac{q}{q-r}$ is sharp, we computed $\rho(W_l(\epsilon)^{-1}A(\epsilon))$ numerically. We found values less than or, when ϵ and h_l are small, close to 1.2. For $\epsilon=1$ (Poisson equation), $f(\zeta)=\zeta^3-4\zeta^2-3\zeta+16$ and so $\overline{\zeta}=\frac{4}{3}+2\mathrm{Re}\left((-\frac{98}{27}+\frac{1}{3}\frac{\sqrt{223}}{\sqrt{3}}i)^{\frac{1}{3}}\right)\approx 3.598$ (and thus $\overline{d}=\frac{\overline{\zeta}+\sqrt{\overline{\zeta}^2-4}}{2}\approx 3.294$). This yields a bound $\frac{q}{q-r}\approx 1.388$. Numerically we found $\rho(W_l^{-1}A_l)\approx 1.178$, 1.179 and 1.180 for $\Omega=(0,1)^2$ and $h_l=\frac{1}{64},\frac{1}{128}$ and $\frac{1}{256}$ respectively.

4 The rotated anisotropic equation for angles $\phi\notin\{0,\frac{\pi}{2},\frac{3\pi}{4}\}$.

For $\phi\notin\{0,\frac{\pi}{2},\frac{3\pi}{4}\}$, 7-point ILU is not exact for $\epsilon=0$; at the boundary the rest does not tend to zero when $\epsilon\downarrow 0$. So, at least a straightforward application of the theory from section 2 does not show robustness of the MGM. However, numerical experiments performed by Oertel and Stüben ([7]) do indicate robustness with 7-point ILU$_\omega$ smoothing if $\phi\in(0,\frac{\pi}{2})\cup(\frac{3\pi}{4},\pi)$ and $\omega=1$ (our ϕ corresponds to $\pi-\phi$ in [7]).

Unmodified ILU turned out to yield an MGM that diverges for many angles. Note that for $\phi\notin\{0,\frac{\pi}{2},\frac{3\pi}{4}\}$ and ϵ small enough, $A_l(\epsilon)$ is not an M-matrix and so $\rho(W_l(\epsilon)^{-1}A_l(\epsilon))<2$ is not guaranteed if $\omega<1$. For $\omega\geq 1$, we have $R_l\geq 0$ which is equivalent to $\rho(W_l^{-1}A_l)\leq 1$ (see [8, proposition 2.8]). We recall that when $\rho(W_l^{-1}A_l)\leq 1$, for proving robustness on can apply theorem 2.3 instead of 2.2 so S_l need not be bounded.

When we repeated the experiments from [7], it appeared, as for $\phi \in \{0, \frac{\pi}{2}, \frac{3\pi}{4}\}$, that for each $\omega \geq 0$ away from the boundary the d_{ij}'s converge to some limit $\overline{d}_\omega$. In the case of ILU$_\omega$, the d_{ij}'s satisfy the recursion

$$d_{ij} = \delta - \frac{\mathit{le}_{ij}^2}{d_{ij+1}} - \frac{\mathit{as}^2}{d_{i-1j}} - \frac{\mathit{lsw}_{ij}^2}{d_{i-1j-1}} + \omega(|r_{i+1j+2}| + |r_{ij}|),$$

with le_{ij}, lsw_{ij} and r_{ij} as in (4). In a way similar to that used in the proof of theorem 3.2 for $\omega = 0$, the resulting fixed point equation for $\overline{d}_\omega$ can be reduced to the equation

$$f_\omega(\zeta, \epsilon) := f(\zeta, \epsilon) - 2\omega|r(\zeta, \epsilon)|(\zeta^2 - 4\mathit{as}^2) = 0$$

for $\zeta = \overline{\zeta}_\omega := \overline{d}_\omega + \frac{\mathit{as}^2}{\overline{d}_\omega}$, where $f(\zeta, \epsilon)$ is as in Theorem 3.2 (there denoted by $f(\zeta)$) and

$$r(\zeta, \epsilon) = \frac{\mathit{ae}\,\mathit{asw}\,\zeta + \mathit{as}(\mathit{asw}^2 + \mathit{ae}^2)}{\zeta^2 - 4\mathit{as}^2} \quad \text{(cf. (7)).}$$

By convergence of d_{ij} to $\overline{d}_\omega$, we have convergence of r_{ij} to $r_\omega := r(\overline{\zeta}_\omega, \epsilon)$. Note that $h_l^2\overline{d}_\omega$, $h_l^2\overline{\zeta}_\omega$ and $h_l^2 r_\omega$ are independent of h_l. For $\phi \in (0, \frac{\pi}{2}) \cup (\frac{3\pi}{4}, \pi)$, but not for $\phi \in (\frac{\pi}{2}, \frac{3\pi}{4})$, we found numerically that $\lim_{\epsilon\downarrow 0} h_l^2 r_\omega = 0$. (However, in contrast to the case $\phi \in \{0, \frac{\pi}{2}, \frac{3\pi}{4}\}$, it turned out that $r_{ij} \leq r_\omega$ was not valid.)

To find a partial explanation for the observed robustness with ILU$_\omega$ when $\phi \in (0, \frac{\pi}{2}) \cup (\frac{3\pi}{4}, \pi)$ and $\omega \geq 1$, we study the size of the "asymptotic rest" $h_l^2 r_\omega$ as a function of ϵ. If the "boundary perturbations" may be neglected, then for $\omega \geq 1$ robustness is proved whenever $\sup_\epsilon\{h_l^2 r_\omega \sup_l h_l^{-2}\rho(K_l(\epsilon)A_l(\epsilon)^{-\alpha})^{\frac{1}{\alpha}}\} < \infty$. We recall that $\sup_l h_l^{-2}\rho(K_l(\epsilon)A_l(\epsilon)^{-\alpha})^{\frac{1}{\alpha}} = \mathcal{O}(\epsilon^{-2})$, and not $\mathcal{O}(\epsilon^{-1})$ as was the case for $\phi \in \{0, \frac{\pi}{2}, \frac{3\pi}{4}\}$.

Except for $\phi = \pi - \arctan(2)$ ($\Leftrightarrow \mathit{asw}|_{\epsilon=0} = \mathit{ae}|_{\epsilon=0}$), for which angle $r(\zeta, 0)$ reduces to $\frac{\mathit{ae}|_{\epsilon=0}\mathit{asw}|_{\epsilon=0}}{\zeta - 2\mathit{as}|_{\epsilon=0}}$ which is never zero, the equations $r(\zeta, 0) = 0$ and $f_\omega(\zeta, 0) = 0$ have a common, and therefore ω-independent, solution $\zeta = \overline{\zeta}^{(0)} := -\frac{\mathit{as}(\mathit{asw}^2 + \mathit{ae}^2)}{\mathit{ae}\,\mathit{asw}}|_{\epsilon=0}$. Via $\zeta = d + \frac{\mathit{as}^2}{d}$, $\overline{\zeta}^{(0)}$ corresponds to the fixed points $-\frac{\mathit{as}\,\mathit{asw}}{\mathit{ae}}|_{\epsilon=0} = h_l^{-2}\cos(\phi)^2$ and $-\frac{\mathit{as}\,\mathit{ae}}{\mathit{asw}}|_{\epsilon=0} = h_l^{-2}(\sin(\phi) + \cos(\phi))^2$ of the recursion for d_{ij} with $\epsilon = 0$. Numerically, we found that indeed $\lim_{\epsilon\downarrow 0} h_l^2\overline{d}_\omega = (\sin(\phi) + \cos(\phi))^2$ if $\phi \in (0, \frac{\pi}{2})$ and $\lim_{\epsilon\downarrow 0} h_l^2\overline{d}_\omega = \cos(\phi)^2$ if $\phi \in (\frac{3\pi}{4}, \pi)$. For $\phi \in (\frac{\pi}{2}, \frac{3\pi}{4})$, $h_l^2\overline{d}_\omega$ converged neither to $(\sin(\phi) + \cos(\phi))^2$ nor to $\cos(\phi)^2$. Note that $\pi - \arctan(2) \in (\frac{\pi}{2}, \frac{3\pi}{4})$.

When we try to analyse $\overline{\zeta}_\omega$ as a function of ϵ, a problem that we encounter is that f_ω is not differentiable to ζ in $(\overline{\zeta}^{(0)}, 0)$. Therefore, now let $\zeta = \overline{\zeta}_\omega^{\pm}$ be the solution of $f_\omega^{\pm}(\zeta, \epsilon) := f(\zeta, \epsilon) \mp 2\omega r(\zeta, \epsilon)(\zeta^2 - 4\mathit{as}^2) = 0$ with $\lim_{\epsilon\downarrow 0}\overline{\zeta}_\omega^{\pm} = \overline{\zeta}^{(0)}$. Since for $\phi \notin \{0, \frac{\pi}{2}, \frac{3\pi}{4}\}$, $\zeta = \overline{\zeta}^{(0)}$ is an isolated root of $f_\omega^{\pm}(\zeta, 0) = 0$, the $\overline{\zeta}_\omega^{\pm}$ are uniquely defined. From our experiments we know that $\lim_{\epsilon\downarrow 0}\overline{\zeta}_\omega = \overline{\zeta}^{(0)}$ if $\phi \in (0, \frac{\pi}{2}) \cup (\frac{3\pi}{4}, \pi)$. Hence, for these angles and ϵ small enough we have $\overline{\zeta}_\omega = \overline{\zeta}_\omega^{+}$

or $\overline{\zeta}_\omega = \overline{\zeta}_\omega^-$. The implicit function theorem tells us that $(\frac{d}{d\epsilon}\overline{\zeta}_\omega^\pm)|_{\epsilon=0} = -\frac{\frac{\partial}{\partial\epsilon}f_\omega^\pm(\overline{\zeta}^{(0)},0)}{\frac{\partial}{\partial\zeta}f_\omega^\pm(\overline{\zeta}^{(0)},0)}$. Some computations show that

$$\frac{\frac{\partial}{\partial\epsilon}f_\omega^\pm(\overline{\zeta}^{(0)},0)}{\frac{\partial}{\partial\zeta}f_\omega^\pm(\overline{\zeta}^{(0)},0)} = \frac{3\cos(\phi)^2\sin(\phi)+\cos(\phi)^4+2\cos(\phi)\sin(\phi)+1}{\cos(\phi)(\cos(\phi)+\sin(\phi))} = \frac{\frac{\partial r}{\partial\epsilon}(\overline{\zeta}^{(0)},0)}{\frac{\partial r}{\partial\zeta}(\overline{\zeta}^{(0)},0)}.$$

We conclude that for $\phi \in (0,\frac{\pi}{2}) \cup (\frac{3\pi}{4},\pi)$,

$$\begin{aligned} h_l^2 r_\omega &= h_l^2 r(\overline{\zeta}_\omega^\pm,\epsilon) \\ &= h_l^2 r(\overline{\zeta}^{(0)},0) + h_l^2\Big(\tfrac{\partial r}{\partial\zeta}(\overline{\zeta}^{(0)},0)(\tfrac{d}{d\epsilon}\overline{\zeta}_\omega^\pm)|_{\epsilon=0} + \tfrac{\partial r}{\partial\epsilon}(\overline{\zeta}^{(0)},0)\Big)\epsilon + \mathcal{O}(\epsilon^2) \\ &= \mathcal{O}(\epsilon^2), \end{aligned}$$

which shows robustness of MG with 7-point ILU_ω smoothing for $\omega \geq 1$ assuming that boundary perturbations can be neglected.

We end by making some remarks about local mode analysis. From $h_l^2 r_\omega = \mathcal{O}(\epsilon^2)$, it is not hard to prove that the "local mode smoothing factor" tends to zero when $\epsilon \downarrow 0$ (cf. [2, table 4.1]). For $\phi \in \{0,\frac{\pi}{2},\frac{3\pi}{4}\}$, it is known that this smoothing factor does not tend to zero when $\epsilon \downarrow 0$ (for example, for $\phi = 0$ and unmodified 7-point ILU it tends to $(3+2\sqrt{2})^{-1}$ ([3, theorem 3.2])). From practice however, it turns out, also for small ϵ, that the MG contraction number depends smoothly on $\phi \in [0,\frac{\pi}{2}] \cup [\frac{3\pi}{4},\pi]$ and that it does not tend to zero uniformly in h_l. We conclude that for anisotropic problems, a local mode analysis of only the smoother can give misleading results. To obtain full insight into the convergence behaviour of the MGM, one also have to perform an analysis of the coarse grid correction.

References

[1] W. Hackbusch. *Multi-Grid Methods and Applications.* Springer-Verlag, Berlin, 1985.

[2] P.W. Hemker. Multigrid methods for problems with a small parameter in the highest derivative. In D.F. Griffiths, editor, *Numerical Analysis,* Proceedings, Dundee 1983, pages 106–121, Berlin, 1984. Lecture Notes in Mathematics 1066, Springer-Verlag.

[3] R. Kettler. Analysis and comparison of relaxation schemes in robust multigrid and preconditioned conjugate gradient methods. In Hackbusch W. and U. Trottenberg, editors, *Multigrid Methods,* Proceedings, Köln-Porz 1981, pages 1–176, Berlin, 1982. Lecture Notes in Mathematics 960, Springer-Verlag.

[4] M. Khalil. Local mode smoothing analysis of various incomplete factorization iterative methods. In Hackbusch W., editor, *Robust Multi-Grid Methods,*

Proceedings of the Fourth GAMM-Seminar, Kiel 1988, pages 155–164, Braunschweig, 1989. NNFM Volume 23, Vieweg.

[5] J. Mandel, S. McCormick, and R. Bank. Variational multigrid theory. In S.F. McCormick, editor, *Multigrid Methods*, chapter 5. SIAM, Philadelphia, Pennsylvania, 1987.

[6] J.A. Meijerink and H.A. van der Vorst. An iterative solution method for linear systems of which the coefficient matrix is a symmetric M-matrix. *Math. Comp.*, 31(137):148–162, January 1977.

[7] K. Oertel and K. Stüben. Multigrid with ILU-smoothing: systematic tests and improvements. In Hackbusch W., editor, *Robust Multi-Grid Methods,* Proceedings of the Fourth GAMM-Seminar, Kiel 1988, pages 188–199, Braunschweig, 1989. NNFM Volume 23, Vieweg.

[8] R.P. Stevenson. Modified ILU as a smoother. Preprint 745, University of Utrecht, September 1992. To appear in Numer. Math.

[9] R.P. Stevenson. New estimates of the contraction number of V-cycle multigrid with applications to anisotropic equations. In Hackbusch W. and G. Wittum, editors, *Incomplete Decompositions (ILU)-Algorithms, Theory, and Applications,* Proceedings of the Eighth GAMM-Seminar, Kiel 1992, pages 159–167, Braunschweig, 1993. NNFM Volume 41, Vieweg.

[10] R.P. Stevenson. Robustness of multi-grid applied to anisotropic equations on convex domains and on domains with re-entrant corners. *Numer. Math.* 66: 373 - 398, 1993.

[11] G. Wittum. Linear iterations as smoothers in multigrid methods: Theory with applications to incomplete decompositions. *IMPACT Comput. Sci. Eng.*, 1:180–215, 1989.

[12] G. Wittum. On the robustness of ILU smoothing. *SIAM J. Sci. Stat. Comput.*, 10(4):699–717, July 1989.

15

Optimal Multigrid Method for Inviscid Flows

Shlomo Ta'asan [1]

ABSTRACT In this paper we describe a novel approach for the solution of inviscid flow problems both for incompressible and subsonic compressible cases. The approach is based on canonical forms of the equations in which subsystems governed by hyperbolic operators are separated from those governed by elliptic ones. The discretizations as well as the iterative techniques for the different subsystems are inherently different. Hyperbolic parts, which describe in general propagation phenomena, are discretized using upwind schemes and are solved by marching techniques. Elliptic parts, which are directionally unbiased, are discretized using h-elliptic central discretizations and are solved by pointwise relaxations together with coarse grid acceleration. The resulting discretization schemes introduce artificial viscosity only for the hyperbolic parts of the system; thus a smaller total artificial viscosity is used, while the multigrid solvers used are much more efficient. Solutions of the subsonic compressible and incompressible Euler equations are achieved at the same efficiency as the full potential and Poisson equations respectively.

1 Introduction

In the past decade a substantial effort has been invested in understanding the Euler equations and their efficient solvers, where multigrid methods play in important role. The two major directions of research in multigrid solution of the Euler equations are the use of coarse grids to accelerate the convergence of the fine grid relaxations [Jameson 83], and the use of defect correction as an outer iteration while use the multigrid method to solve for the low order operator involved [Hemker 86]. Extensive research has been conducted in both directions, and refer-

[1]This research was made possible in part by funds granted to the author through a fellowship program sponsored by the Charles H. Revson Foundation and in part by the National Aeronautics and Space Administration under NASA Contract No. NAS1-19480 and NAS1-18605 while the author was in residence at ICASE, NASA Langley Research Center, Hampton, Va 23681.
The Weizmann Institute of Science, Rehovot 76100, Israel, and
Institute for Computer Applications in Science and Engineering
NASA Langley Research Center, Hampton VA 23681.

ences can be found, for example, in [Wesseling 92]. Both approaches can be shown to be limited in their potential. For hyperbolic equations, methods that are based on defect correction have h-dependent convergence rates, while the other methods have p-dependent convergence rates where p is the order of the scheme involved. This unacceptable situation has led to the research outlined in this paper. More extensive discussion of the different issues will be presented elsewhere.

The poor behavior of coarse grid acceleration for hyperbolic equations, and even worse behavior for high order discretizations, has led to the conclusion that coarse grids should not be used to accelerate the convergence for hyperbolic problems. Rather the relaxation should converge all components of such problems. This is possible since hyperbolic problems describe propagation phenomena and marching techniques in the appropriate directions are very effective for solving them. For elliptic problems, on the other hand, local relaxation with good smoothing properties can be accelerated by coarse grid corrections leading to a very fast solver. Moreover, these problems cannot be solved efficiently by any local relaxation alone, and coarse grids acceleration is necessary. Thus hyperbolic equations do not need coarse grids acceleration, while for elliptic equations such acceleration is necessary.

These observations have motivated a study concerning the separation of the different parts, hyperbolic and elliptic, in inviscid flow problems. The result was a canonical form for the inviscid equation where the hyperbolic and elliptic parts reside in different blocks of an upper triangular form of the system [Ta'asan 93]. These forms are the analog of the decomposition of the time dependent one dimensional Euler equations into characteristic directions and Riemann invariants. The insight gained by the use of the canonical variables enables one to construct genuinely multidimensional schemes for the equations. It unifies the treatment of the compressible subsonic case with the incompressible case, although these two cases have been studied by different methods up to now. Canonical boundary conditions are also obtained [Ta'asan 93] and enable the proper numerical treatment of general boundary conditions.

Schemes that are based on the canonical forms have been developed. These schemes use upwind discretization only for the hyperbolic variables and central h-elliptic discretization for the elliptic ones. The resulting schemes are also compatible with the uniqueness and non-uniqueness of the inviscid equations under different geometries and boundary conditions. In particular, the non-uniqueness of solutions for exterior flows around smooth bodies is evident for these schemes, and only the addition of a global condition, e.g., circulation, ensures uniqueness. In existing schemes this issue is obscure, since there seems to be no direct analog of the physical behavior.

The relaxations based on the canonical forms use a marching technique in the stream-direction for the hyperbolic quantities. These are the total pressure $P = p + q^2/2$ for incompressible flows, and entropy s and total enthalpy H for compressible subsonic flows. The velocity components are relaxed by a Kacmarz relaxation using preconditioned residuals.

All schemes presented here use staggered grids and are free of spurious oscillations even as the Mach number approach zero. Their formulation is done by conservative finite volumes and is constructed for structured as well as unstructured meshes. The unknown variables for the incompressible case are the normal velocity components on the cell faces and the total pressure. The pressure and tangential velocity components are calculated from the other components. For the compressible case the variables are the normal velocity components, the total enthalpy, and the entropy. Other quantities, such as pressure and density, are calculated from these by well known algebraic relations, while the tangential velocities are calculated from the normal velocities at the neighboring cells.

Numerical results are given for a two dimensional flow around a cylinder both for incompressible and subsonic compressible Euler equations. These problems already include the major difficulties in real problems and serve as a good test for the method proposed. Second order schemes are used for both cases and the solutions are obtained with the efficiency of the full potential and the Poisson equations respectively.

2 Canonical Forms and Discretization Rules

The discretization and efficient solution of elliptic systems of partial differential equations is quite well understood. One of the important concepts here is h-ellipticity [Brandt 84]. It guarantees that the stability of high frequencies for the discrete problem is in correspondence to that of the differential system. For the later, an ellipticity is defined in terms of the symbol $\hat{P}(\omega)$ as

$$\det \hat{P}(\omega \geq C|\omega|^{2m} \tag{2.1}$$

while h-ellipticiy is defined as

$$\det \hat{P}^h(\theta) \geq C|\theta|^{2m} \qquad |\theta| \leq \pi \tag{2.2}$$

Discretizations which are $h - elliptic$ admit local relaxation methods with good smoothing properties. This, together with efficient coarse grid acceleration for smooth components, makes standard multigrid methods very efficient for such discretizations. Although other types of discretization also admit fast multigrid solvers, we restrict our focus to $h - elliptic$ discretization for elliptic problems.

The discretization of hyperbolic equations which in general describe propagation phenomena can be done naturally using upwind biased schemes. The application of the above ideas to the steady state inviscid incompressible and subsonic compressible equations is not straightforward, since these equations are neither elliptic nor hyperbolic, but rather mixed hyperbolic-elliptic. The optimal treatment of the problem should therefore include an identification of these two parts, which have inherently different behavior and call for different numerical treatment both on the level of the discretization and the solver. The device for this is a canonical form of the equations, described in details in [Ta'asan 93].

For a two dimensional flow the canonical form of the incompressible Euler equations, in terms of velocities u, v and total pressure $P = p + (u^2 + v^2)/2)$, is

$$\left(\begin{array}{cc|c} D_x & D_y & 0 \\ qD_y & -qD_x & \frac{1}{q}D_0 \\ \hline 0 & 0 & Q \end{array}\right) \begin{pmatrix} u \\ v \\ P \end{pmatrix} = \begin{pmatrix} 0 \\ 0 \\ 0 \end{pmatrix}, \tag{2.3}$$

while in the compressible case it is

$$\left(\begin{array}{cc|cc} D_1 & D_2 & 0 & 0 \\ qD_y & -qD_x & -\frac{c^2}{\gamma(\gamma-1)}D_0 & \frac{1}{q}D_0 \\ \hline 0 & 0 & -T\rho Q & 0 \\ 0 & 0 & 0 & \rho Q \end{array}\right) \begin{pmatrix} u \\ v \\ s \\ H \end{pmatrix} = \begin{pmatrix} 0 \\ 0 \\ 0 \\ 0 \end{pmatrix} \tag{2.4}$$

where:

$$\begin{array}{l} D_1 = \rho/c^2((c^2 - u^2)D_x - uvD_y) \\ D_2 = \rho/c^2((c^2 - v^2)D_y - uvD_x) \\ D_0 = vD_x - uD_y \end{array} \tag{2.5}$$

In view of these forms we can use the following discretization rule for the inviscid equations:

(a). *Use central (unbiased) h-elliptic discretizations for elliptic subsystems*

(b). *Use upwind biased schemes for hyperbolic subsystems*

3 Discretization

Let a domain $\Omega \in \mathbb{R}^2$ be divided into arbitrary cells. Let the vertices, edges and cells be V,E and C respectively. The well known Euler formula

$$\#V + \#C + \#holes = \#E + 1 \tag{3.1}$$

suggests several possibilities for discretization of different systems on structured and unstructured meshes.

3.1 Incompressible Euler

For the discretization schemes for the incompressible Euler equations we rewrite the Euler formula (3.1) as

$$\#V + \#V + \#C + \#holes = \#V + \#E + 1 \tag{3.2}$$

where the left hand side will be related to the equations and the right hand side to the unknowns.

The unknowns used are the normal velocity components on edges, and the total pressure P on vertices. A continuity equation is discretized on each cell and two momentum equations on each vertex. Thus, we obtain the following diagram

$$\begin{array}{lcl} \mathbf{V}\cdot\mathbf{n} & \Longleftrightarrow & \#E \\ P & \Longleftrightarrow & \#V \\ \mathrm{div}\mathbf{V}=0 & \Longleftrightarrow & \#C \\ -\mathbf{V}\times\mathrm{curl}\mathbf{V}+\nabla P=0 & \Longleftrightarrow & 2\#V \\ \int_{\Gamma_{hole}}\mathbf{V}\cdot\mathbf{t}d\sigma & \Longleftrightarrow & \#holes \end{array} \tag{3.3}$$

As can be seen form equation (3.2), a compatibility condition has to be specified. When specifying as boundary conditions the normal velocity component at every boundary point, it is clear from the continuity equation that the required compatibility condition is $\int \mathbf{V}\cdot\mathbf{n}ds = 0$. Moreover, (3.2) also suggests that an extra condition per hole is required as is shown in (3.3). This is in agreement with the uniqueness of the differential equation.

The above argument for the location of variables and equations must be implemented for general grids. Thus an integral form of the equation will be used, namely:

$$\begin{gathered} \int_{\gamma}\mathbf{V}\cdot\mathbf{n}ds=0 \\ \int_{\bar{\gamma}}((\mathbf{V}\cdot\mathbf{n})\mathbf{V}-\tfrac{1}{2}(\mathbf{V}\cdot\mathbf{V})\mathbf{n}+P\mathbf{n})ds=0 \end{gathered} \tag{3.4}$$

Control volumes for the continuity and momentum equations are shown for an unstructured grid case in Figure 1. Note that the control volumes for the momentum equations consist of cells of the dual grid.

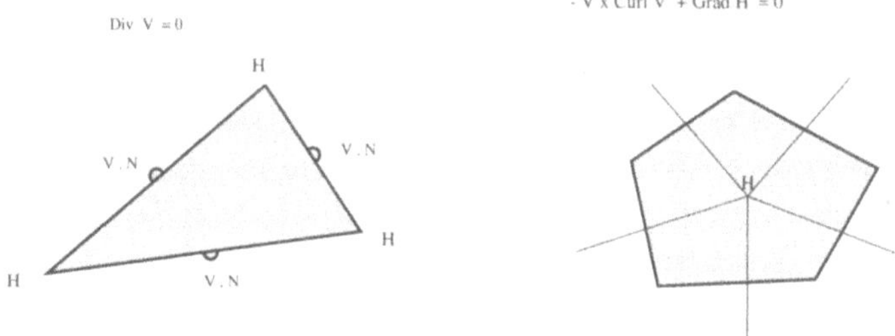

FIGURE 1

The discretization of the continuity equation is straightforward, since all the required quantities for that equation are in the appropriate location. In discretizing the momentum equation, we consider first the canonical form to decide which quantities are to be upwinded and which ones are to be discretized using central differencing. As can be seen the total pressure, which is one of our unknowns, is propagating along streamlines. Its discretization in the third equation in the canonical form is to be done using an upwind biased scheme, since upwinding a term that propagates along certain lines introduces only a small error in its discretization. For that reason other terms involving derivatives of P in the vorticity equations should not be upwinded. Note that only the total pressure may need an upwind biased scheme, while the velocity components require a central discretization everywhere.

The above arguments can now be transformed into the conservative form. To this end we introduce the following decomposition of the derivatives of P

$$P_x = \frac{u}{q^2}(uP_x + vP_y) + \frac{v}{q^2}(vP_x - uP_y) \tag{3.5}$$

and a similar decomposition for P_y

The first term in the above formula represents a derivative of the total pressure in the direction of the flow, and the second one involves the derivative in a direction perpendicular to the flow. Upwinding only part of the terms involving P is done as follows: Let $\mathbf{e}_1 = (u/q, v/q)$ and $\mathbf{e}_2 = (v/q, -u/q)$. The normal vector to an edge is written as

$$\mathbf{n} = (\mathbf{n} \cdot \mathbf{e}_1)\mathbf{e}_1 + (\mathbf{n} \cdot \mathbf{e}_2)\mathbf{e}_2 \tag{3.6}$$

and is used for the discretization of the P terms, namely,

$$\textstyle\int_{\bar{\gamma}} P\mathbf{n}ds \approx \mathbf{d}P = \sum_{l\in\bar{\gamma}} [P_l^{up}(\mathbf{n}_l \cdot \mathbf{e}_1^l)\mathbf{e}_1^l + P_l^c(\mathbf{n}_l \cdot \mathbf{e}_2^l)\mathbf{e}_2^l] ds_l \tag{3.7}$$

where P_l^{up}, P_l^c are upwind and central approximations to P on the edges. All other terms are done in a directionally unbiased way, leading to a scheme which involve artificial viscosity only for the total pressure, and also there, only for the derivatives in the streamwise direction. The full discretization is

$$\begin{gathered}\textstyle\sum_{j\in\gamma} \mathbf{V}_j \cdot \mathbf{n}_j ds_j = 0 \\ \textstyle\sum_{l\in\bar{\gamma}} [(\mathbf{V}_l \cdot \mathbf{n}_l)\mathbf{V}_l - \frac{1}{2}(\mathbf{V}_l \cdot \mathbf{V}_l)\mathbf{n}_l] ds_l + \mathbf{d}P = 0\end{gathered} \tag{3.8}$$

where $\gamma, \bar{\gamma}$ are control volumes for the continuity and momentum equations respectively.

On a uniform rectangular grid some simplification is obtained. The discretization of the total pressure term reduces to

$$\begin{aligned} hP_x^h = \; & u\{u[P^{up}_{\alpha+\frac{1}{2},\beta} - P^{up}_{\alpha-\frac{1}{2},\beta}] + v[P^{up}_{\alpha,\beta+\frac{1}{2}} - P^{up}_{\alpha,\beta-\frac{1}{2}}]\}/q^2 + \\ & v\{v[P^{c}_{\alpha+\frac{1}{2},\beta} - P^{c}_{\alpha-\frac{1}{2},\beta}] - u[P^{c}_{\alpha,\beta+\frac{1}{2}} - P^{c}_{\alpha,\beta-\frac{1}{2}}]\}/q^2 \end{aligned} \tag{3.9}$$

where u, v, q^2 are approximated at the point (α, β) using symmetric formulas, and P^c, P^{up} are central and upwind biased approximations given, for example, by

$$\begin{aligned} P^{c}_{\alpha+\frac{1}{2},\beta} &= \tfrac{1}{2}(P_{\alpha,\beta} + P_{\alpha+1,\beta}) \\ P^{up}_{\alpha+\frac{1}{2},\beta} &= sgn(u)(1.5P_{\alpha,\beta} - .5P_{\alpha+1,\beta}) + \\ & \quad (1 - sgn(u))(1.5P_{\alpha+1,\beta} - .5P_{\alpha+2,\beta}) \end{aligned} \tag{3.10}$$

Similar equations are constructed for extrapolating in the other direction.

It can be shown that the resulting scheme does not admit spurious oscillations. If on the other hand, a central approximation to P were used everywhere the resulting scheme would admit spurious oscillations for the total pressure as well as for the pressure.

3.2 2D Compressible Euler Equations

As we have emphasized before, the structure of the incompressible Euler and the subsonic compressible Euler are very similar, although the number of variables is different. The numerical treatment of the two cases is similar.

Rewriting the Euler formula as

$$\#V + \#V + \#C + \#C + \#holes = \#C + \#V + \#E + 1 \tag{3.11}$$

one obtain the following choice of discretization. Let H be associated with the cell centers, the normal velocity components be at the edges as before, and the entropy be at the vertices. With each cell we associate one continuity equation and one energy equation while the two momentum equations are associated with each vertex. Quantities other than the above are calculated by well known algebraic relations for the thermodynamical quantities and by averaging for the tangential velocity components. The following diagram is obtained,

$$\begin{array}{lcl} \mathbf{V}\cdot\mathbf{n} & \Longleftrightarrow & \#E \\ s & \Longleftrightarrow & \#V \\ H & \Longleftrightarrow & \#C \\ \mathrm{div}\rho\mathbf{V} = 0 & \Longleftrightarrow & \#C \\ -\mathbf{V}\times\mathrm{curl}\mathbf{V} + \nabla P = 0 & \Longleftrightarrow & 2\#V \\ \mathrm{div}\rho\mathbf{V}H = 0 & \Longleftrightarrow & \#C \\ \int_{\Gamma_{hole}} \mathbf{V}\cdot\mathbf{t}d\sigma & \Longleftrightarrow & \#holes \end{array} \tag{3.12}$$

Control volume for an unstructured mesh, for the continuity, energy, and momentum equations are shown in Figure 2.

The canonical form for the compressible equations suggests that only the entropy and the total enthalpy will be discretized using upwind biased schemes, and only in the appropriate terms; that is, only in those in which a derivative in the streamwise direction is involved. Other derivatives involving these quantities should be discretized using central differencing. Decomposing the unit normal vectors to the edges as before we get

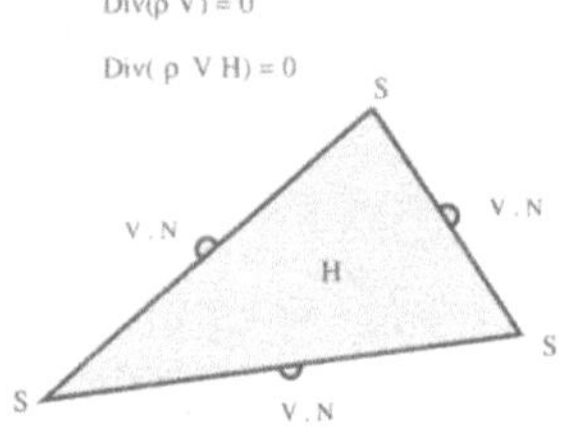

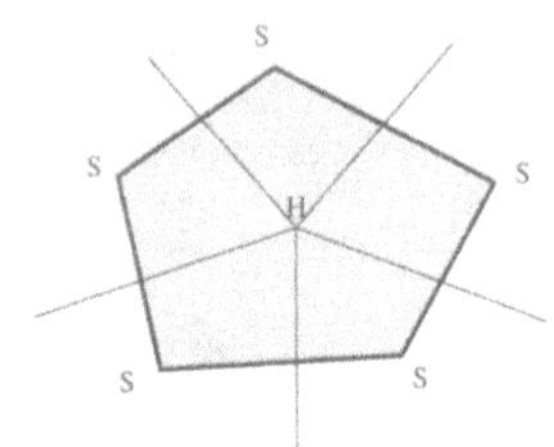

FIGURE 2

$$\mathbf{d}p = \sum_{l \in \bar{\gamma}} [p_l^{up}(\mathbf{n}_l \cdot \mathbf{e}_1^l)\mathbf{e}_1^l + p_l^c(\mathbf{n}_l \cdot \mathbf{e}_2^l)\mathbf{e}_2^l] ds_l \tag{3.13}$$

$$\tag{3.14}$$

where

$$p^{up} = p(S^{up}, H^c, q^2) \tag{3.15}$$

$$p^c = p(S^c, H^c, q^2). \tag{3.16}$$

The full discretization is then

$$\sum_{j \in \gamma} \bar{\rho}_j \mathbf{V}_j \cdot \mathbf{n}_j ds_j = 0 \tag{3.17}$$

$$\sum_{l \in \bar{\gamma}} ((\bar{\rho}_l \mathbf{V}_l \cdot \mathbf{n}_l) \mathbf{V}_l ds_l + \mathbf{d}p = 0 \tag{3.18}$$

$$\sum_{j \in \gamma} \bar{\rho}_j \mathbf{V}_j \cdot \mathbf{n}_j H_j^{up} ds_j = 0 \tag{3.19}$$

where $\bar{\rho}$ is computed using a symmetric formula for all the quantities involved, (i.e., velocities, total enthalpy, and entropy), and γ and $\bar{\gamma}$ denote control volumes for the different equations.

4 Multigrid Algorithm

The multigrid solver, like the discretization, is based on the canonical forms mentioned in section 2. Its main ingredient which differs from other methods is the relaxation method. Other elements of the multigrid method are standard and will be mentioned only briefly.

As can be seen from the canonical form, the hyperbolic and elliptic parts for subsonic flows are separated. Since these subsystems are of very different nature, it is unlikely that the same numerical process will be optimal for both. Indeed, it can be shown that coarse grids are inefficient in accelerating certain smooth components for hyperbolic problems. For these problems convergence rate is roughly $(2^p - 1)/2^p$ for a $p - order$ method. The better the scheme, the less coarse grids help. This behavior suggests that coarse grids are not appropriate for accelerating convergence for hyperbolic problems. The relaxation should therefore converge all components in the problem. While for elliptic problems relaxation cannot be efficient for smooth components, for hyperbolic problems the situation is different. Marching in the direction of the physical flow is very efficient in converging all components and eliminates the need for coarse grid acceleration. For elliptic problems, on the other hand, relaxation techniques with good smoothing properties combined with coarse grid acceleration yield optimal solvers. The separation of the different subsystems presented by the canonical form allows one to construct an optimal solver for the full system. Marching techniques will be used for the hyperbolic quantities, while local relaxation with good smoothing will be used for the elliptic parts.

4.1 Relaxation: Incompressible Euler

The discretization of the equation is done in conservative form and some transformations are required between that form and the canonical form in order to relax the equations. Denoting the residual of the incompressible Euler equations in conservative form by (R^p, R^u, R^v) and the residual of the canonical form of the equations by (r_i^1, r_i^2, r_i^3), the following relation holds

$$\begin{pmatrix} r_i^1 \\ r_i^2 \\ r_i^3 \end{pmatrix} = \begin{pmatrix} 1 & 0 & 0 \\ -q^2 A & -v & u \\ -q^2 A & u & v \end{pmatrix} \begin{pmatrix} R^p \\ R^u \\ R^v \end{pmatrix} \tag{4.1}$$

where A is an averaging operator needed since R^p and R^v are not located at the same points.

The relaxation of the incompressible Euler equation is done as follows. The total pressure P is relaxed in the streamwise direction using the residual r_i^3 and the discretization of the operator $uD_x + vD_y$. Thus we get

$$P_{\bar{\gamma}} \longleftarrow P_{\bar{\gamma}} + (r_i^3)_{\bar{\gamma}} / c_{\bar{\gamma}}^P \tag{4.2}$$

where $c_{\bar{\gamma}}^P$ is the diagonal term multiplying $P_{\bar{\gamma}}$ in the preconditioned residual r_i^3. This is followed by relaxation of the normal components of the velocity vector using a Kacmarz relaxation for the resulting preconditioned equations, which are similar the Cauchy-Riemann equations, giving the relaxation

$$\mathbf{V}_j \cdot \mathbf{n}_j \longleftarrow \mathbf{V}_j \cdot \mathbf{n}_j + (r_i^1)_\gamma ds_j / \sum_{k \in \gamma} (ds_k)^2 \qquad j \in \gamma \tag{4.3}$$

for the continuity equation and

$$\mathbf{V}_l \cdot \mathbf{t}_l \longleftarrow \mathbf{V}_l \cdot \mathbf{t}_l + (r_i^2/q)_{\bar{\gamma}} ds_l / \sum_{k\in\bar{\gamma}} (ds_k)^2 \qquad l \in \bar{\gamma} \tag{4.4}$$

for the vorticity equation. Here we assumed for simplicity of exposition that the grid and its dual are orthogonal, although one can work with the general case as well. In that case the variables $\mathbf{V}_l \cdot \mathbf{t}_l$ coincide with the normal velocity component of one of the neighboring cells.

4.2 Relaxation: Compressible Euler

Let the residual of the compressible Euler equations be denoted by $(R^\rho, R^{\rho u}, R^{\rho v}, R^H)$ and the ones for the canonical form by $(r_c^1, r_c^2, r_c^3, r_c^4)$. Then the following relation prevails

$$\begin{pmatrix} r_c^1 \\ r_c^2 \\ r_c^3 \\ r_c^4 \end{pmatrix} = \begin{pmatrix} 1 & 0 & 0 & 0 \\ -q^2 A & -v & u & 0 \\ -q^2 A & u & v & 0 \\ -H & 0 & 0 & 1 \end{pmatrix} \begin{pmatrix} R^\rho \\ R^{\rho u} \\ R^{\rho v} \\ R^H \end{pmatrix} \tag{4.5}$$

The relaxation for the compressible Euler is done in a similar way to that of the incompressible equations. The total enthalpy is relaxed first, using the preconditioned residual r_c^4

$$H_\gamma \longleftarrow H_\gamma + (r_c^4)_\gamma / c_\gamma^H \tag{4.6}$$

followed by relaxing the entropy using

$$s_{\bar{\gamma}} \longleftarrow s_{\bar{\gamma}} + (r_c^3)_\gamma / c_{\bar{\gamma}}^s \tag{4.7}$$

where c_γ^H and $c_{\bar{\gamma}}^s$ are the diagonal coefficient in the discrete version of $uD_x + vD_y$ and $-\rho T(uD_x + vD_y)$ respectively. This is followed by relaxing the continuity and vorticity equations. The relaxation for the continuity equation is done by keeping the values for the density frozen, i.e.,

$$\mathbf{V}_j \cdot \mathbf{n}_j \longleftarrow \mathbf{V}_j \cdot \mathbf{n}_j + (r_c^1)_\gamma \bar{\rho}_j ds_j / \sum_{k\in\gamma} (\bar{\rho}_k ds_k)^2 \qquad j \in \gamma \tag{4.8}$$

and the vorticity equation is relaxed as

$$\mathbf{V}_l \cdot \mathbf{t}_l \longleftarrow \mathbf{V}_l \cdot \mathbf{t}_l + (r_c^2/(q\rho))_{\bar{\gamma}} ds_l / \sum_{k\in\bar{\gamma}} (ds_k)^2 \qquad l \in \bar{\gamma} \tag{4.9}$$

Note that the preconditioning of the discrete system does not result in an exact upper triangular form. Lower diagonal terms exist but these are of order $O(h^2)$ and do not affect the design of the relaxation and other numerical processes.

The coarsening part of the multigrid method involved is standard and its details are omitted. Coarse grids are created by combining neighboring fine grid cells into a coarse grid cell. Linear interpolation of corrections and full weighting of residual and functions are used in an FMG-FAS formulation [Brandt 84].

5 Numerical Results

We present here numerical results for flow over a cylinder, which already presents all the difficulties encountered in incompressible flows and subsonic compressible flows. For both cases an O-mesh body fitted grid was used. The grid extended to a distance of about 10 chords and the aspect ratio for it was around 1. The examples are given to show the convergence (toward physical solution) of the schemes presented here. All solutions were obtained with residual on the order of 10^{-8}. Numerical experiments are shown for levels 4,5 and 6 which correspond to 32 x 16, 64 x 32 and 128 x 64 grids. Plots of the pressure coefficient C_p are given in Figure 3 and are in agreement with the exact analytical solution for the incompressible case and the Prandl-Gaure approximation for the compressible case, i.e.,

$$\frac{p^{inc}-p_\infty}{\frac{1}{2}\rho_\infty U_\infty^2} = 1.-4\sin^2\theta \qquad \frac{p^{c}-p_\infty}{\frac{1}{2}\rho_\infty U_\infty^2} = (1.-4\sin^2\theta)/\sqrt(1-M_\infty^2) \tag{5.10}$$

The boundary conditions used for the incompressible case were

$$\begin{array}{ll} \mathbf{V}\cdot\mathbf{n}=0 & \text{wall} \\ \mathbf{V}\cdot\mathbf{n}=(U_\infty,0)\cdot\mathbf{n} & \text{far}-\text{field} \\ P=p_\infty+\frac{1}{2}U_\infty^2 & \text{inflow} \end{array} \tag{5.11}$$

The implementation of the boundary condition for the velocity components was straightforward since the unknown variables are the normal velocity components to the edges. The implementation of the inflow boundary condition for P was done by prescribing it at the vertex inflow boundary points. Outflow condition for this quantity was a second order extrapolation.

The boundary conditions used for the compressible Euler were

$$\begin{array}{ll} \rho\mathbf{V}\cdot\mathbf{n}=0 & \text{wall} \\ \rho\mathbf{V}\cdot\mathbf{n}=\rho_\infty(U_\infty,0)\cdot\mathbf{n} & \text{far}-\text{field} \\ H=H_\infty & \text{inflow} \\ s=s_\infty & \text{inflow} \end{array} \tag{5.12}$$

with $M_\infty = .1$ The implementation of the boundary condition for the velocity components was also natural for this case, though slightly more involved since it was nonlinear as the density depends on the velocity field. The implementation of the inflow boundary condition for the total enthalpy was done by introducing the inflow value at the inflow edges. The entropy at the inflow vertex points was also prescribed. Both the entropy and total enthalpy were extrapolated at the outflow farfield boundary, as well as on the wall.

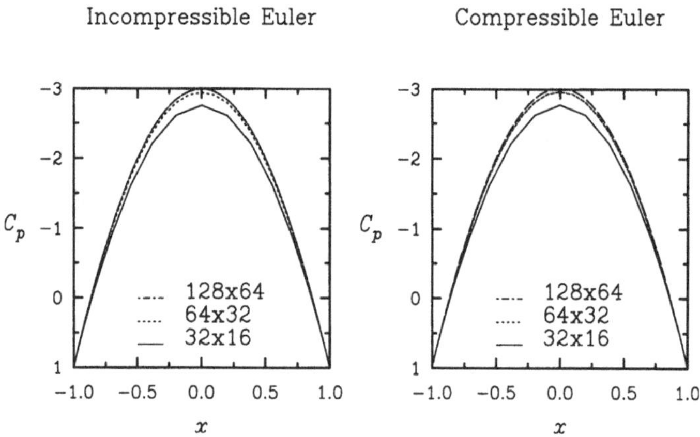

FIGURE 3: Flow around a Cylinder

Both cases had an asymptotic convergence rate of about .18 for a W(2,2) cycle, and the resulting pressure coefficients shown in Figure 3 correspond to solutions converged to residuals level of 10^{-8}.

References

[Brandt 84] A. Brandt: Multigrid Techniques 1984 Guide with Applications to Fluid Dynamics. GMD Studien Nr 85.

[Hemker 86] Hemker, P.W.: Defect Correction and Higher Order Schemes for the Multigrid Solution of the Steady Euler Equations, *Multigrid Methods* II, W. Hackbush and U. Trottenberg (Eds), (*Lecture Notes in Mathematics*) Springer Verlag, Berlin, 149-165.

[Jameson 83] A. Jameson: Solution of the Euler Equations for Two Dimensional Transonic Flow by Multigrid Method, *Appl. Math. and Computat.*, 13, 327-355.

[Ta'asan 93] S. Ta'asan: Canonical forms of Multidimensionl Inviscid Flows. ICASE Report No. 93-34

[Wesseling 92] Wesseling P. *An Introduction to Multigrid Methods* John Wiley and Sons 1992.

16

Multigrid Techniques for Simple Discretely Divergence–free Finite Element Spaces

Stefan Turek[1]

ABSTRACT We derive some basic properties for a class of discretely divergence free finite elements. These make possible a new proof of the *smoothing property* in a standard multigrid algorithm for the Stokes equations. In addition with appropriate divergence-free grid transfer routines of second order accuracy we get the full multigrid convergence. We demonstrate how to develop and implement efficiently these operators and confirm our theoretical results by numerical tests.

1. The simple nonconforming finite element spaces

We consider the usual weak formulation of the steady Stokes problem with bilinear forms $a(\mathbf{u},\mathbf{v}) := (\nabla\mathbf{u},\nabla\mathbf{v})$ and $b(p,\mathbf{v}) := -(p,\nabla\cdot\mathbf{v})$:

$$\begin{aligned}&\text{Find a pair } \{\mathbf{u},p\} \in \mathbf{H}_0^1(\Omega)\times \mathrm{L}_0^2(\Omega)\ ,\ \text{such that}\\ &\quad a(\mathbf{u},\mathbf{v}) + b(p,\mathbf{v}) + b(q,\mathbf{u}) = (\mathbf{f},\mathbf{v})\quad ,\ \forall\,\{\mathbf{v},q\}\in \mathbf{H}_0^1(\Omega)\times \mathrm{L}_0^2(\Omega).\end{aligned} \tag{V}$$

For the discretization let $\mathbf{T}_h$ be a regular decomposition of the domain Ω into triangles or quadrilaterals denoted by T , where $h > 0$ is a measure on the maximum diameter of the elements of $\mathbf{T}_h$. To obtain the fine mesh $\mathbf{T}_h$ from a coarse mesh $\mathbf{T}_{2h}$ we simply connect opposing midpoints (true domain boundaries are respected). In the new grid $\mathbf{T}_h$ coarse midpoints become vertices. For the approximation of problem (V) by the finite element method we introduce discrete spaces $\mathbf{H}_h \approx \mathbf{H}_0^1(\Omega)$ and $\mathrm{L}_h \approx \mathrm{L}_0^2(\Omega)$. In the quadrilateral case we use the reference element $\hat{T} = [-1,1]^2$ and define for each $T \in \mathbf{T}_h$ the one-to-one (bilinear) transformation $\psi_T : \hat{T} \to T$. Then we set (*rotated bilinear elements*, see [5])

$$\hat{Q}_1(T) := \{q\circ\psi_T^{-1}\,|\,q \in \mathrm{span}\langle x^2-y^2, x, y, 1\rangle\}\,, \tag{1}$$

while in the triangular case $P_1(T)$ is used. The degrees of freedom are determined

[1]Institut für Angewandte Mathematik, Universität Heidelberg, Im Neuenheimer Feld 294, D-69120 Heidelberg, Germany

by the nodal functionals $\{F_\Gamma^{(a/b)}(\cdot)\,,\,\Gamma$ edges of $\mathbf{T}_h\}$, with m_Γ midpoint of Γ ,

$$F_\Gamma^{(a)}(v) := |\Gamma|^{-1}\oint_\Gamma v\,d\gamma \quad \text{or} \quad F_\Gamma^{(b)}(v) := v(m_\Gamma). \tag{2}$$

Either choice is unisolvent with respect to $P_1(T)$ and $\hat{Q}_1(T)$, but in the quadrilateral case each leads to a different finite element space (since the applied midpoint rule is only exact for linear functions). Then, the corresponding (*parametric*) finite element spaces $\mathbf{H}_h = \mathbf{H}_h^{(a/b)}$ and L_h are defined as (Γ_i inner edges, Γ_b boundary edges)

$$\mathrm{L}_h := \{q_h \in \mathrm{L}_0^2(\Omega)\,|\,q_{h|T} = \text{const.}, \forall T \in \mathbf{T}_h\} \quad , \quad \mathbf{H}_h^{(a/b)} := \mathrm{S}_h^{(a/b)} \times \mathrm{S}_h^{(a/b)}, \tag{3}$$

$$\mathrm{S}_h^{(a/b)} := \left\{ \begin{array}{l} v_h \in \mathrm{L}^2(\Omega)\,|\,v_{h|T} \in \hat{Q}_1(T), \forall T \in \mathbf{T}_h, v_h \text{ continuous w.r.t.} \\ \\ \text{all nodal functionals } F_{\Gamma_i}^{(a/b)}(\cdot), \forall \Gamma_i\,, F_{\Gamma_b}^{(a/b)}(v_h) = 0, \forall \Gamma_b \end{array} \right\}. \tag{4}$$

In the triangular case, $\hat{Q}_1(T)$ is replaced by $P_1(T)$. Our definitions lead to piecewise constant pressure approximations and edge oriented velocity approximations with midpoints or integral mean values as degrees of freedom. Since the spaces $\mathbf{H}_h^{(a/b)}$ are nonconforming, i.e., $\mathbf{H}_h^{(a/b)} \not\subset \mathbf{H}_0^1(\Omega)$, we have to work with elementwise defined discrete bilinear forms and corresponding energy norms. Let

$$a_h(\mathbf{u}_h, \mathbf{v}_h) := \sum_{T\in T_h} \int_T \nabla \mathbf{u}_h \cdot \nabla \mathbf{v}_h\,dx \quad , \quad \|\mathbf{v}_h\|_h := (a_h(\mathbf{v}_h, \mathbf{v}_h))^{1/2}, \tag{5}$$

$$b_h(q_h, \mathbf{v}_h) := -\sum_{T\in T_h} q_{h|T}\,Q_T(\mathbf{v}_h) \quad , \quad Q_T(\mathbf{v}_h) := \sum_{\Gamma\subset\partial T} |\Gamma|\,F_\Gamma^{(a/b)}(\mathbf{v}_h)\cdot \mathbf{n}_\Gamma \tag{6}$$

Furthermore, let $i_h^{(a/b)} : \mathbf{H}_0^1(\Omega) \to \mathbf{H}_h^{(a/b)}$ be the global interpolation operator in $\mathbf{H}_h^{(a/b)}$, which is determined by

$$F_\Gamma(i_h^{(a/b)}\mathbf{v}) = F_\Gamma(\mathbf{v}) \quad , \forall \Gamma \subset \partial\mathbf{T}_h. \tag{7}$$

With some additional regularity assumptions on the domain and the used mesh in the quadrilateral case (see [5],[6]) we can state:

Lemma 1 *For the interpolation operators* $i_h = i_h^{(a/b)}$ *we have*

$$\|v - i_h v\|_0 + h\,\|v - i_h v\|_h \le ch^2\,\|v\|_2 \quad , \forall v \in \mathrm{H}_0^1(\Omega) \cap \mathrm{H}^2(\Omega)\,. \tag{8}$$

Lemma 2 *There exist unique solutions* $\{\mathbf{u}_h, p_h\} \in \mathbf{H}_h^{(a/b)} \times \mathrm{L}_h$ *such that*

$$\|\mathbf{u} - \mathbf{u}_h\|_0 + h\,\|\mathbf{u} - \mathbf{u}_h\|_h + h\,\|p - p_h\|_0 \le ch^2\{\|\mathbf{u}\|_2 + \|p\|_1\}. \tag{9}$$

For the explicit construction of the divergence–free subspace $\mathbf{H}_h^d \subset \mathbf{H}_h$ we make the following definition.

Definition 1 *A function* $\mathbf{v}_h \in \mathbf{H}_h$ *is called* discretely divergence–free *if*

$$b_h(q_h, \mathbf{v}_h) = 0\,,\ \forall\, q_h \in \mathrm{L}_h. \tag{10}$$

Because the pressure space is piecewise constant an equivalent criterion is

$$Q_T(\mathbf{v}_h) = 0\,,\ \forall T \in \mathbf{T}_h. \tag{11}$$

With these modifications we can introduce a subspace $\mathbf{H}_h^d \subset \mathbf{H}_h$, and our discrete problem for the velocity only is reduced to:

Find $\mathbf{u}_h^d \in \mathbf{H}_h^d$, such that

$$a_h(\mathbf{u}_h^d, \mathbf{v}_h^d) = (\mathbf{f}, \mathbf{v}_h^d) \quad , \forall\, \mathbf{v}_h^d \in \mathbf{H}_h^d. \tag{$\mathrm{V}_\mathrm{h}^\mathrm{d}$}$$

Finally, the corresponding pressure $p_h \in \mathrm{L}_h$ is determined by the condition

$$b_h(p_h, \mathbf{v}_h^r) = (\mathbf{f}, \mathbf{v}_h^r) - a_h(\mathbf{u}_h^d, \mathbf{v}_h^r) \quad , \forall\, \mathbf{v}_h^r \in \mathbf{H}_h^r, \tag{12}$$

where the functions $\mathbf{v}_h^r$ are in the curl–free part of the complete space $\mathbf{H}_h$. In our configuration this is performed by a marching process from element to element, without solving any linear system of equations.

2. The divergence–free subspaces and their properties

Consider a general quadrilateral $T \in \mathbf{T}_h$ (see Figure 1) with vertices a^i, midpoints m^j, edges Γ^j, unit tangential vectors $\mathbf{t}^j$ and normal unit vectors $\mathbf{n}^j$. Let $\varphi_h^j \in \mathrm{S}_h$ be the usual nodal basis functions of the finite element space $\mathrm{S}_h = \mathrm{S}_h^{(a/b)}$, restricted to the element T, satisfying $F_{\Gamma^i}(\varphi_h^j) = \delta_{ij}$, $i, j = 1, \ldots, 4$. Then, the first group

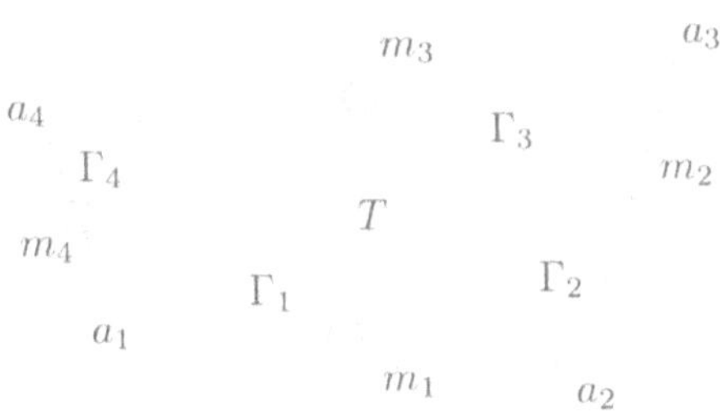

FIGURE 1: General quadrilateral T

of trial functions $\{\mathbf{v}_h^{i,t}\}$ of $\mathbf{H}_h^d$, corresponding to the edges of $\mathbf{T}_h$, is given by the local definition

$$\mathbf{v}_{h|T}^{i,t} \in \{\varphi_h^j \mathbf{t}^j \,,\, j = 1,\ldots,4\}. \tag{13}$$

The second group $\{\mathbf{v}_h^{i,\psi}\}$, corresponding to the vertices, is locally determined by

$$\mathbf{v}_{h|T}^{i,\psi} \in \{\frac{\varphi_h^k \mathbf{n}^k}{|\Gamma^k|} - \frac{\varphi_h^j \mathbf{n}^j}{|\Gamma^j|} \,,\, j = 1,\ldots,4 \,,\, k = (j+2) \bmod 4 + 1\}. \tag{14}$$

Thus, we get approximations for the tangential velocities on the edges, and for the streamfunction values at the nodes. The full space $\mathbf{H}_h^d$ is the direct sum of these two subspaces. Defining the inner product $(\cdot,\cdot)_{d,h}$ on $\mathbf{H}_h^d$ by

$$(\mathbf{u}_h, \mathbf{v}_h)_{d,h} := \frac{1}{4} \sum_{T \in T_h} |T| \sum_{\Gamma \in \partial T} F_\Gamma(\mathbf{u}_h) \cdot F_\Gamma(\mathbf{v}_h) \,, \tag{15}$$

the induced norm $\|\cdot\|_{d,h}$ is equivalent to the L^2-norm, and both groups of trial functions are orthogonal relative to this form. If we eliminate one of the functions $\{\mathbf{v}_h^{i,\psi}\}$ by prescribing the value in one (boundary) point, we get a basis for the discretely divergence–free subspace $\mathbf{H}_h^d$, assuming that our problem has only one boundary component. This is a simple consequence of the orthogonality relation corresponding to $\|\cdot\|_{d,h}$ and the fact that for $\mathbf{v}_h = \sum \Psi^i \mathbf{v}_h^{i,\psi}$:

$$\|\mathbf{v}_h\|_{d,h}^2 \sim \sum_{T \in T_h} |T| \sum_{\Gamma^j \in \partial T} \frac{|\Psi^{j+1} - \Psi^j|^2}{|\Gamma^j|^2} \sim \sum_{T \in T_h} \sum_{k=1}^{4} |\Psi^{k+1} - \Psi^k|^2 \,. \tag{16}$$

This means, the mass matrix is spectrally equivalent to a stiffness matrix corresponding to the discretization of the Poisson equation with natural boundary conditions by conforming bilinear elements. Let S_h^l be this usual conforming finite element space with nodal basis, satisfying $\varphi_h^{j,l}(a_i) = \delta_{ij}$ for all vertices a_i of $\mathbf{T}_h$. By S_l we denote the corresponding positive definite stiffness matrix with

$$S_l^{(i,j)} := \sum_{T \in T_h} \int_T \nabla \varphi_h^{i,l} \cdot \nabla \varphi_h^{j,l} \, dx \,. \tag{17}$$

Additionally we define the Stokes stiffness matrix S_d in $\mathbf{H}_h^d$,

$$S_d^{(i,j)} := a_h(\mathbf{v}_h^{i,d}, \mathbf{v}_h^{j,d}) \,, \tag{18}$$

and we set for functions $\mathbf{v}_h = \sum X^{d,i} \mathbf{v}_h^{i,d} \in \mathbf{H}_h^d$ the discrete norm scale $|||\cdot|||_s$,

$$|||\mathbf{v}_h|||_s := ((X^d)^T S_d^s X^d)^{1/2} \,, \tag{19}$$

where $s = 0$ corresponds to the euclidian vector norm $\|\cdot\|_E$, and $s = 1$ to the energy norm $\|\cdot\|_h$. The following Lemma is necessary for the multigrid convergence proof, when the *smoothing property* will be shown.

Lemma 3 *There holds for* $\mathbf{v}_h = \sum X^{d,i}\mathbf{v}_h^{i,d} \in \mathbf{H}_h^d$: $|||\mathbf{v}_h|||_{1/2} \leq ch^{-1}\|\mathbf{v}_h\|_0$.

Proof.
We show some estimates for $s = 0$ and $s = 1$, and then prove the final result for $s = 1/2$ using some interpolation arguments (see also [4]). First, we get

$$|||\mathbf{v}_h|||_1^2 \leq c\,(\Psi^T S_l\, S_l \Psi + U_t^T S_p U_t) \leq c\,(\Psi^T S_l\, S_l \Psi + U_t^T U_t)\,, \tag{20}$$

and by definition we additionally have

$$|||\mathbf{v}_h|||_0^2 = \Psi^T \Psi + U_t^T U_t\,. \tag{21}$$

Then, a first result in matrix–vector notation reads (with I_p identity matrix)

$$|||\mathbf{v}_h|||_s^2 \leq c \begin{bmatrix} \Psi \\ U_t \end{bmatrix}^T \begin{bmatrix} S_l\, S_l & 0 \\ 0 & I_p \end{bmatrix}^s \begin{bmatrix} \Psi \\ U_t \end{bmatrix}, \tag{22}$$

and with some interpolation arguments for norm scales we achieve for $s = 1/2$

$$|||\mathbf{v}_h|||_{1/2}^2 \leq c\,(\Psi^T S_l\, \Psi + U_t^T U_t)\,. \tag{23}$$

By a basic finite element estimate we finally reach: $|||\mathbf{v}_h|||_{1/2} \leq ch^{-1}\|\mathbf{v}_h\|_0$. □

Furthermore, we can state the following estimates for the condition numbers:

Lemma 4 *Let* $\mathbf{v}_h \in \mathbf{H}_h^d$ *, then:*

1) For the stiffness matrix S_d *:* $ch^2|||\mathbf{v}_h|||_0 \leq |||\mathbf{v}_h|||_1 \leq c\,|||\mathbf{v}_h|||_0$

2) For the mass matrix M_d *:* $ch^2|||\mathbf{v}_h|||_0 \leq \|\mathbf{v}_h\|_0 \leq ch\,|||\mathbf{v}_h|||_0$.

3. The multigrid algorithm and its analysis

Let $\{\mathbf{T}_{h_l}\}_{h_l}$ be a family of regular subdivisions which are achieved using the refinement process from Section 1. The discrete Stokes problem on level k reads:

Find $\mathbf{u}_k^d \in \mathbf{H}_k^d$, such that

$$a_k(\mathbf{u}_k^d, \mathbf{v}_k^d) = (\mathbf{f}, \mathbf{v}_k^d) \quad , \forall\, \mathbf{v}_k^d \in \mathbf{H}_k^d\,. \tag{$\mathrm{V_k^d}$}$$

As before write $\mathbf{v}_k \in \mathbf{H}_k^d$ as

$$\mathbf{v}_k = \sum_l X^{d,l}\mathbf{v}_k^{l,d} = \sum_i \Psi^i(h_k \cdot \mathbf{v}_k^{i,\psi}) + \sum_j U_t^j \mathbf{v}_k^{j,t}, \tag{24}$$

and introduce, corresponding to the euclidian scalar product $< \cdot, \cdot >_E$ for vectors, the discrete scalar product $(\cdot, \cdot)_k$ where

$$(\mathbf{v}_k, \mathbf{w}_k)_k := \sum_i \Psi_v^i \Psi_w^i + \sum_j U_{t,v}^j U_{t,w}^j \quad , \quad (\mathbf{v}_k, \mathbf{w}_k)_k =< X_v^d, X_w^d >_E\,. \tag{25}$$

The prolongation operator $I_{k-1}^k : \mathbf{H}_{k-1}^d \to \mathbf{H}_k^d$ and its adjoint restriction operator $I_k^{k-1} : \mathbf{H}_k^d \to \mathbf{H}_{k-1}^d$ are defined through

$$(I_k^{k-1}\mathbf{w}_k, \mathbf{v}_{k-1})_{k-1} = (\mathbf{w}_k, I_{k-1}^k\mathbf{v}_{k-1})_k \quad , \forall\, \mathbf{v}_{k-1} \in \mathbf{H}_{k-1}^d \, , \forall\, \mathbf{w}_k \in \mathbf{H}_k^d \, . \tag{26}$$

Further, we define the positive definite discrete operator $A_k : \mathbf{H}_k^d \to \mathbf{H}_k^d$ by

$$(A_k\mathbf{v}_k, \mathbf{w}_k)_k = a_k(\mathbf{v}_k, \mathbf{w}_k) \quad , \forall\, \mathbf{v}_k, \mathbf{w}_k \in \mathbf{H}_k^d \, , \tag{27}$$

such that the eigenvalues λ_k^i of A_k satisfy the relation

$$O(h_k^4) \le \lambda_k^1 \le \ldots \le \lambda_k^{max} \le c \, , \tag{28}$$

where c is a constant independent of h_k . Finally we introduce the operator P_k^{k-1} : $\mathbf{H}_k^d \to \mathbf{H}_{k-1}^d$, which is the adjoint of I_{k-1}^k relative to $a_k(\cdot,\cdot)$,

$$a_{k-1}(P_k^{k-1}\mathbf{w}_k, \mathbf{v}_{k-1}) \;=\; a_k(\mathbf{w}_k, I_{k-1}^k\mathbf{v}_{k-1}) \, , \forall\, \mathbf{v}_{k-1} \in \mathbf{H}_{k-1}^d \, , \forall\, \mathbf{w}_k \in \mathbf{H}_k^d \tag{29}$$

such that $\quad P_k^{k-1} = A_{k-1}^{-1} I_k^{k-1} A_k$,

and again we introduce the mesh–dependent norm scale $|||\cdot|||_{s,k}$ on $\mathbf{H}_k^d$ where

$$|||\mathbf{v}_k|||_{s,k} := (A_k^s\mathbf{v}_k, \mathbf{v}_k)_k^{1/2} \, . \tag{30}$$

The k-level iteration $MG(k, \mathbf{u}_k^0, \mathbf{g}_k)$ for solving $A_k\mathbf{u}_k = \mathbf{g}_k$

For $k = 1$, $MG(1, \mathbf{u}_1^0, \mathbf{g}_1)$ is the exact solution: $MG(1, \mathbf{u}_1^0, \mathbf{g}_1) = A_1^{-1}\mathbf{g}_1$.

For $k > 1$, there are four steps:

1) m-Presmoothing steps using the damped Jacobi–iteration

$$\mathbf{u}_k^l = \mathbf{u}_k^{l-1} + \omega_k(\mathbf{g}_k - A_k\mathbf{u}_k^{l-1}) \, , \, l = 1, \ldots, m \, , \tag{31}$$

where ω_k has to be smaller than the inverse of the largest eigenvalue λ_k^{max} .

2) Correction step
Calculate the restricted defect

$$\mathbf{g}_{k-1} = I_k^{k-1}(\mathbf{g}_k - A_k\mathbf{u}_k^m), \tag{32}$$

and let $\mathbf{u}_{k-1}^i \in \mathbf{H}_{k-1}^d$ $(1 \le i \le p \, , \, p \ge 2)$ be defined recursively by

$$\mathbf{u}_{k-1}^i = MG(k-1, \mathbf{u}_{k-1}^{i-1}, \mathbf{g}_{k-1}) \, , \, 1 \le i \le p \quad , \quad \mathbf{u}_{k-1}^0 = \mathbf{0} \, . \tag{33}$$

3) Step size control
Calculate $\mathbf{u}_k^{m+1}$ by

$$\mathbf{u}_k^{m+1} = \mathbf{u}_k^m + \alpha_k I_{k-1}^k \mathbf{u}_{k-1}^p , \tag{34}$$

where the parameter α_k may be a fixed value or chosen adaptively so as to minimize the error $\mathbf{u}_k^{m+1} - \mathbf{u}_k$ in the energy norm,

$$\alpha_k = \frac{(\mathbf{g}_k - A_k \mathbf{u}_k^m, I_{k-1}^k \mathbf{u}_{k-1}^p)_k}{(A_k I_{k-1}^k \mathbf{u}_{k-1}^p, I_{k-1}^k \mathbf{u}_{k-1}^p)_k} . \tag{35}$$

4) m-Postsmoothing steps
Analogously to step 1) apply m smoothing steps on $\mathbf{u}_k^{m+1}$ and obtain $\mathbf{u}_k^{2m+1}$.

For the convergence analysis we restrict to the case of a two level method ($k = 2$) without postsmoothing and step length control ($\alpha_k = 1$), and show the usual *smoothing* and *approximation property* for the damped Jacobi–method. The essential new approach is that the *smoothing property* may be shown using only the properties of the finite element spaces. This is in contrast to Brenner [2], where the relation between linear divergence–free finite elements and the Morley element was used.

Lemma 5 *(***Smoothing property***)*
For the error $\mathbf{e}_h^m := \mathbf{u}_h - \mathbf{u}_h^m$ *there holds:* $|||\mathbf{e}_h^m|||_1 \leq cm^{-1/4} h^{-1} \|\mathbf{e}_h^0\|_0 .$

Proof.
Applying m damped Jacobi–steps to $\mathbf{e}_h^0 = \mathbf{u}_h - \mathbf{u}_h^0$ yields

$$\mathbf{e}_h^m = (I_h - \omega_h A_h)^m \mathbf{e}_h^0 . \tag{36}$$

Hence,
$$\begin{aligned} A_h^{1/2} \mathbf{e}_h^m &= A_h^{1/2} (I_h - \omega_h A_h)^m \mathbf{e}_h^0 \\ &= \omega_h^{-1/4} A_h^{1/4} \omega_h^{1/4} A_h^{1/4} (I_h - \omega_h A_h)^m A_h^{-1/4} A_h^{1/4} \mathbf{e}_h^0 \end{aligned}$$

$$\begin{aligned} |||A_h^{1/2} \mathbf{e}_h^m|||_0^2 &\leq \omega_h^{-1/2} |||A_h^{1/4} \omega_h^{1/4} A_h^{1/4} (I_h - \omega_h A_h)^m A_h^{-1/4}|||_0^2 \, |||A_h^{1/4} \mathbf{e}_h^0|||_0^2 \\ &\leq c|||(\omega_h A_h)^{1/4} (I_h - \omega_h A_h)^m|||_0^2 \, |||A_h^{1/4} \mathbf{e}_h^0|||_0^2 , \end{aligned}$$

respectively, $|||\mathbf{e}_h^m|||_1^2 \leq c|||(\omega_h A_h)^{1/4} (I_h - \omega_h A_h)^m|||_0^2 \, |||\mathbf{e}_h^0|||_{1/2}^2$.

By standard arguments for positive definite operators (see, e.g., [1]) we get

$$|||(\omega_h A_h)^{1/4} (I_h - \omega_h A_h)^m|||_0 \leq cm^{-1/4} , \tag{37}$$

and, by Lemma 3, $|||\mathbf{e}_h^m|||_1 \leq cm^{-1/4} |||\mathbf{e}_h^0|||_{1/2} \leq cm^{-1/4} h^{-1} \|\mathbf{e}_h^0\|_0 .$ □

The main work was already done in the preceding section, by proving Lemma 3. Now we show the appropriate *approximation property* following the ideas of Brenner [2]. The main difference is that the evaluation for the quadrilateral case is done without using explicitely that the finite elements are divergence–free in a pointwise sense (which they are not). Since in the 2–level iteration the correction equation is solved exactly, the coarse grid solution $\mathbf{u}_{2h} := \mathbf{u}_{2h}^p$ satisfies:

$$\mathbf{u}_{2h} = A_{2h}^{-1}\mathbf{g}_{2h} = A_{2h}^{-1}[I_h^{2h}(\mathbf{g}_h - A_h\mathbf{u}_h^m)] = A_{2h}^{-1}I_h^{2h}A_h\mathbf{e}_h^m = P_h^{2h}\mathbf{e}_h^m \,. \tag{38}$$

Then, for the proof we require the following assumptions on I_{2h}^h :

<u>**Condition I**</u>

1) $\|I_{2h}^h\mathbf{v}_{2h} - \mathbf{v}_{2h}\|_0 \leq ch\|\mathbf{v}_{2h}\|_{2h}\,,\ \forall\,\mathbf{v}_{2h} \in \mathbf{H}_{2h}^d$.

2) $\exists\,\Pi_{h/2h} : \mathbf{V}(\Omega)\cap\mathbf{H}^2(\Omega) \to \mathbf{H}_{h/2h}^d\,,\ \mathbf{V}(\Omega) = \{\mathbf{v}\in\mathbf{H}_0^1(\Omega) : \nabla\cdot\mathbf{v} = 0\}$:

$$\begin{aligned} \|\mathbf{v} - \Pi_{h/2h}\mathbf{v}\|_0 + h\|\mathbf{v} - \Pi_{h/2h}\mathbf{v}\|_{h/2h} &\leq ch^2\|\mathbf{v}\|_2\,,\ \forall\,\mathbf{v}\in\mathbf{V}(\Omega)\cap\mathbf{H}^2(\Omega)\,, \\ \|\Pi_h\mathbf{v} - I_{2h}^h\Pi_{2h}\mathbf{v}\|_0 + h\|\Pi_h\mathbf{v} - I_{2h}^h\Pi_{2h}\mathbf{v}\|_h &\leq ch^2\|\mathbf{v}\|_2\,,\ \forall\,\mathbf{v}\in\mathbf{V}(\Omega)\cap\mathbf{H}^2(\Omega)\,. \end{aligned}$$

Lemma 6 *(***Approximation property***)*
There holds for $\mathbf{v}_h \in \mathbf{H}_h^d$ *and* $\hat{\mathbf{v}}_h = (I_h - I_{2h}^hP_h^{2h})\mathbf{v}_h \in \mathbf{H}_h^d$ *, with condition I :*

$$\|\hat{\mathbf{v}}_h\|_0 \leq ch|||\mathbf{v}_h|||_1\,.$$

We can sum up both lemmas in the following Theorem.

Theorem 1 *(***Convergence of the 2–level scheme***)*
Let $\mathbf{e}_h^{m+1}$ *be the error after one 2-level step with m damped Jacobi–smoothing steps from an initial error* $\mathbf{e}_h^0$ *. Using the grid transfer routines* I_{2h}^h *, fulfilling condition I, we obtain the error reduction, with* $\varrho(m) := m^{-1/4}$ *,*

$$\|\mathbf{e}_h^{m+1}\|_0 \leq C\varrho(m)\|\mathbf{e}_h^0\|_0\,.$$

In the following we will construct transfer operators $I_{2h}^h : \mathbf{H}_{2h}^d \to \mathbf{H}_h^d$ satisfying condition I. The first, a natural choice, analogous to scalar nonconforming finite elements (see [3],[7]), is the L^2–projection $I_{2h}^{h,P}$ from $\mathbf{H}_{2h}^d$ into $\mathbf{H}_h^d$ with

$$(I_{2h}^{h,P}\mathbf{v}_{2h}, \mathbf{v}_h) = (\mathbf{v}_{2h}, \mathbf{v}_h)\,,\ \forall\,\mathbf{v}_{2h}\in\mathbf{H}_{2h}^d\,,\ \forall\,\mathbf{v}_h\in\mathbf{H}_h^d\,. \tag{39}$$

In matrix–vector notation with coefficient vectors X_{2h} and $X_h = I_{2h}^{h,P}X_{2h}$ we get

$$X_h = M_{h,h}^{-1}N_{h,2h}X_{2h} \quad , \quad M_{h,h}^{(i,j)} = (\mathbf{v}_h^{i,d}, \mathbf{v}_h^{j,d}) \quad , \quad N_{h,2h}^{(i,j)} = (\mathbf{v}_h^{i,d}, \mathbf{v}_{2h}^{j,d})\,, \tag{40}$$

where $M_{h,h}$ is the mass matrix on level h , and $N_{h,2h}$ the *transfer* matrix.

Lemma 7 *The transfer operator* $I_{2h}^h = I_{2h}^{h,P}$ *satisfies condition I .*

The problem with this transfer operator lies in the fact that the mass matrix corresponds to a second order problem, and the part for the streamfunction values is equivalent to a conformingly discretized Laplacian operator (see (16)). Consequently, we need fast Poisson–solvers for each prolongation and restriction, typically a second (standard) multigrid algorithm. However, even with fast Poisson–solvers the numerical amount is very large. This led us to look for simpler transfer operators for which one obtains the same convergence rates with much less numerical amount. For this we present two operators which work on the macro elements of level $2h$, interpolating directly into the divergence–free subspace. Then, the problem is to show that the approximation properties are sufficiently good. Consider the following macro element with streamfunction values Ψ_i and tangential components Ut_j . One regular refinement leads to Figure 2. We have to define five

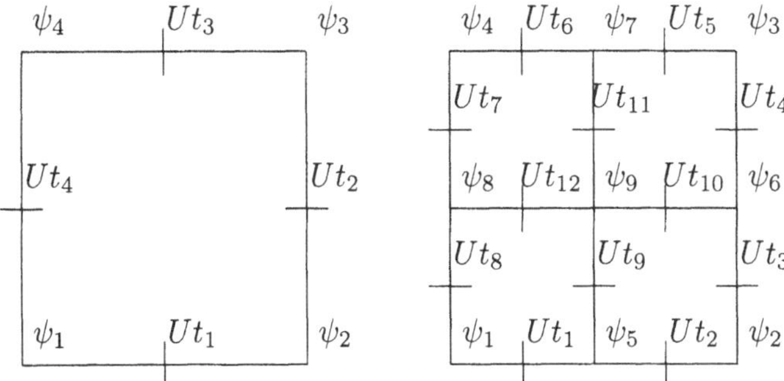

FIGURE 2: Macro and refined elements

new streamfunction values at the vertices, and tangential components on all new edges. In vertices belonging to the macro triangulation the values are preserved.

The macro elementwise interpolation algorithm

(a). Transfer the divergence–free coefficient vector (Ψ_{2h}, Ut_{2h}) into the primitive coefficient vector (U_{2h}, V_{2h}) .

(b). Interpolate "fully" (see below) on the macro elements to get (U_h, V_h) .

(c). Compute Ut_h and Un_h on all fine grid edges.

(d). Set $\Psi_h = \Psi_{2h}$ in the macro nodes and calculate in the new vertices the values for Ψ_h by integrating Un_h .

(e). Take the average for Ψ_h and Ut_h , which lie on macro edges.

For the following we denote the "full" interpolation (step b.) of the primitive nonconforming finite elements by I_{2h} . The problem for the analysis is that the values for the inner normal velocities are only implicitely given, while the tangential ones are directly defined. Let us start using the full interpolation for I_{2h} , this means using linear/rotated bilinear interpolation on each macro element. Then we get the following function values (Figure 3) on the new edges (see also [7]). For

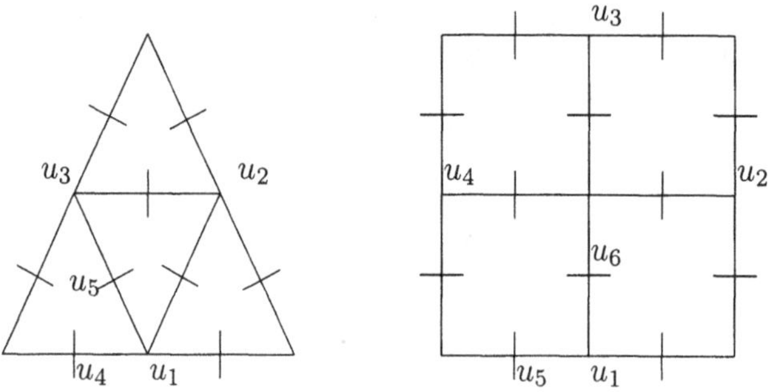

FIGURE 3: Configuration for interpolation

the trial space $\mathbf{H}_h^{(a)}$, for instance, we can calculate

$$u_5 = u_1 - \frac{1}{2}u_2 + \frac{1}{2}u_4 \quad , \quad u_6 = \frac{5}{8}u_1 + \frac{1}{8}(u_2 + u_3 + u_4)\,. \tag{41}$$

This choice for I_{2h} is denoted by I_{2h}^L . Another possibility is to use a constant interpolation I_{2h}^K on each macro element, which results in

$$u_5 = u_1 \quad , \quad u_6 = u_1. \tag{42}$$

Before giving some theoretical results we make some remarks concerning an efficient implementation. The procedure 1) – 5), previously described, looks complicated. However, the essential idea is to rewrite this procedure using local matrices, resulting in elementwise defined 21×8 matrices. This has to be done very carefully, but the gain is a discretely divergence–free interpolation operator with numerical amount comparable to corresponding operators for scalar Poisson–equations ([6]).

Lemma 8 *The operator $I_{2h}^{h,L}$ (with I_{2h}^L) satisfies both estimates of condition I, while the $I_{2h}^{h,K}$ satisfies only the first relation.*

Nevertheless $I_{2h}^{h,K}$ will be used in our following test calculations, too. Another approach for developing this operator is the following one. Consider the discretization of the generalized Stokes problem with $\alpha > 0\,,\, \varepsilon \geq 0$,

$$\alpha\mathbf{u} - \varepsilon\Delta\mathbf{u} + \nabla p = \mathbf{f}\,, \nabla\cdot\mathbf{u} = 0\,. \tag{43}$$

As $\varepsilon \to 0$ the influence of the Stokes operator weakens, and for $\varepsilon = 0$ we only have to solve a linear system with mass matrix M_d . Since this matrix is spectrally equivalent to a matrix with conformingly discretized Laplacian part (for the streamfunction (16)) it seems to be natural to solve this system using conforming multigrid routines. This procedure, however, is exactly the same like the proposed one for $I_{2h}^{h,K}$, if for the tangential part on the edges the operator I_{2h}^{K} is taken.

4. Numerical results

As smoothing operator we restrict ourselves to the Gauß–Seidel method which has the same numerical amount as the Jacobi–iteration on scalar workstations like the SUN 4/260 used. The case of the ILU–method is studied in [6], in which we give an overview of renumbering strategies. In the following table we present the number of unknowns (NEQ), the convergence rate κ and the efficiency rate γ,

$$\kappa = \sqrt[8]{|r^{(8)}|/|r^{(0)}|} \quad , \quad \gamma = -\frac{1000T_8}{8NEQ\log\kappa} . \tag{44}$$

Here, $r^{(8)}$ denotes the residue after 8 iterations, and T_8 the corresponding time. The efficiency rate γ measures the time needed to gain one digit per unknown. The numbers are generated using an F–cycle, since the V–cycle seems to be unstable sometimes, while the W–cycle shows no advantages. As finite element we use the space $\mathbf{H}_h = \mathbf{H}_h^{(b)}$, and we set $m = 2$. We calculate the standard problem of the *Stokes Driven Cavity* on the unit square, with the following coarse grids as typical representants of possible meshes (see Figure 4). The finer subdivisions are

FIGURE 4: Used coarse grids

achieved by the regular refinement process from the previous section. The tables show that the projection method led to very good convergence rates but had the largest numerical amount. The constant operator $I_{2h}^{h,K}$ led to surprisingly good results if the grid was regular. For irregular grids, $I_{2h}^{h,K}$ was much less robust than $I_{2h}^{h,L}$.

From these tests we conclude that $I_{2h}^{h,L}$ is the best since it satisfied the requirements: Small numerical amount, good convergence rates, robust against grid and parameter variations and theoretically analysable. In connection with Gauß–Seidel iteration as smoother in our proposed algorithm we seem to have found a good candidate as a *Black Box* solver for linear systems in a fully nonstationary Navier–Stokes code, as can be seen in [6],[7].

TABLE 1: Rates for *Stokes Driven Cavity*

	κ			γ		
Grid I	3201	12545	49665	3201	12545	49665
$I_{2h}^{h,K}$	0.193	0.210	0.298	0.597	0.612	0.840
$I_{2h}^{h,L}$	0.104	0.092	0.106	0.622	0.548	0.619
$I_{2h}^{h,P}$	0.066	0.058	0.050	1.080	1.010	1.250
Grid II	2417	9441	37313	2417	9441	37313
$I_{2h}^{h,K}$	0.356	0.379	0.440	0.856	0.947	1.200
$I_{2h}^{h,L}$	0.136	0.143	0.143	0.605	0.667	0.792
$I_{2h}^{h,P}$	0.175	0.196	0.220	1.880	1.940	2.890
Grid III	2433	9473	37377	2433	9473	37377
$I_{2h}^{h,K}$	0.286	0.321	0.545	0.811	0.998	1.950
$I_{2h}^{h,L}$	0.175	0.193	0.211	0.964	1.100	1.200
$I_{2h}^{h,P}$	0.151	0.184	0.210	3.660	3.980	5.250

References

[1] Bank, R.E., Dupont, T.: *An optimal order process for solving finite element equations*, Math. Comp., 36, 35–51 (1981)

[2] Brenner, S.C.: *A nonconforming multigrid method for the stationary Stokes equations*, Math. Comp., 55, 411–437 (1990)

[3] Braess, D., Verfürth, R.: *Multi-grid methods for non-conforming finite element methods*, Technical Report 453, SFB 123, University Heidelberg, 1988

[4] Peisker, P.: *Zwei numerische Verfahren zur Lösung der biharmonischen Gleichung unter besonderer Berücksichtigung der Mehrgitteridee*, Thesis, Bochum 1985

[5] Rannacher, R., Turek, S.: *A simple nonconforming quadrilateral Stokes element*, Numer. Meth. Part. Diff. Equ., 8, 97–111, 1992

[6] Turek, S.: *Ein robustes und effizientes Mehrgitterverfahren zur Lösung der instationären, inkompressiblen 2-D Navier-Stokes-Gleichungen mit diskret divergenzfreien finiten Elementen*, Thesis, Heidelberg 1991

[7] Turek, S.: *Tools for simulating nonstationary incompressible flow via discretely divergence-free finite element models*, submitted to International Journal for Numerical Methods in Fluids

[8] Turek, S.: *Multigrid techniques for a class of discretely divergence-free finite element spaces*, Technical Report, University Heidelberg, 1994 (to appear)

17

Grid-independent Convergence Based on Preconditioning Techniques

A. van der Ploeg, E.F.F. Botta and F.W. Wubs[1]

1 Introduction

Today numerical calculations are no longer restricted to a class of simple problems, but cope with complicated simulations and complex geometries. In many situations the accuracy of the numerical solution is determined by the limited amount of computer power and memory. Therefore much attention has been given to the development of numerical methods for solving the large sparse system of equations $Ax = b$ obtained by discretising some partial differential equation. Since direct methods require much computer storage and CPU-time, a large variety of iterative methods has been derived. In this paper we will focus on iterative methods like MICCG and algebraic multigrid. Gustafsson [1] has shown that for several problems the CPU-time using MICCG is $O(N^{5/4})$ in 2 dimensions and $O(N^{7/6})$ for 3D-problems, where N is the total number of unknowns. Multigrid methods perform even better and for a large class of problems they have an optimal order of convergence: the amount of work and storage is proportial to the number of unknowns N. However, due to the required proper smoothers and the restriction and prolongation operators at each level, the implementation of multigrid for practical problems is much more complicated than that of MICCG. Here we look for a combination of these properties: an incomplete LU-decomposition such that the preconditioned system can be solved with the optimal computational complexity $O(N)$ by a conjugate gradient-like method. The basic idea behind this preconditioning technique is the same as in multigrid methods. In Section 2 a preconditioning technique is described which uses a partition of the unknowns based on the sequence of grids in multigrid. After a renumbering of the unknowns according to this partition, L and U are obtained from an incomplete decomposition based on a drop tolerance [2]. The construction of L and U makes no restriction with respect to the sparsity pattern of A and the computational complexity for

[1]University of Groningen, Department of mathematics, P.O. Box 800, 9700 AV Groningen

the building of the incomplete decomposition is $O(N)$.
In Section 3 results are presented from the above method applied to some representative test problems described in the literature. These results show the robustness of the method. A comparison with some existing techniques is given in Section 4.

2 The preconditioning technique

Before an incomplete decomposition of A is made, a renumbering of the unknowns based on the multigrid idea is performed. Consider a sequence of nested grids $\Omega_1, \Omega_2, \ldots, \Omega_\gamma$, where $\Omega_\gamma \subset \Omega_{\gamma-1} \cdots \subset \Omega_1$. If all grids are uniform Ω_m has mesh size $2^{m-1}h$, where h is the mesh size of the finest grid Ω_1. The set of unknowns at the m-th level is now defined by $W_m = \Omega_m \backslash \Omega_{m+1}$, where $\Omega_{\gamma+1} = \emptyset$. For Dirichlet boundary conditions and a lexicographical numbering within the levels, we obtain for the inner grid points of a uniform rectangular 8×8 grid:

$$\begin{array}{cccccc} 1 & 2 & 3 & 4 & 5 & 6 \\ 7 & \mathit{28} & 8 & \mathit{29} & 9 & \mathit{30} \\ 10 & 11 & 12 & 13 & 14 & 15 \\ 16 & \mathit{31} & 17 & \mathit{36} & 18 & \mathit{32} \\ 19 & 20 & 21 & 22 & 23 & 24 \\ 25 & \mathit{33} & 26 & \mathit{34} & 27 & \mathit{35} \end{array}$$

The points with numbers 1 to 27 belong to the first level W_1. Similarly the two sets of points 28 to 35 and 36 belong to W_2 and W_3 respectively. In [3] an algorithm for the generation of such a numbering is given in the more general case where the mesh is not uniform. Numbering the unknowns as described above results in a system of linear equations which can be written as

$$\begin{bmatrix} A_{11} & A_{12} \\ A_{21} & A_{22} \end{bmatrix} \begin{bmatrix} x_1 \\ x_2 \end{bmatrix} = \begin{bmatrix} b_1 \\ b_2 \end{bmatrix}$$

where x_1 is the vector containing the unknowns of the first level W_1 and x_2 those of the second grid Ω_2. This partitioning of the matrix can be repeated for the matrix in the lower-right corner until we arrive at the coarsest grid.
The preconditioning technique consists now of making a splitting $A = LU + R$ in which the elements r_{ij} all satisfy $|r_{ij}| \leq \varepsilon_{ij}$. Herein ε_{ij} represents a drop tolerance which should be chosen carefully in order to obtain a proper incomplete decomposition of A. We will show that it is advantageous to choose this drop tolerance small for the block in the lower-right corner. Suppose that ε_{ij} can be chosen such that LU has the block structure

$$\begin{bmatrix} LU & A_{12} \\ A_{21} & A_{22} \end{bmatrix} = \begin{bmatrix} L & 0 \\ A_{21}U^{-1} & A_{22} - A_{21}U^{-1}L^{-1}A_{12} \end{bmatrix} \begin{bmatrix} U & L^{-1}A_{12} \\ 0 & I \end{bmatrix} \quad (1)$$

This implies that the residual matrix R has the block structure

$$\begin{bmatrix} LU - A_{11} & 0 \\ 0 & 0 \end{bmatrix}$$

The vector Rx only contains components of the first level. With this type of preconditioner all low-frequency errors are eliminated immediately and the iterative method only has to remove high-frequency errors with a wavelength in the order of the mesh size. In [3] it is shown that for a discretised Laplace operator the choice (1) leads to a preconditioned matrix with a condition number which is bounded by 2. Of course it is not realistic to use (1) as a preconditioner, since this requires the inverse of the block $A_{22} - A_{21}U^{-1}L^{-1}A_{12}$, but it is possible to choose ε_{ij} such that one obtains a residual matrix with small elements r_{ij} in the lower-right corner and a limited amount of fill-in.

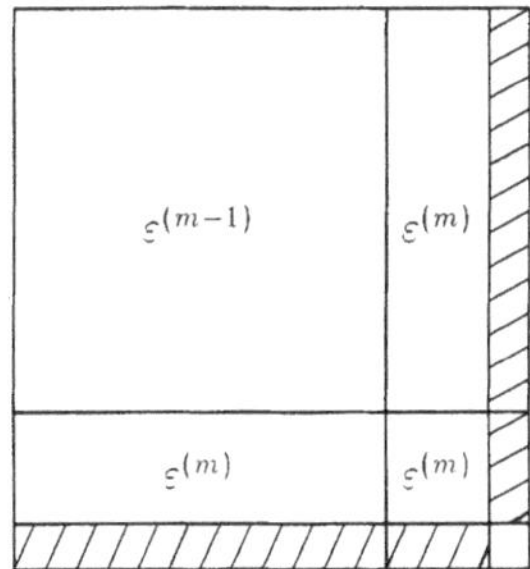

FIGURE 1: The drop tolerance for isotropic problems.

In the following we describe our choice for ε_{ij}. Suppose that A is obtained from a standard discretisation of a Poisson equation in two dimensions on a uniform rectangular grid. After a renumbering of the unknowns, as mentioned before, we consider the corresponding block partitioning of the matrix A. The drop tolerance ε_{ij} is kept constant within each of the diagonal blocks and starting with $\varepsilon = \varepsilon^{(1)}$ in the first diagonal block, corresponding with level W_1, we let the drop tolerance decrease by multiplying with a factor $c < 1$ at each new level. In the lower-triangular part ε_{ij} is chosen equal to ε_{ii} and in the upper-triangular part it follows from symmetry. Fig. 1 shows the drop tolerance near the diagonal blocks corresponding with W_{m-1} and W_m. Herein $\varepsilon^{(m)} = c\varepsilon^{(m-1)} = c^{m-1}\varepsilon^{(1)}$. For most problems the choice of $\varepsilon^{(1)}$ and c is not very critical. In 2D-problems $c = 0.2$ is a reasonable choice, but in 3D-problems the optimal value for c is smaller. The above choice can also be used very well for different mesh sizes h and k in horizontal and vertical direction, respectively, as long as h and k have the same order of magnitude. In [3] it is shown that when $h \ll k$ or $k \ll h$ it is much better to choose ε_{ij} as

$$\varepsilon_{ij} = \varepsilon \frac{h^2 k^2}{(h^2 + k^2)\rho_{ij}^2} \tag{2}$$

where ρ_{ij} is the distance between the two grid points with numbers i and j, and ε is a parameter which must be chosen in advance (the numerical experiments will demonstrate that good results are obtained with $\varepsilon \approx 0.2$). When a non-uniform grid is used, the drop tolerance can be adapted to the local mesh size [3].

3 Numerical experiments

In order to demonstrate the preconditioning technique described in the previous section, we show the results of 3 different situations. Example 3 was taken from [4]. All numerical experiments have been carried out in double precision on an HP-720 workstation and with the iterative method applied on the preconditioned system $L^{-1}AU^{-1}\tilde{x} = L^{-1}b$, where $\tilde{x} = Ux$. In all cases, the initial solution was some random vector.

It appeared to be advantageous to number the unknowns in the separate groups according to a red-black ordering. This numbering has also the advantage that the resulting method can be implemented more efficiently on supercomputers. This means that the best ordering for scalar computers is also optimal for vector and parallel computers. When the sparsity pattern of the factors L and U is chosen such that all elements of the residual matrix are in absolute value smaller than ε, this is indicated with ILU(ε) or MILU(ε). When the technique described in the previous section is used this is indicated with NGIC(ε) or NGILU(ε) were NGIC stands for Nested Grids Incomplete Choleski. When NGIC(ε) is combined with the conjugate gradient method, this is indicated with NGICCG(ε). In all examples the parameter c was chosen equal to 0.2.

When the sparsity pattern of the matrix $L + U$ is taken the same as that of A, this is indicated with the word standard. The efficient Eisenstat implementation [5] was used whenever it was possible.

Example 1. The first example shows the results of solving a Poisson equation on the unit square $[0,1] \times [0,1]$ with Neumann boundary conditions everywhere, discretised on a uniform grid. This problem is of interest for calculating the pressure in an incompressible fluid. Since the level of the solution is not fixed, the coefficient matrix is singular. Therefore, the conjugate gradient method is implemented as described in [6]. Fig. 2 shows the number of flops per unknown necessary for the conjugate gradient method to fulfil the stopping criterion

$$\|L^{-1}(b - AU^{-1}\tilde{x}^n)\|_2 < 10^{-6}\|L^{-1}(b - AU^{-1}\tilde{x}^0)\|_2$$

versus the number of unknowns. In order to improve the efficiency of MICCG we applied small perturbations of the main diagonal: before the modified incomplete Choleski-decomposition was made all diagonal elements were multiplied with a factor $1 + 10h^2$. These perturbations decrease the number of MICCG-iterations considerably. The results of Fig. 2 clearly show the effect of choosing different incomplete Choleski-decompositions. With standard ICCG the number of flops per

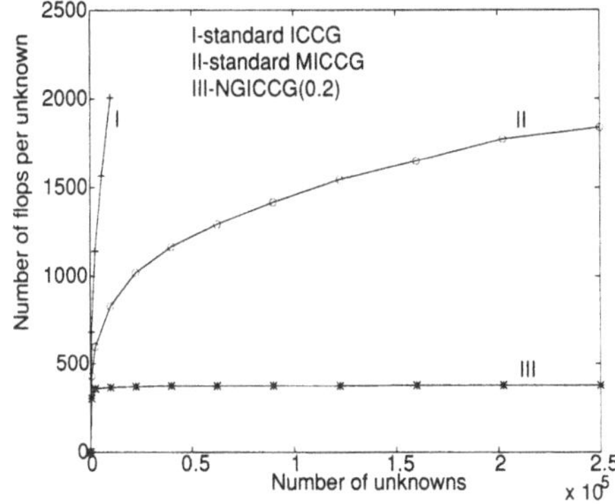

FIGURE 2: Numerical results for a discretised Poisson problem.

unknown grows very strongly with the number of unknowns. Standard MICCG performs much better but the amount of work per unknown still grows with mesh-refinement. With NGICCG as described in Section 2, the number of flops per unknown is approximately constant.

Fig. 3 gives an impression of the sparsity pattern of $L+L^T$ in case of 100 unknowns. The size of a dot represents the absolute value of the corresponding matrix entry.

Example 2. Our second example is the discretised convection-diffusion equation

$$-\Delta u(x,y) + 10^4 \left\{ (\tfrac{1}{2} - x)^3 \frac{\partial}{\partial x} u(x,y) + (\tfrac{1}{2} - y)^3 \frac{\partial}{\partial y} u(x,y) \right\} = f(x,y)$$

on the square $[0,1] \times [0,1]$ with Neumann boundary conditions everywhere. For the discretisation we used a rectangular grid with constant mesh size $1/(M-1)$ in both directions and central differences for all derivatives. The coefficient matrix is not necessarily an M-matrix, because the mesh-Péclet numbers can be larger than 2. Since the system is non-symmetric, it was solved with preconditioned Bi-CGSTAB with the same stopping criterion as in Example 1. We have compared the preconditioning of standard ILU with NGILU(0.2). The results are summarised in Fig. 4 giving the number of flops per unknown against the number of unknowns. The results of the standard MILU-preconditioning cannot be shown because even for $M = 256$ ($\approx 0.65 \times 10^5$ unknowns) the construction of the incomplete factorisation breaks down due to the generation of small elements on the main diagonal. For $M = 512$ ($\approx 2.62 \times 10^5$ unknowns) the construction of the MILU-factorisation can be completed and leads combined with Bi-CGSTAB to 49 iterations and approximately 2150 flops per unknown.

We observed that the convergence behaviour of Bi-CGSTAB with standard ILU as preconditioner was very irregular, whereas with NGILU(0.2) this behaviour is much smoother. In case of a very irregular convergence behaviour the accuracy of the calculated solution may be spoiled by cancellation effects ([4]). Indeed, when using standard ILU, the 2-norm of the real residual was about 100 times larger than the 2-norm of the calculated residual on which the stopping criterion was

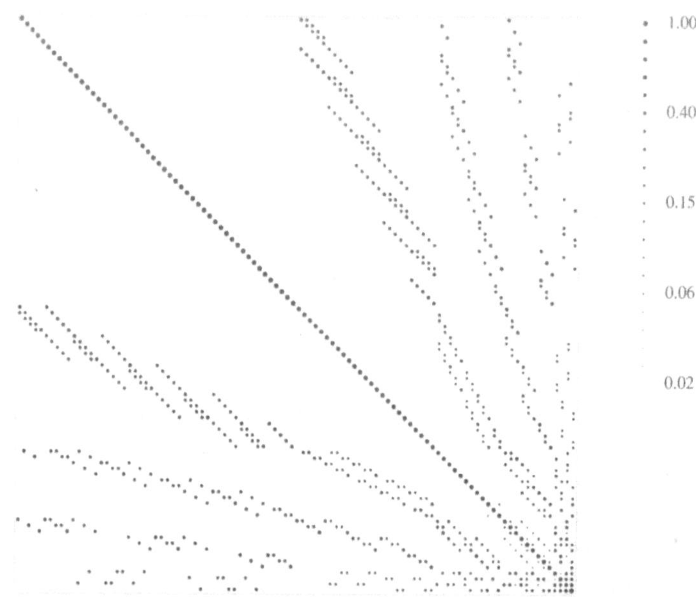

FIGURE 3: Sparsity pattern of $L + L^T$

based. When using NGILU(0.2) both norms were approximately the same.

Example 3. The following test problem is a simplified aquifer problem and is taken from [4]. The non-symmetric system of linear equations stems from discretisation of the convection-diffusion equation

$$-\frac{\partial}{\partial x}(A\frac{\partial u}{\partial x}) - \frac{\partial}{\partial y}(A\frac{\partial u}{\partial y}) + 2e^{2(x^2+y^2)}\frac{\partial u}{\partial x} = F(x,y)$$

on the square $[0,1] \times [0,1]$. The diffusion coefficient function A is given in Fig. 5 in which the dashed area indicates the region in which $A = 10000$. Further $F(x,y) = 0$ everywhere, except for the small square in the center, where $F(x,y) = 100$. We have Dirichlet boundary conditions along all boundaries as shown in Fig. 5. The partial differential equation was discretised on a rectangular grid with mesh size $1/M$, and central differences for all derivatives. Fig. 6 shows the number of flops per unknown necessary for Bi-CGSTAB versus the number of unknowns. As stopping criterion for the iteration we used

$$\|L^{-1}(b - AU^{-1}\tilde{x}^n)\|_2 < 10^{-6}$$

Again the standard MILU-factorisation breaks down and therefore no results for this preconditioning technique can be shown. With MILU(0.02) the construction of the preconditioner does not break down. The results of Fig. 6 clearly show the

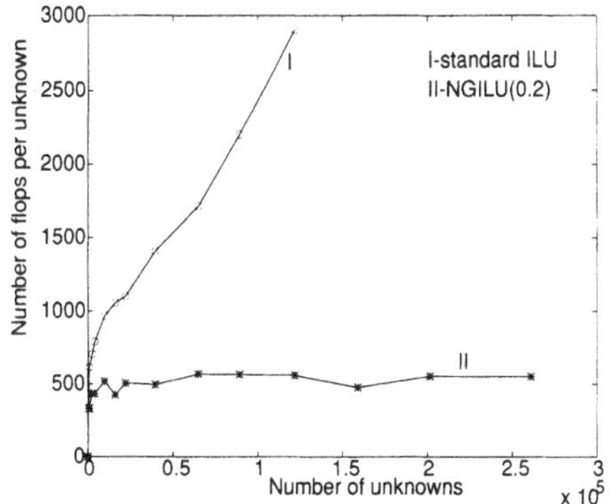

FIGURE 4: Number of flops per unknown for example 2 on various grids.

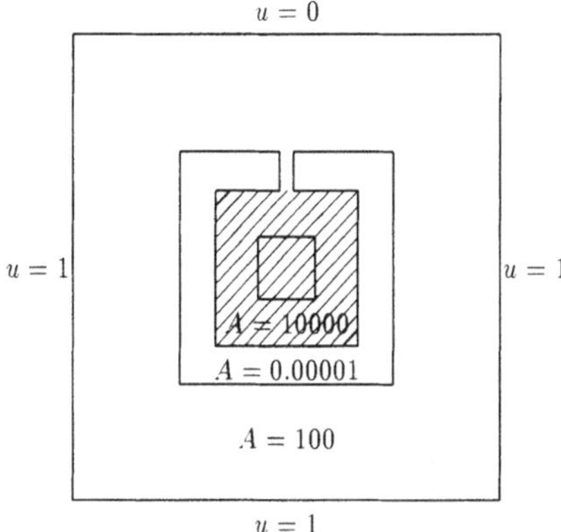

FIGURE 5: The diffusion coefficient for example 3.

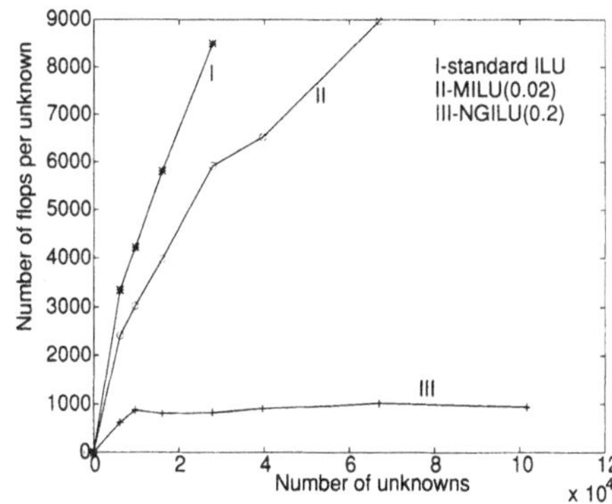

FIGURE 6: Numerical results for example 3.

effect of the choice of the preconditioner. With standard ILU and with MILU(0.02) the number of flops per unknown increases very strongly with mesh-refinement, whereas with NGILU(0.2) the amount of work per unknown is approximately constant. When the mesh size was 1/320 in both directions, the average number of entries in one row of the matrix $L+U$ was 12.4, and not more than 11 iterations of Bi-CGSTAB were necessary in order to fulfil the stopping criterion.
Fig. 7 shows the convergence behaviour of preconditioned Bi-CGSTAB when the mesh size was 1/200, and we note the relatively smooth convergence behaviour of the iterative method combined with NGILU(0.2).

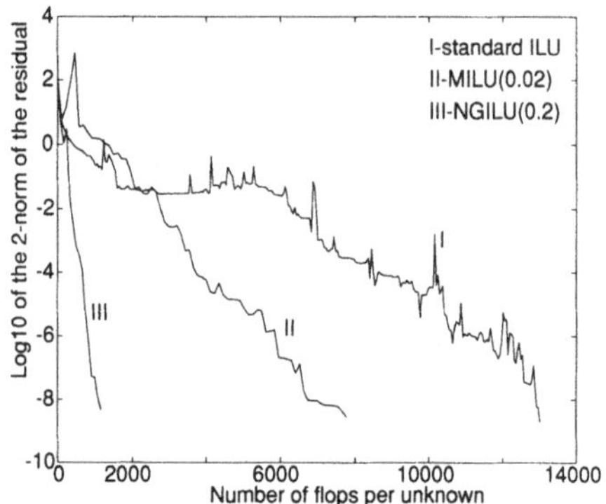

FIGURE 7: Convergence behaviour of Bi-CGSTAB for example 3 on a 201×201-grid.

4 Comparison with other techniques

In this section we take a closer look at the condition number of the preconditioned matrix in case the coefficient matrix A stems from a standard discretisation of a Poisson equation on a rectangular grid with constant mesh size h. We want to construct an incomplete Choleski-decomposition of A such that the preconditioned matrix $L^{-1}AL^{-T}$ has a condition number as small as possible. Although we only consider the case of A being symmetric and positive (semi-)definite, the results of Section 3 show that the preconditioning technique is of interest for a much broader class of problems.
We will see that for a certain choice of the drop tolerance parameter ε_{ij} the condition number of $L^{-1}AL^{-T}$ grows with the mesh size h as $O(h^{-\alpha})$ with $\alpha < \frac{1}{2}$ and that this particular choice of ε_{ij} is far from optimal. In [3] it is shown that even if the largest part of L is chosen very sparse, we still can construct a preconditioner for which cond$(L^{-1}AL^{-T}) \leq 2$. The grid points are divided into 4 parts: the red and black points of the first level, and the red and black points of the second level.

For example:

$$
\begin{array}{cccccccccc}
* & \circ & * & \circ & * & \circ & * & \circ & * & \circ \\
\circ & \star & \circ & \bullet & \circ & \star & \circ & \bullet & \circ & \star \\
* & \circ & * & \circ & * & \circ & * & \circ & * & \circ \\
\circ & \bullet & \circ & \star & \circ & \bullet & \circ & \star & \circ & \bullet \\
* & \circ & * & \circ & * & \circ & * & \circ & * & \circ \\
\circ & \star & \circ & \bullet & \circ & \star & \circ & \bullet & \circ & \star
\end{array}
$$

The points are numbered in the sequence '$\circ$','$*$', '$\bullet$', '$\star$'. Suppose that the main diagonal of A is scaled to unity, then according to this partition, A has the block structure

$$
\begin{bmatrix} I_1 & A_{12} & A_{13} & A_{14} \\ A_{21} & I_2 & 0 & 0 \\ A_{31} & 0 & I_3 & 0 \\ A_{41} & 0 & 0 & I_4 \end{bmatrix} \tag{3}
$$

where $A_{ij}^T = A_{ji}$ and I_j is an identity matrix. We want to make a complete Choleski-decomposition LL^T of the matrix $A - R$, such that L is sparse. The residual matrix R is taken so small that LL^T resembles A. We consider a choice of LL^T which leads to a residual matrix with the block structure

$$
R = \begin{bmatrix} 0 & 0 & 0 & 0 \\ 0 & R_{22} & 0 & 0 \\ 0 & 0 & R_{33} & 0 \\ 0 & 0 & 0 & R_{44} \end{bmatrix} \tag{4}
$$

After eliminating all '$\circ$'-points we obtain the Schur-complement

$$
S_2 = \begin{bmatrix} I_2 - A_{21}A_{12} & -A_{21}A_{13} & -A_{21}A_{14} \\ -A_{31}A_{12} & I_3 - A_{31}A_{13} & -A_{31}A_{14} \\ -A_{41}A_{12} & -A_{41}A_{13} & I_4 - A_{41}A_{14} \end{bmatrix}
$$

By lumping all off-diagonal elements onto the diagonal, the first diagonal block $I_2 - A_{21}A_{12}$ is approximated by the diagonal $D_2 = \frac{1}{2}I_2$. This implies that $R_{22} = \frac{1}{2}I_2 - A_{21}A_{12}$ with row sums zero. We continue in this manner by eliminating all '$*$'-points and by approximating the block in the upper-left corner by $D_3 = \frac{1}{2}I_3$. One can show that this leads to the block $R_{33} = \frac{1}{4}I_3 - 2A_{31}A_{12}A_{21}A_{13}$ of the residual matrix. After eliminating all '$\bullet$'-points it follows that R_{44} is the residual matrix in the approximation of the Schur-complement

$$
S_4 = \tfrac{3}{4}I_4 - 2A_{41}A_{12}A_{21}A_{14} - 8A_{41}A_{13}A_{31}A_{14}
$$

It can be shown that this block is a sparse M-matrix with a nine-point stencil. One way to proceed is to make a block decomposition of S_4 in the same way as in (3), and then make an incomplete decomposition as described above for A. This is exactly the nested recursive two-level factorisation method described by Axelsson and Eijkhout in [7]. They prove that if the incomplete Choleski-decomposition is made in this way, the condition number of $L^{-1}AL^{-T}$ grows with the mesh size h as $O(h^{-0.69})$.

The so-called RRB-method described by Brand in [8] follows the same strategy for the first ${}^2\log M$ levels. When the level number exceeds ${}^2\log M$ the drop tolerance is set to zero. In [8] it is proven that when A stems from the discretisation of a Poisson equation on a uniform rectangular grid with Dirichlet boundary conditions, this strategy results in an incomplete Choleski-decomposition such that the condition number of $L^{-1}AL^{-T}$ grows with the mesh size h as $O(h^{-\alpha})$, with $\alpha < \frac{1}{2}$. With this preconditioning technique one does not obtain grid-independent convergence. In [7] the nested two-level factorisation method is combined with nested polynomial approximations in order to obtain a method of optimal order of computational complexity. In this paper, we follow a different approach: we make a more accurate incomplete Choleski-decomposition of S_4 by choosing a smaller drop-tolerance during the construction of an MIC-decomposition of S_4 as described in Section 2.

Numerical verification of the condition number.

To obtain insight in the condition number of $L^{-1}AL^{-T}$ we have computed the maximum eigenvalue of this matrix, which is equal to the spectral condition number, by an iterative method. The coefficient matrix results from a standard five-point discretisation of the Poisson equation on a rectangular grid with constant mesh size $1/(M+1)$, hence A has dimension M^2. We have used Dirichlet boundary conditions everywhere. At every new level the drop tolerance ε is decreased by multiplying with a factor $c = 0.2$. The results are summarised in Fig. 8 which shows the computed largest eigenvalue versus M. For comparison we also give the results of the RRB-method described in [8]. From these results we conclude that with

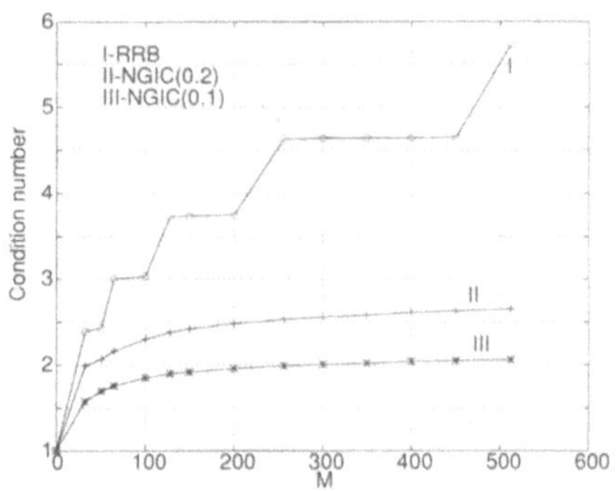

FIGURE 8: Condition number of $L^{-1}AL^{-T}$ versus M.

the RRB-method the condition number of $L^{-1}AL^{-T}$ increases only very slightly with mesh-refinement. With NGIC(0.2) and NGIC(0.1) the results are even better: the condition number hardly increases with mesh-refinement. When we choose a smaller value for the parameter c, the condition number of $L^{-1}AL^{-T}$ appears to be bounded by 2. Table 1 shows a number of calculated largest eigenvalues of $L^{-1}AL^{-T}$ for RRB and for NGIC with some choices of ϵ and c. The average number of entries in one row of L is given in brackets. Remark that when $c = 0.1$

TABLE 1: Calculated condition numbers of $L^{-1}AL^{-T}$.

M	RRB	NGIC	NGIC	NGIC
		$\varepsilon = c = 0.2$	$\varepsilon = 0.1,\ c = 0.2$	$\varepsilon = 0.2,\ c = 0.1$
32	2.39(5.0)	1.990(5.2)	1.577(6.9)	1.98992(6.4)
64	3.00(5.1)	2.162(5.5)	1.762(7.5)	1.99753(7.1)
128	3.73(5.2)	2.378(5.7)	1.898(7.8)	1.99939(7.7)
256	4.63(5.2)	2.532(5.9)	1.990(8.0)	2.000(8.1)
512	5.73(5.2)	2.647(6.0)	2.057(8.2)	2.000(8.3)

and $\varepsilon = 0.2$ the difference between 2 and the largest eigenvalue behaves like $O(h^2)$. From the difference between the last two columns of this table we conclude that the choice of the parameters ε and c is not very critical.

5 Conclusions and discussion

In this paper, a new preconditioning technique is described which shows grid-independent convergence when combined with any conjugate gradient-like method. This technique is relatively easy to implement: we only have to make an incomplete LU-decomposition of A. Essential in the method is the choice of a drop tolerance controlling the size of $R = A - LU$ and the ordering of the unknowns. The ordering is similar to that in multigrid approaches and makes it possible to construct an incomplete LU-decomposition which can be used in eliminating effectively both high- and low-frequency errors. The method is demonstrated for a Poisson equation, a convection-diffusion equation and an aquifer problem. In all cases, the method is much cheaper than standard (M)ICCG. This difference is more pronounced for the really difficult problems and increases with the dimension. The convergence behaviour is, in contrast to that of standard (M)ICCG, always smooth, which is advantageous for the construction of stopping criteria when the linear solver is used as an inner-iteration method within some inexact Newton method.
The computational work consists of three parts: the construction of the preconditioning matrix, its application, and the number of iterations. From the numerical experiments we conclude that the construction of the preconditioner grows linearly with the number of unknowns. Its application is linear with the fill-in. This fill-in is about a factor two larger than that of A, which is modest and comparable to that of a standard ILU-decomposition. With the present method an iteration step is significantly cheaper than one multigrid step, due to additional smoothing operations needed in the latter method. Some preliminary results show that

the number of iterations is comparable and hence the computational work of the present method is less than that of multigrid.
In our view, the results in this paper show the potential of the introduced method. Of course, more analysis (e.g. rigourous convergence proof) and further optimisation (e.g. choice of the drop tolerance and ordering of the unknowns) is needed. Moreover, implementations on modern computers which exploit the parallelism present in the algorithm must be studied. A different topic is the application of these ideas to problems which are far from elliptic, such as the Navier-Stokes equations. Research on these subjects is in progress.

ACKNOWLEDGEMENT

The authors wish to thank Prof. Dr. Henk A. Van der Vorst for providing the FORTRAN-code that generates the matrix for the third test problem.

References

[1] I. Gustafsson. A class of 1:st order factorization methods. *BIT*, 18:142–156, 1978.

[2] A. van der Ploeg. Preconditioning techniques for non-symmetric matrices with application to temperature calculation of cooled concrete. *Int. J. for Num. Methods in Engineering*, 35(6):1311–1328, 1992.

[3] A. van der Ploeg, E.F.F. Botta, and F.W. Wubs. Grid-independent convergence based on preconditioning techniques. Technical Report W-9310, University of Groningen, Department of Mathematics, 1993.

[4] H.A. van der Vorst. Bi-CGSTAB: A fast and smoothly converging variant of Bi-CG for the solution of nonsymmetric linear systems. *SIAM J. Sci. Stat. Comput.*, 13(2):631–644, 1992.

[5] S.C. Eisenstat. Efficient implementation of a class of preconditioned conjugate gradient methods. *SIAM J. Sci. Stat. Comput.*, 2:1–4, 1981.

[6] E.F. Kaasschieter. Preconditioned conjugate gradients for solving singular systems. In A. Hadjidimos, editor, *Iterative methods for the solution of linear systems.* Elsevier, North Holland, 1988. Reprinted from the Journal of Computational and Applied Mathematics, Volume 24, Numbers 1 and 2.

[7] O. Axelsson and V. Eijkhout. The nested recursive two-level factorization method for nine-point difference matrices. *SIAM J. Sci. Stat. Comput.*, 12:1373–1400, 1991.

[8] Cl.W. Brand. An incomplete-factorization preconditioning using red-black ordering. *Num. Math.*, 61:433–454, 1992.

18

A New Residual Smoothing Method for Multigrid Acceleration Applied to the Navier-Stokes Equations

Zhong Wen Zhu, Chris Lacor and Charles Hirsch[1]

ABSTRACT A new residual smoothing method, based on a decomposition into forward and backward sweeping steps, is investigated by means of a Fourier analysis on the two-dimensional convection-diffusion equation. Both central and upwind space discretizations are considered together with explicit multi-stage Runge-Kutta time-stepping. The Fourier analysis shows improved high-frequency-damping, which plays an important role in the multigrid acceleration. The numerical results of 2D flat plate laminar flow calculations confirm the efficiency of the new residual smoothing approach

1 Introduction

Explicit multi-stage time stepping schemes are widely applied to the Euler/Navier-Stokes equation solvers because of their simplicity [1],[2],[3]. In order to obtain fast convergence with a high CFL number, the implicit residual smoothing approach with a central form was introduced by Jameson et al [4] in early 1980's. Later, several authors [5],[6],[7] slightly adapted the residual smoothing strategy in order to deal with high-aspect-ratio meshes in viscous calculations. In 1991, Blazek et al [8] proposed an upwind-biased residual smoothing method which shows improved high-frequency-damping if combined with upwind schemes for hypersonic flow. The drawback seems to be an increased complexity and computational effort because the smoothed residual is based on the characteristic variables. In addition, since both of methods require the solution of tri-diagonal systems, they are difficult to vectorize.

A new approach, defined as a forward-backward residual smoothing, was pro-

[1]Department of Fluid Mechanics, Vrije Universiteit Brussel, Pleinlaan 2, 1050 Brussels, Belgium

posed by Zhu, Lacor and Hirsch [9] recently. It can be combined with either central or upwind schemes. The one-dimensional Fourier analysis shows that it offers improved smoothing properties for all the frequencies and that it is also computationally more efficient since it vectorizes completely. The application of the forward-backward residual smoothing to the two-dimensional Navier-Stokes equations is investigated in detail in this paper. In the viscous calculation, the high-aspect-ratio meshes can not be avoided near the wall. The effect of the high-aspect-ratio on the convergence is studied by means of a Fourier analysis on the two-dimensional convection diffusion equation. Both central and upwind space discretizations are considered together with explicit multi-stage Runge-Kutta time-stepping. The analysis shows that the smoothing properties can be improved by implicit residual smoothing for reasonable aspect-ratios. In this case the forward-backward residual smoothing shows better smoothing properties than previous residual smoothing approaches. Finally, the 2D flat plate laminar flow test case is selected to confirm the efficiency of the new approach.

2 Two Dimensional Model Problem

Consider the following linear, scalar, two-dimensional convection-diffusion equation

$$\frac{\partial u}{\partial t} + a\frac{\partial u}{\partial x} + b\frac{\partial u}{\partial y} = \alpha(\frac{\partial^2 u}{\partial x^2} + \frac{\partial^2 u}{\partial y^2}) \tag{1}$$

where a, b, $\alpha \geq 0$. The convective term can be represented in the κ form as follows:

$$\begin{aligned} u_x \Delta x = \delta_x^- u_{i,j} = (u_{i,j} - u_{i-1,j}) + \psi \left[\frac{1-\kappa}{4}(u_{i,j} - 2u_{i-1,j} + u_{i-2,j}) + \right. \\ \left. \frac{1+\kappa}{4}(u_{i+1,j} - 2u_{i,j} + u_{i-1,j}) \right] \end{aligned} \tag{2}$$

$$\begin{aligned} u_y \Delta y = \delta_y^- u_{i,j} = (u_{i,j} - u_{i,j-1}) + \psi \left[\frac{1-\kappa}{4}(u_{i,j} - 2u_{i,j-1} + u_{i,j-2}) + \right. \\ \left. \frac{1+\kappa}{4}(u_{i,j+1} - 2u_{i,j} + u_{i,j-1}) \right] \end{aligned} \tag{3}$$

where $\psi = 0$ for a first order upwind scheme. For $\psi = 1$, the parameter κ controls the upwind biasing which yields different schemes for the following values of κ: -1 (second-order upwind), 0 (Fromm scheme), 1/3 (third-order accurate), 1/2 (Quick scheme). When $\kappa = 1$, the central scheme is recovered and the following artificial viscosity should be added for stability.

$$\begin{aligned} \mu \Delta x^4 \frac{\partial^4 u}{\partial x^4} &= \mu(u_{i+2,j} - 4u_{i+1,j} + 6u_{i,j} - 4u_{i-1,j} + u_{i-2,j}) \\ \mu \Delta y^4 \frac{\partial^4 u}{\partial y^4} &= \mu(u_{i,j+2} - 4u_{i,j+1} + 6u_{i,j} - 4u_{i,j-1} + u_{i,j-2}) \end{aligned} \tag{4}$$

where μ is the artificial viscosity coefficient. The viscous terms in equation (1) are discretized in a central form:

$$\begin{aligned} u_{xx}\Delta x^2 &= \delta_x^2 u_{i,j} &= u_{i+1,j} - 2u_{i,j} + u_{i-1,j} \\ u_{yy}\Delta y^2 &= \delta_y^2 u_{i,j} &= u_{i,j+1} - 2u_{i,j} + u_{i,j-1} \end{aligned} \tag{5}$$

After introducing the following parameters:

$$\sigma_x \equiv a\frac{\Delta t}{\Delta x}, \qquad \theta \equiv \frac{b}{a}, \qquad \lambda \equiv \frac{\Delta x}{\Delta y}, \qquad \frac{1}{Re} \equiv \frac{\alpha}{a\Delta x} \tag{6}$$

the semi-discretized equation in time is given as

$$\Delta t\frac{du_{i,j}}{dt} - R_{i,j} - R_{i,j}^{av} = 0 \tag{7}$$

where $R_{i,j}$ is the residual of the physical flux and $R_{i.j}^{av}$ is the residual of the artificial viscosity which corresponds only to the central scheme. It is zero for $\kappa \neq 1$. They have the following forms:

$$\begin{aligned} R_{i,j} &= -\sigma_x[\delta_x^- + \theta\lambda\delta_y^- - \frac{1}{Re}(\delta_x^2 + \lambda^2\delta_y^2)]u_{i,j} \\ R_{i,j}^{av} &= -\sigma_x\mu(\delta_x^4 + \theta\lambda\delta_y^4)u_{i,j} \end{aligned} \tag{8}$$

After applying the m-stage explicit time-stepping to equation (8), one obtains

$$\begin{aligned} u_{i,j}^{[0]} &= u_{i,j}^n \\ u_{i,j}^{[1]} &= u_{i,j}^{[0]} + \alpha_1[R_{i,j} + R_{i,j}^{av}] \\ u_{i,j}^{[2]} &= u_{i,j}^{[0]} + \alpha_2[R_{i,j} + R_{i,j}^{av}] \\ &\vdots \\ u_{i,j}^{[k]} &= u_{i,j}^{[0]} + \alpha_k[R_{i,j} + R_{i,j}^{av}] \\ &\vdots \\ u_{i,j}^{[m]} &= u_{i,j}^{[0]} + \alpha_m[R_{i,j} + R_{i,j}^{av}] \\ u_{i,j}^{n+1} &= u_{i,j}^{[m]} \end{aligned} \tag{9}$$

where α_j is the multi-stage time-stepping coefficient of stage j. Then the amplification factor can be written by

$$G(z) = \frac{\widehat{u}^{n+1}}{\widehat{u}^n} = 1 + z + \sum_{i=2}^{n}\beta_i z^i \tag{10}$$

where $\widehat{u}$ denotes the amplitude of Fourier symbol u, z is the Fourier symbol of the residual, and $\beta_i = \alpha_m\alpha_{m-1}\cdots\alpha_{m-i-1}$. m is the number of the explicit time-stepping stages.

3 Forward-Backward Residual Smoothing

One-dimensional Smoothing Model

The forward-backward residual smoothing approach called hereafter FB-RS was proposed in [9]. For one-dimensional problems, it is defined by

$$(1+\varepsilon_f\delta^-)(1-\varepsilon_b\delta^+)\overline{R}_i \;=\; R_i \tag{11}$$

where ε_f, ε_b are forward or backward smoothing parameters, and δ^-, δ^+ are forward or backward operators respectively. It contains both central (C-RS, $\varepsilon_f=\varepsilon_b$) and upwind residual smoothing methods (U-RS, $\varepsilon_b=0$). According to the one-dimensional analysis, the stability region of FB-RS is shown in figs. 1 and 2.

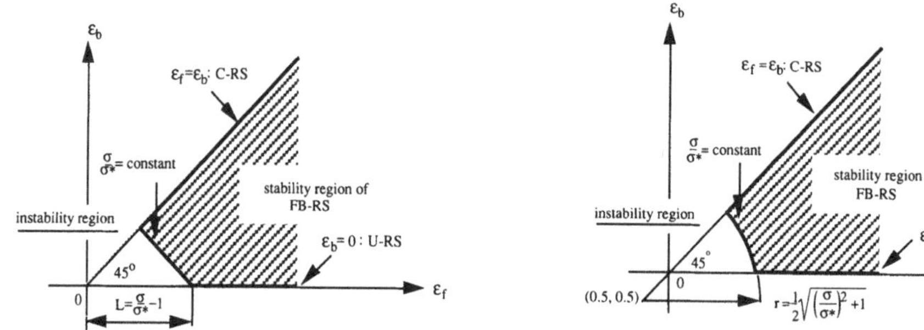

Fig. 1 Stability limit for central schemes Fig. 2 Stability limit for upwind schemes

The maximum ratio of the CFL numbers of the smoothed to unsmoothed multistage scheme was estimated for central schemes as

$$\frac{\sigma}{\sigma^*} = 1+\varepsilon_f+\varepsilon_b \qquad\qquad \varepsilon_f \geq \varepsilon_b \tag{12}$$

and for upwind schemes:

$$\frac{\sigma}{\sigma^*} = \sqrt{1+4\varepsilon_f(1+\varepsilon_f)+4\varepsilon_b(1+\varepsilon_b)} \qquad\qquad \varepsilon_f \geq \varepsilon_b \tag{13}$$

The relation between two parameters (ε_f, ε_b) decides the effect of the implicit residual smoothing. Theoretically, it should be taken according to the high-frequency-damping properties of the smoothed scheme. Since it seems difficult to find an analytical relation, even from the one dimensional linear equation, the following simple linear relation is assumed.

$$\varepsilon_b \;=\; C_{fb}\,\varepsilon_f \tag{14}$$

where $0 \leq C_{fb} \leq 1$. In addition, relation (12) can replace relation (13) for simplicity, since it is more severe. For all the schemes (central, 1st-order and 2nd-order upwind), the FB-RS will be given by:

$$\varepsilon_b = C_{fb}\,\varepsilon_f, \qquad \varepsilon_f = \frac{1}{1+C_{fb}}\left(\frac{\sigma}{\sigma^*} - 1\right) \tag{15}$$

Two-dimensional Smoothing Model

For two-dimensional problems, several sweeping forms are proposed in [9]. Here only Alternative Direction Residual Smoothing is selected since it will be implemented in the 3D code which also allows to run 2D testcases.

$$(1+\varepsilon_{fx}\delta_x^-)(1-\varepsilon_{bx}\delta_x^+)(1+\varepsilon_{fy}\delta_y^-)(1-\varepsilon_{by}\delta_y^+)\overline{R}_{i,j} = R_{i,j} \tag{16}$$

where $\varepsilon_{fx}, \varepsilon_{bx}, \varepsilon_{fy}, \varepsilon_{by}$ are smoothing parameters. $\overline{R}_{i,j}$ is the smoothed residual. The Fourier symbol of the smoothed residual can be described as:

$$z_R = \frac{z}{z_{p,x}z_{p,y}} \tag{17}$$

with

$$\begin{aligned} z_{p,x} &= 1+\varepsilon_{fb,x}(1-\cos\phi_x) + I\eta_{fb,x}\sin\phi_x \\ z_{p,y} &= 1+\varepsilon_{fb,y}(1-\cos\phi_y) + I\eta_{fb,y}\sin\phi_y \end{aligned}$$

where $I = \sqrt{-1}$, ϕ_x, ϕ_y is a phase angle, and

$$\begin{aligned} \varepsilon_{fb,x} &= \varepsilon_{f,x}+\varepsilon_{b,x}+2\varepsilon_{f,x}\varepsilon_{b,x} \quad & \eta_{fb,x} &= \varepsilon_{f,x}-\varepsilon_{b,x} \\ \varepsilon_{fb,y} &= \varepsilon_{f,y}+\varepsilon_{b,y}+2\varepsilon_{f,y}\varepsilon_{b,y} \quad & \eta_{fb,y} &= \varepsilon_{f,y}-\varepsilon_{b,y} \end{aligned} \tag{18}$$

Fourier Analysis of High-Frequency-Damping

Referring to equation (10), the stability region of the multi-stage schemes can be drawn in the complex plane (shown fig. 3). The Fourier symbol of smoothed residual z_R should be inside this region to make the scheme stable. Further when the Fourier symbol of the residual in the high frequency region (see fig. 4) is taken as z_{HF}, better smoothing properties can be obtained if $G(z_{HF})$ is less than a small constant value.

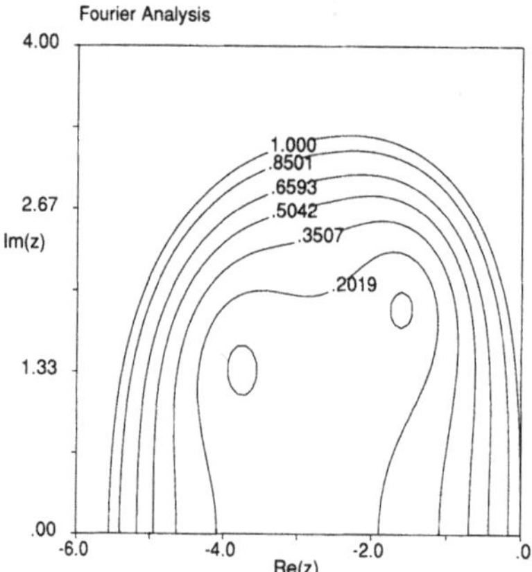

Fig. 3 Stablity limit of time-stepping

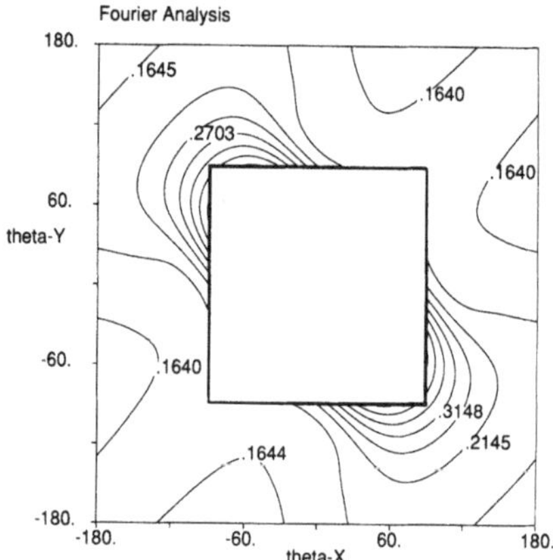

Fig. 4 Two-dimensional high frequency region

2D Convection Equation

The residual of the 2D convection equation ($1/Re$=0.0) can be written as (see equation (8)):

$$R_{i,j} = -\sigma_x(\delta_x^- + \theta\lambda\delta_y^-)u_{i,j} \tag{19}$$

For the second-order upwind scheme (ψ=1, κ=–1), the Fourier symbol of the unsmoothed residual is:

$$\begin{aligned} Re(z) &= -\sigma_x[(1-\cos\phi_x)^2 + \theta\lambda(1-\cos\phi_y)^2] \\ Im(z) &= -\sigma_x[\sin\phi_x(2-\cos\phi_x) + \theta\lambda\sin\phi_y(2-\cos\phi_y)] \end{aligned} \tag{20}$$

If the high frequency point ($\phi_x = \phi_y = \pi$) is considered, the Fourier symbol of the smoothed residual has the following form:

$$z_R = -\frac{1+\varepsilon_{f,x}+\varepsilon_{b,x}}{(1+2\varepsilon_{fb,x})(1+2\varepsilon_{fb,y})}4\sigma_x^*(1+\theta\lambda) = -f(\varepsilon_f,\varepsilon_b)4\sigma_x^*(1+\theta\lambda) \tag{21}$$

where σ_x^* is a CFL number of the unsmoothed scheme (see equation (12)). Since the function $f(\varepsilon_f,\varepsilon_b)$ is less than one, then $|z_R| \le 4\sigma_x^*(1+\theta\lambda) = |z|$. If the multi-stage coefficients are optimized based on the unsmoothed residual z, the residual smoothing method will increase the smoothing factor because it makes the real part of the residual away from the original optimial value and close to zero. In addition, considering $\varepsilon_f+\varepsilon_b$=const. (see equation (15)), the maximum of $f(\varepsilon_f,\varepsilon_b)$ is obtained at $\varepsilon_b = 0$, while its minimum corresponds to $\varepsilon_f = \varepsilon_b$. Hence the following relation can be easily found

$$|Re(z_R)|_{C-RS} \le |Re(z_R)|_{FB-RS} \le |Re(z_R)|_{U-RS} \tag{22}$$

It means that U-RS will bring the real part of z_R close to the original optimal value and show good high-frequency-damping. C-RS gives the poorest high-frequency-damping since it brings the real part of the residual to zero. FB-RS is between them. Further, under the stability condition, in order to obtain the function f as large as possible, the same ratio $C_{fb} = \varepsilon_b/\varepsilon_f$ should be taken along x- and y-direction.

It should be noted that under certain condition, e.g. $f|_{\varepsilon_b=0} \geq 1.0$, U-RS will increase the amplification factor, even cause the scheme unstable since it brings the real part of the residual too much to the stability limit. In this case, FB-RS exhibits its advantage. For other kinds of schemes, the same conclusion can be drawn.

2D Convection-Diffusion Equation with High-Aspect-Ratio

In order to numerical simulate boundary layer flow, a mesh with high-aspect-ratio has to be used. That is

$$\Delta x = constant, \qquad \Delta y \to 0 \qquad \lambda = \frac{\Delta x}{\Delta y} \to \infty \tag{23}$$

Since normally the flow angle is also very small in the boundary, the product $\theta\lambda \to o(1)$. Then the unsmoothed residual becomes (referring to equation (8)):

$$R_{i,j} = -\frac{\sigma_x}{Re}\lambda^2\delta_y^2 u_{i,j} \tag{24}$$

and its Fourier symbol is:

$$z = -2\frac{\sigma_x}{Re}\lambda^2(1-\cos\phi_y) \tag{25}$$

For the m-stage Runge-Kutta scheme, the stability condition is

$$0 \leq |Re(z)| \leq C_{smax} \qquad C_{smax} = \{max[Re(C_s)] : \ |G(C_s)| = 1\} \tag{26}$$

Combining equation (25) and (26), the stability condition for 2D convection-diffusion equation in the case of high-aspect-ratio is:

$$0 \leq \sigma_x \leq \frac{Re}{4\lambda^2}C_{smax} \tag{27}$$

when λ increases, the CFL number has to be reduced to keep the scheme stable.

For the high-frequency-damping, since the Fourier symbol of the residual z is only function of ϕ_y, z is always zero in the high frequency region ($\pi/2 \leq \phi_x \leq \pi$,$\phi_y$=0.). It is impossible to reduce the amplification factor by an implicit residual smoothing method. However, for other high frequencies, such as (ϕ_x=0,ϕ_y=π), the improvement of amplification factor can be obtained. In this case, the Fourier symbol of the residual is:

$$z_R = -\frac{4\sigma_x}{Re}\lambda^2 \frac{1}{1+2\varepsilon_{fb,y}} \tag{28}$$

From above expression, it is seen that the same or large CFL number can be taken in the high-aspect-ratio case with residual smoothing methods. In a similar way, the same relation as (22) can be easily derived. The FB-RS can give better smoothing properties.

Numerical Investigation

In the numerical analysis, the smoothing factor, defined as the maximum of amplification factor in the high-frequency region (seeing fig. 4), is used to measure the high-frequency-damping. For the FB-RS, the same ratio $C_{fb} = \varepsilon_b/\varepsilon_f$ is taken for both directions according to previous theoretical analysis and numerical calculation of the smoothing factor.

The central scheme with 4th-order artificial viscosity is used with five-stage Runge-Kutta time-stepping and optimal coefficients (1/4, 1/6, 3/8, 1/2, 1). The artificial viscosity term is re-calculated only at the first two-stages.

$\kappa = 1,\ 1/Re = 0.0,\ \lambda = 1.0,\ \sigma/\sigma^* = 2,\ \mu = 1/32$

θ	No-RS	C-RS	U-RS	FB-RS	$(C_{fb,x},\ C_{fb,y})$
1.0	0.8914	0.8724	unstable	0.8137	(0.40, 0.40)
0.5	0.8879	0.9144	unstable	0.8716	(0.30, 0.30)
0.1	0.9770	0.9568	unstable	0.9567	(0.60, 0.60)

Table 1 Smoothing factor of central scheme with 5-stage Runge-Kutta optimal coefficients (σ^*=1.85)

The smoothing factor of the central scheme at σ/σ^*=2.0 with the different residual smoothing approaches is shown in the table 1. With the increment of flow angle (θ), the smoothing factor with C-RS or FB-RS decreases. FB-RS shows better smoothing properties than C-RS. U-RS causes the instability of the central scheme. At very small flow angle, the smoothing factor is close to one for all the methods.

The first-order upwind scheme with four-stage Runge-Kutta explicit time-stepping is considered first. The optimal coefficients (0.0796, 0.2026, 0.4285, 1.0) are taken from [10].
Table 2 shows the comparison of smoothing factor among the different residual smoothing methods with the increment of flow angle (θ). The smoothing factor of upwind scheme with C-RS is reduced by increasing θ. The further improvement of smoothing properties is obtained by using U-RS instead of C-RS. FB-RS is between them. In addition, in the case of $\theta = 0.5, 1.0$, the smoothing factor with C-RS is larger than without residual smoothing, but the smoothing factor with

$\psi = 0,\ 1/Re = 0.0,\ \lambda = 1.0,\ \sigma/\sigma^* = 2$

θ	No-RS	C-RS	U-RS	FB-RS	$(C_{fb,x},\ C_{fb,y})$
1.0	0.2369	0.5112	0.1560	0.1942	(0.20, 0.20)
0.5	0.5207	0.5654	0.3533	0.3865	(0.20, 0.20)
0.1	0.8791	0.7793	0.7730	0.7733	(0.20, 0.20)

Table 2 Smoothing factor of 1st-order upwind scheme with 4-stage Runge-Kutta optimal coefficients (σ^*=1.29)

U-RS is even smaller than without residual smoothing. This confirms the previous analysis.

For the second-order upwind scheme, the optimal coefficients (0.0934, 0.2331, 0.4788, 1.0) are also taken from [10]. Based on the table 3, the same conclusion as the first-order upwind scheme can be obtained. Again U-RS shows the best smoothing properties, and C-RS is the poorest. FB-RS is between them.

$\psi = 1,\ \kappa = -1,\ 1/Re = 0.0,\ \lambda = 1.0,\ \sigma/\sigma^* = 2$

θ	No-RS	C-RS	U-RS	FB-RS	$(C_{fb,x},\ C_{fb,y})$
1.0	0.5095	0.7007	0.4768	0.5316	(0.20, 0.20)
0.5	0.7300	0.7907	0.6717	0.6972	(0.20, 0.20)
0.1	0.9405	0.9013	0.8905	0.8887	(0.80, 0.80)

Table 3 Smoothing factor of 2nd-order upwind scheme with 4-stage Runge-Kutta optimal coefficients (σ^*=0.62)

4 Application to The Navier-Stokes Equations

The FB-RS approach can be applied to the Navier-Stokes equations after the following modifications are made. In order to satisfy the stability condition ($\varepsilon_f \geq \varepsilon_b$), the local smoothing parameters are given according to the sign of velocity.

$$\begin{aligned} \varepsilon_{f\xi,L} &= \frac{1}{2}\left\{[1+sgn(u_\xi)]\,\varepsilon_{f\xi} + [1-sgn(u_\xi)]\,\varepsilon_{b\xi}\right\} \\ \varepsilon_{b\xi,L} &= \frac{1}{2}\left\{[1-sgn(u_\xi)]\,\varepsilon_{f\xi} + [1+sgn(u_\xi)]\,\varepsilon_{b\xi}\right\} \end{aligned} \tag{29}$$

where $u_\xi = \vec{v}\cdot\vec{n}_\xi$ with $\vec{n}_\xi$ as a surface normal. It is same for other directions. In order to deal with the high-aspect-ratio, Martinelli and Jameson [5] proposed the variable coefficient model:

$$\begin{aligned} \frac{\sigma}{\sigma^*}\varphi(r) &= 1+\varepsilon_{f\xi}+\varepsilon_{b\xi} \\ \frac{\sigma}{\sigma^*}\varphi(\frac{1}{r}) &= 1+\varepsilon_{f\eta}+\varepsilon_{b\eta} \end{aligned} \tag{30}$$

with

$$\varphi(r) = \frac{1+r^\alpha}{1+r} \tag{31}$$

where α is a parameter ($0 \leq \alpha \leq 1$), and r is given by

$$r_\xi = \frac{\lambda_\xi}{\lambda_\eta} \qquad \lambda_\xi = |\vec{v} \cdot \vec{n}_\xi| + c|\vec{n}_\xi|, \ \lambda_\eta = |\vec{v} \cdot \vec{n}_\eta| + c|\vec{n}_\eta| \tag{32}$$

where $\vec{v}$ is a velocity vector.

The FB-RS approach was developed within the framework of a 3D Multiblock/ Multigrid Navier-Stokes solver, called EURANUS (European Aerodynamic Numerical Simulator) [3]. It comprises a variety of upwind/central schemes, and uses both explicit solvers and implicit solvers. The multigrid method is the FAS scheme and V-, W- or sawtooth-cycles can be chosen. The grid loop is outside the block loop, so that all blocks are treated in phase.

5 Numerical Results and Discussions

Two-dimensional flat plate laminar calculation with 49x41x2 mesh points is used to illustrate the FB-RS efficiency. The inlet Machnumber, Reynolds number is 0.52 and 1.0×10^6 respectively. Both central and upwind schemes are chosen. Convergence is accelerated with 3-level V-cycle multigrid and with the FB-RS method.

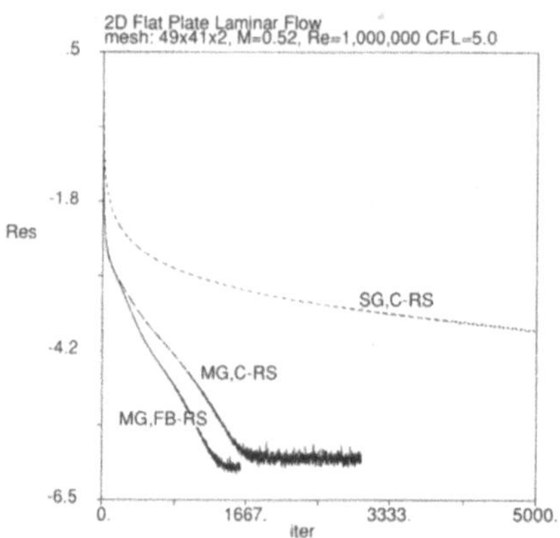

Fig. 5 Central scheme (X-momentum)

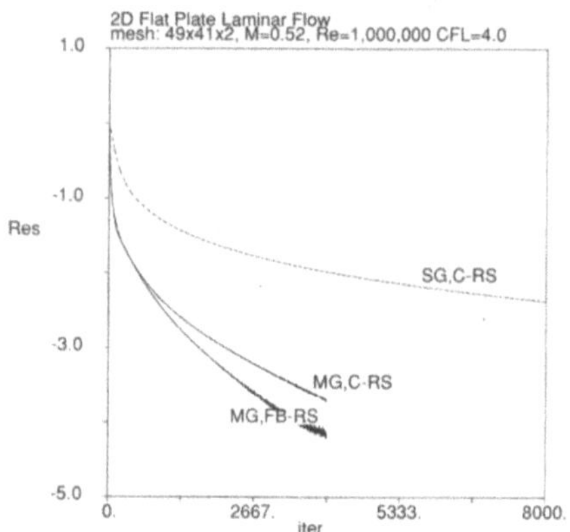

Fig. 6 Upwind scheme (X-momentum)

For the central scheme, artificial viscosity (2nd-order (0.05), 4th-order(0.05)) which is only re-calculated at first- and third stage of five stage Runge-Kutta scheme is used. The optimal coefficients for the explicit time-stepping are taken from [1]. Fig. 5 shows that the number of iterations is reduced around 25% by replacing C-RS with FB-RS ($C_{fb,x} = 0.75, C_{fb,y} = 0.90$, referring to equation(15)).

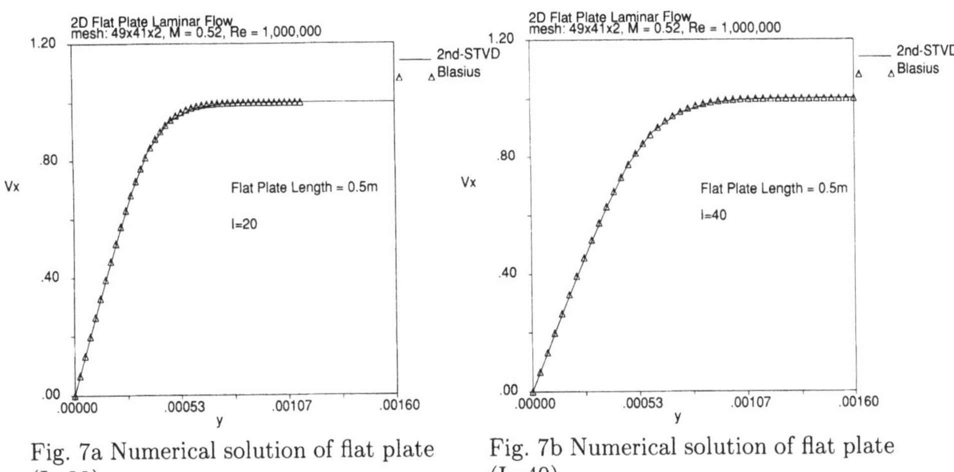

Fig. 7a Numerical solution of flat plate (I=20)

Fig. 7b Numerical solution of flat plate (I=40)

For the upwind scheme, a new second-order symmetric TVD scheme based on an effective ratio is used [11] [12]. The optimal coefficients for the four-stage Runge-Kutta time-stepping are taken from [10]. Fig. 6 shows that the number of iterations is reduced by around 30% with FB-RS ($C_{fb,x} = 0.60, C_{fb,y} = 0.75$) instead of C-RS. In addition, the high-accuracy solutions are shown in figs. 7a and 7b.

6 Conclusions

The application of the forward-backward residual smoothing to the Navier-Stokes equations is presented. According to the Fourier analysis of the smoothing model on the 2D convection-diffusion equation, it shows the improvement of smoothing properties with FB-RS compared to C-RS, U-RS in the case of reasonable aspect ratio. The numerical results from 2D flat plate laminar calculation confirm the efficiency of the new residual smoothing approach.

Future work will concentrate on the optimization of the smoothing parameters combined with the coefficients of multi-stage Runge-Kutta schemes.

References

[1] Jameson A. Multigrid Algorithms for Compressible Flow Calculation. MAE Report 1743, Text of Lecture given at 2nd European Conference on Multigrid Methods, 1985.

[2] Vatsa V.N. Turkel E. and Abolnassani J.S. Extension of Multigid Methodology to Supersonic/Hypersonic 3-D Viscous Flows. ICASE Report No–91–66, 1991.

[3] Lacor C. Hirsch Ch. Eliasson P. Lindblad I. and Rizzi A. Hypersonic Navier-Stokes Computations about Complex Configurations. Proceedings of the First European Computational Fluid Dynamics Conference, pp.1089-1096, 1992.

[4] Jameson A. Schmidt W. and Turkel E. Numerical Solutions of the Euler Equations by Finite Volume Methods Using Runge-Kutta Time-stepping Schemes. AIAA Paper 81–1259, 1981.

[5] Martinelli L. and Jameson A. Validation of a Multigrid Method for the Reynolds Averaged Equations. AIAA Paper 88–0414, 1988.

[6] Swanson R.C. and Turkel E. Artificial Dissipation and Central Difference Schemes for the Euler and Navier-Stokes Equations. AIAA Paper 87–1107–CP, 1987.

[7] Radespiel R. and Kroll N. Multigrid Schemes with Semi-coarsening for Accurate Computations of Hypersonic Viscous Flows. DLR IB 129-90/19, 1991.

[8] Blazek J. Kroll N. Radespiel R. and Rossow C.-C. Upwind Implicit Residual Smoothing Method for Multi-stage Schemes. AIAA Paper 91–1533–CP, 1991.

[9] Zhu Z.W. Lacor C. and Hirsch Ch. A New Residual Smoothing Method for Multigrid, Multi-stage Schemes. AIAA Paper 93–3356–CP, 1993.

[10] Catalano L. A. and Deconinck H. Two-dimensional Optimization of Smoothing Properties of Multi-stage Schemes Applied to Hyperbolic Equations. 3rd European Multigrid Methods, Bonn, Oct. 1–4, 1990.

[11] Lacor C. and Zhu Z.W. Global Formulation for Upwind/Symmetric TVD Schemes. VUB-STRO-CFD Report 93-01, 1993.

[12] Zhu Z.W. Simple, Efficient and High-accuracy Symmetric TVD Scheme. VUB-STRO-CFD Report 93-07, 1993.

Contents of the volume "Contributions to Multigrid"

"Contributions to Multigrid" is a second volume with contributions to the Fourth European Multigrid Conference. It is published by CWI, Amsterdam, in the Mathematical Tract Series, number 103, ISBN 90-6196-439-3.

1) *Alfio Borzì*
Burgers Equation and Multi-Grid Techniques

2) *Alfio Borzì*
On the Extension of the Twolevel Method for Operator Equations in Hilbert Space

3) *K.M.J. de Cock*
Multigrid Convergence Acceleration for the 2D Euler Equations Applied to High-lift Systems

4) *O. Dorok, F. Schieweck and L. Tobiska*
A Multigrid Method for Solving the Boussinesq Approximation of the Navier-Stokes Equations

5) *Jürgen Fuhrmann and Klaus Gärtner*
On matrix data structures and the stability of multigrid algorithms

6) *Wilhelm Heinrichs*
Spectral Multigrid Methods for the Reformulated Stokes Equations

7) *Jan Janssen and Stefan Vandewalle*
Multigrid Waveform Relaxation on Spatial Finite Element Meshes

8) *Edgar Katzer*
A Subspace Decomposition Twogrid Method for Hyperbolic Equations

9) *Ch. Kloppmann, D. Schwamborn and J.P. Singh*
Multigrid Solution of the 2-D Navier-Stokes Equations for Transonic Internal and External Flows

10) *B. Koobus, M-H. Lallemand, G. Carré and A. Dervieux*
Unstructured Multigrid by Volume Agglomeration for diffusion problems

11) *Hans Kuerten and Bernard Geurts*
A Multigrid Multiblock Solver for Compressible Turbulent Flow

12) *Yvonne Luh*
A Study on Narrow Stencil Discretizations and Multigrid Solvers for the Generalized Stokes Equations

13) *Lesley R. Matheson and Robert E. Tarjan*
A Critical Analysis of Multigrid Methods on Massively Parallel Computers

14) *C.W. Oosterlee and P. Wesseling*
A Multigrid Waveform Relaxation Method for Time Dependent Incompressible Navier-Stokes Equations

15) *Kris Riemslagh and Erik Dick*
Implementation Aspects of the Multigrid Formulation of Finite Volume Algorithms on Unstructured Grids

16) *C. Srinivasa*
Solution of Full Potential Equation on an Airfoil by Multigrid Technique

17) *S. Zeng and P. Wesseling*
Multigrid for the Incompressible Navier-Stokes Equations on Staggered Grids in General Coordinates

Zeitfracht Medien GmbH
Ferdinand-Jühlke-Straße 7
99095 Erfurt, Deutschland
produktsicherheit@kolibri360.de